Through 32 separate revisions and
reprintings since 1942, this manual
has helped more than 150,000 mechanics
prepare for their FAA ratings.

STUDY TEXT AND GUIDE
TO THE FAA RATING

A ZWENG MANUAL

PAN AMERICAN NAVIGATION SERVICE, INC.

12021 VENTURA BOULEVARD • NORTH HOLLYWOOD • CALIFORNIA   91604

AIRFRAME
MECHANICS'
MANUAL

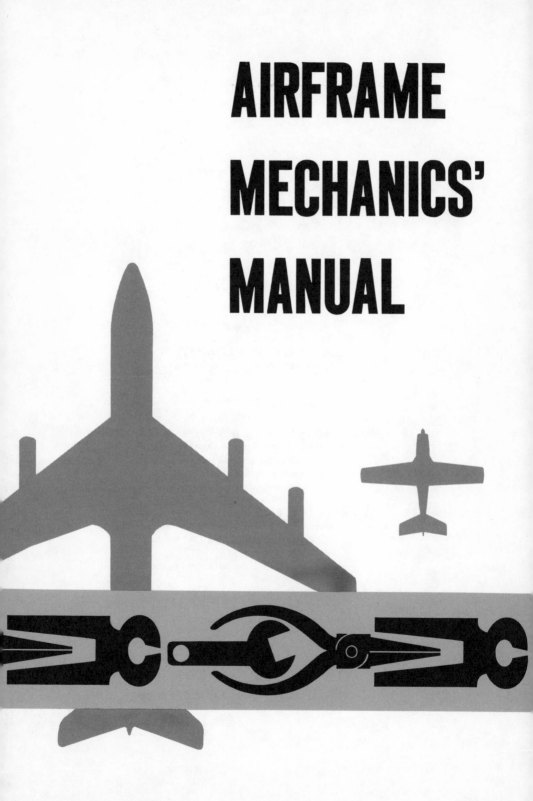

AIRFRAME MECHANICS' MANUAL:
A STUDY TEXT AND GUIDE TO THE FAA RATING

First published in this form in 1971.

Manufactured in the U.S.A.

Library of Congress Card Catalog Number: 79-138652

International Standard Book Number: 0-87219-017-X

A WORD ABOUT THIS MANUAL

ALTHOUGH THIS is a completely new manual from beginning to end, it has a publishing history that goes back to 1942 when the aircraft mechanic's job was still a comparatively simple one. Through 32 separate revisions and reprintings, the contents of the manual were changed to meet the changing (and more demanding) requirements of the FAA airframe and powerplant mechanic ratings.

Today each of the two ratings calls for highly specialized and unique skills, and a single manual can hardly do justice to the wide range of knowledge that the FAA expects of licensed mechanics—either airframe *or* powerplant. For this reason, Pan American Navigation Service, publishers nearly 30 years ago of the first *Aircraft and Engine Mechanics' Manual* (as it was then called), decided it was necessary to provide the airframe mechanic with a separate, informed, and thoroughly up-to-date means of preparing for the FAA tests on the craft he has chosen to practise. The present manual is the result. It is scheduled to be followed in due course by a completely new manual on the powerplant mechanic rating, which has become every bit as demanding as the airframe.

This is not a test guide in the sense that it merely offers some multiple-choice questions based on the FAA airframe exams with answers to be memorized in order to "guarantee" a passing grade. Nor is it a mechanical or audio-visual short-cut to easy, simplified learning of what is neither an easy nor a simple craft. The turbine-powered aircraft alone has taken care of that. *Airframe Mechanics' Manual* is, as the sub-title says it is, a study text and guide. That is, it provides—in the clearest, most concise manner possible—the information the FAA requires of a licensed airframe mechanic. The sample examinations follow the text in order that the student can test himself on whether that information has really been absorbed. If it has, then he can feel confident of meeting the FAA's standards.

This is the time-honored process of learning. No other means yet devised has made it obsolete.

For his contributions to this wholly new version of *Airframe Mechanics' Manual* the publishers would especially like to acknowledge their thanks to Ernest Brooks, instructor of aircraft mechanics for 30 years.

CONTENTS

AIRFRAME MECHANICS' MANUAL

ELIGIBILITY FOR AIRFRAME MECHANIC CERTIFICATE

The following mechanic certification requirements for the applicant seeking an airframe rating also include the requirements for the powerplant rating. There is a close relationship between the eligibility requirements of both mechanic ratings, and in many instances the applicant for the airframe rating will also be eligible for the powerplant rating.

Certificate requirements are in the areas of general eligibility, knowledge, experience, and skills. This section deals mainly with the general eligibility and experience requirements, with knowledge and skill requirements also included.

GENERAL ELIGIBILITY REQUIREMENTS

Age: A person must be at least 18 years of age to be issued a mechanic certificate. An applicant who meets all the requirements except age may apply for a certificate and complete the required written examinations. However, he must wait until age 18 to complete the oral and practical phases of the examinations, at which time he will be issued a mechanic certificate with appropriate rating/s.

Language: An applicant must be able to read, write, speak, and understand the English language, or, in the case of an applicant who does not meet this requirement and who is employed outside the United States by a U.S. air carrier, have his certificate endorsed "Valid only outside the United States."

Tests: All the tests (written, oral, and practical) prescribed by Federal Aviation Regulations must be satisfactorily passed within a period of 24 months.

Ratings: The Federal Aviation Regulations provide for two mechanic ratings—*airframe* and *powerplant.* The requirements for at least one rating must be met to be eligible for a mechanic certificate. A certificated mechanic applying for an additional rating must meet the experience, knowledge, and skill requirements for the rating sought, and pass satisfactorily all of the tests prescribed within a period of 24 months.

11

EXPERIENCE REQUIREMENTS. Each applicant for a mechanic certificate or rating must present either an appropriate graduation certificate from a certificated mechanic school or documentary evidence, satisfactory to the Administrator, of:

(a) At least 18 months of practical experience with the procedures, practices, materials, tools, machine tools, and equipment generally used in constructing, maintaining, or altering airframes, or powerplants appropriate to the rating sought, or:

(b) At least 30 months of practical experience concurrently performing the duties appropriate to both the airframe and powerplant ratings.

The documentary evidence specified above may take any of various forms such as letters from present or former employers, military service records, business records, etc., as long as the FAA inspector or advisor who reviews it is able to determine that the type and amount of experience meets the experience requirements. *Note:* Any recognized field of aviation involving the construction, maintenance, or alteration of airframes or powerplants offers possibilities for acquiring the required experience. Experience should be obtained by performing or assisting in the performance of basic mechanic work involved in aircraft construction, repair, maintenance, or alteration. The basic principles covering the installation and maintenance of propellers are included in the powerplant rating tests.

KNOWLEDGE REQUIREMENTS. Each applicant for a mechanic certificate or rating must, after meeting the applicable experience requirements, satisfactorily pass a written test covering aircraft construction and maintenance appropriate to the rating he seeks. In addition, the written tests include regulations, procedures, and practices pertinent to the repair, alterations, and inspection of aircraft.

The applicant must pass each section of the written test before applying for the oral and practical tests. On satisfactory completion of the oral and practical tests the applicant is issued a temporary certificate with the appropriate rating/s. The temporary certificate is valid for 90 days. During this 90-day period the applicant will receive a permanent certificate from the Administrator. The permanent certificate will be properly endorsed and numbered, and will be valid until revoked or suspended by the FAA or surrendered to the FAA voluntarily.

SKILL REQUIREMENTS. Each applicant for a mechanic certificate or rating must pass an oral and a practical test on the rating (airframe or powerplant) he seeks. The oral and practical tests cover the applicant's basic skill in performing practical projects on the subjects covered by the written examination for the particular rating.

These tests (oral and practical) are given by an FAA Inspector or by a qualified person designated by the FAA as a mechanic examiner. In either case an appointment must be made by the mechanic applicant as to the time and place of the tests.

It is suggested that the mechanic applicant using this book refer to and read Federal Aviation Regulations, Part 65, *"Certification: Airmen Other Than Flight Crew-members,"* located inside the rear cover of this manual. A section of FAR Part 65 refers specifically to mechanic applicants.

THE WRITTEN TESTS

The applicable written tests for a mechanic certificate or rating can be taken at FAA Flight Standards district offices and also at most Flight Service Stations (FSS). The tests are administered by, or under the supervision of, specifically designated FAA employees. Some Flight Standards district offices give written tests at other locations by prior arrangement. Flight Service Stations often administer tests after the normal workday and on weekends.

Though not a requirement, most FAA offices that administer written tests recommend that an appointment be made prior to the intended date of examination so that the appropriate personnel will be on hand to give the tests and to assure that testing space be available.

An applicant is expected to begin the written test early enough to complete it within the time allotted and before office closing time. Five hours each are allowed for the airframe and powerplant tests. Applicants who have not previously had their experience evaluated to determine whether they are eligible to take the test should allow additional time for this determination to be made by the appropriate FAA personnel.

When eligibility to take the test is confirmed, the FAA office will issue a test booklet, blank answer sheet, and all other material necessary to take the test. The applicant is not required to furnish any test or reference materials, nor permitted to use or take notes during the test.

Airframe and powerplant tests are made up of multiple-choice questions. Each question presents a problem and lists four possible answers. Only one of the answers will solve the problem completely and correctly. The tests do not contain questions designed to trick or mislead the applicant.

After completion of the test, the answer sheet will be forwarded by the FAA office to a central location for grading. The minimum passing grade for FAA tests is 70%. Written test grades will be mailed directly to the applicant using the address given by the applicant on his answers sheet at the time he takes the test. Five working days (exclusive of time enroute in the mail) are normally required for answer sheets to be processed and grades reported.

Both airframe and powerplant mechanic written tests are separated into three sections—in the former: (1) General, (2) Airframe Structures, and (3) Airframe Systems and Components; and in the latter: (1) General, (2) Powerplant Theory and Maintenance, and (3) Powerplant Systems and

Components.

An applicant will not be required to take the General section of an airframe or powerplant test if he can show that he has previously passed it. Proof of passing may be in the form of a mechanic certificate with the alternate rating or an Airman Written Examination Report for the alternate rating that shows a passing grade on the General section. If an Airman Written Examination Report is used, the passing credit must have been earned within the preceding 24 months.

RETESTING AFTER FAILURE. An applicant who fails a written, oral, or practical test for a certificate and rating, or for an additional rating, may apply for retesting: (a) after 30 days following the date he failed the test, *or* (b) upon presenting a statement from whichever of the following persons is applicable, certifying that he has given the applicant at least five hours of additional instruction in each of the subjects failed and now considers that the applicant is ready for retesting. *Note:* The persons qualified to write this statement are: (a) a certificated and appropriately rated mechanic, (b) a certificated and appropriately rated ground instructor, or (c) a certificated repairman experienced in the subject/s failed.

MECHANIC TEST CONTENTS

This listing of the subject material covered by the questions in the mechanic tests shows what the applicant should know and be able to do. Each section of the listing contains major headings (identified by letters A, B, C, etc.) under which are listed one or more action lines. Each action line is made up of 3 elements: (1) the action, (2) the subject, and (3) the level. For instance, the major heading "C. Weight and Balance" in the General section has two action lines:

(a) Weigh aircraft—Level 2.

(b) Perform complete weight and balance check and record data—Level 3.

The action lines tell what the end result or objective of the applicant's study and experience should be. Many action lines show more than one action and more than one subject.

The purpose of the level indicated for each action line is to help limit the amount of study that must be done or the skill that must be developed to pass the mechanic tests. Three levels have been identified. The higher the level, the more comprehensive the knowledge and skill required in that subject area. A Level 1 action line requires a less extensive knowledge of the subject and no skill demonstration to pass the mechanic tests. A Level 2 action line requires a fairly good understanding of the subject indicated, the theories and principles associated with it, and the ability to perform basic operations. Level 3, the highest level, requires a thorough knowledge of the subject and an understanding of how it relates to the total operation and maintenance of aircraft. The operations necessary to complete Level 3 items must be performed skillfully enough so that, if performed on an aircraft, the aircraft could be returned to service.

A detailed description of the meaning of each level is:

Level 1:

Know—basic facts and principles.

Be able to—find information and follow directions and written instructions.

No skill demonstration is required.

Level 2:

Know and understand—principles, theories and concepts.

Be able to—find and interpret information and perform basic operations.

A high level of skill is not required.

Level 3:

Know, understand, and apply—facts, principles, theories, and concepts. Understand how they relate to the total operation and maintenance of aircraft.

Be able to—make independent and accurate airworthiness judgments. Perform all operations to a return-to-service standard.

A fairly high skill level is required.

Following each action line is a list of statements that describe the specific subjects of mechanic test questions. They are intended to be used during study and while experience is being gained, to help direct the applicant's attention toward areas covered by the tests.

15

GENERAL

A. BASIC ELECTRICITY.

Measure capacitance and inductance.—Level 1:

The effect of inductive reactance in an electrical circuit.

The interrelationship of capacitive and inductive reactance.

The term that describes the combined resistive forces in an a.c. electrical circuit.

The unit of measurement for capacitance and inductance.

Calculate and measure conductivity and electrical power.—Level 2:

Determine the power furnished by a generator to an electrical system consisting of various electrical units having specific load ratings.

Determine the power required by an electric motor that is operating at a specified efficiency and load.

Measure voltage, current, resistance, continuity, and leakage.—Level 3:

Use an ohmmeter to check for open or short circuits.

The test instruments used to check continuity.

The basic operating principle of d.c. electrical instruments.

The basic operating principles of a galvanometer.

Connect voltmeters and ammeters into an electrical circuit.

The purpose of a shunt resistor when used with an ammeter.

The meaning of prefixes such as micro, mega, kilo, and milli as used in expressing electrical quantities.

Determine the relationship of voltage, current, and resistance in electrical circuits.—Level 3:

Determine the current flow in an electrical circuit using variable resistance and voltage values.

Determine the power requirements of an electrical circuit when the voltage and resistance values are specified.

The current relationship in a parallel electrical circuit.

The unit of measurement used to express electrical power.

The principles of electromagnetic induction.

The characteristics of magnets and magnetic lines of force.

The factors that affect the voltage drop in an electrical conductor.

Determine the resistance of an electrical device from the wattage and voltage values specified.

Calculate the voltage drop across a resistor.

Read and interpret electrical circuit diagrams.—Level 3:

Trace electrical circuits using aircraft wiring diagrams.

Identify electrical system malfunctions by reference to circuit diagrams.

Identify the commonly used aircraft electrical symbols.

Inspect and service batteries.—Level 3:

Remove spilled electrolyte and treat all adjacent surfaces.

Remove and install a battery in an aircraft with a single-wire electrical system.

Connect batteries to a constant-current battery charger.

Determine the specific gravity of the battery electrolyte.

Perform a high-rate-discharge condition test of batteries.

The design factors that affect battery voltage and capacity.

The factors that determine battery charging rate on a constant voltage source.

The indications of a shorted battery cell.

The significance of battery capacity ratings.

The effects of increased internal resistance on battery operation.

The effects of connecting battery cells in series or parallel.

The relative advantages of lead-acid and nickel-cadmium batteries for use in aircraft.

The principles of battery construction.

Check battery electrolyte levels.

The relationship between battery state of charge and the temperature at which the electrolyte will freeze.

The purpose of and requirements for ventilating batteries and battery compartments in civil aircraft.

The effect of excessive charging rates on batteries.

B. AIRCRAFT DRAWINGS.

Use drawings, symbols, and schematic diagrams.—Level 2:

Interpret the various types of lines employed in blueprints and schematics.

Use schematic diagrams to analyze system malfunctions.

Extract a specific electrical circuit from a system drawing.

Know why dimensions are used and how they are shown on aircraft drawings.

Use installation diagrams to locate and identify components.

Draw sketches of repairs and alterations. —Level 3:

Illustrate a major repair or alteration.

Use dividers, compass, ruler, T-square, etc., in the development of sketches of repairs and alterations.

Use standard drafting procedures.

Use blueprint information.—Level 3:

The information presented in blueprint title blocks.

The common symbols used on aircraft blueprints.

Install and modify component parts by reference to blueprints.

Identify the changes made to a blueprint.

Use graphs and charts.—Level 3:

Determine electric cable size and current-carrying capacity.

Determine engine power requirements.

C. WEIGHT AND BALANCE.

Weigh aircraft.—Level 2:

Use aircraft specifications for weighing purposes.

Locate jacks and scales in the correct position.

Prepare aircraft for weighing.

Perform complete weight-and-balance check and record data.—Level 3:

Determine that the forward or rearward center-of-gravity (c.g.) limit is not exceeded on a specified aircraft.

The point of reference for all weight-and-balance measurements.

The procedure for computing "minimum fuel."

Locate the information that should be known to compute weight and balance.

The method of expressing additions or removals of equipment for weight-and-balance purposes.

Determine the fully loaded center of gravity of an aircraft.

Determine the "maximum authorized weight" of an aircraft.

The method of determining aircraft empty weight, when engine oil and hydraulic fluid are contained in supply tanks.

The effect on weight and balance of replacing a component with another of different weight and location.

Calculate the maximum cargo or baggage weight that can be carried by an aircraft.

The record requirements for weight-and-balance data.

The hazards of exceeding aircraft fore and aft center-of-gravity limits.

The critical conditions of helicopter load and balance.

Determine aircraft empty weight and empty weight center of gravity.

Define maximum gross weight.

Determine the moment of an item of equipment.

Account for tare weight when weighing an aircraft.

D. FLUID LINES AND FITTINGS.

Fabricate and install rigid and flexible fluid lines and fittings.—Level 3:

Single- and double-flare tubing.

Install Military Standard (MS) flareless fittings.

The significance of the identification stripes that appear on aircraft hose.

Fabricate and install beaded tubing.

Use lubricants and sealants in the assembly of lines and fittings.

Identify flexible hydraulic lines.

Install hose clamps.

Determine the bend radii for rigid tubing.

Fabricate aluminum tubing using standard AN flared tube fittings.

Route fluid lines in entryways and passenger, crew, or baggage compartments.

Repair metal tube lines.

Route fluid lines adjacent to electrical power cables.

Install rigid tubing.

Select tube-flaring tools.

Identify AN fitting materials from color designators.

The maximum reduction in original outside diameter allowed when bending aluminum alloy hydraulic lines.

The procedure to follow if scratches are detected on an aluminum alloy tube.

The storage requirements for hydraulic hose.

Install flexible hydraulic hose.

The lubricant used when assembling oxygen fittings.

E. MATERIALS AND PROCESSES.

Identify and select appropriate nondestructive testing methods.—Level 1:

The use of radiography in aircraft and component inspection.

The use of ultrasonic inspection methods for detecting cracks.

The applicability of magnetic particle inspection methods to engine crankshafts.

The method for detecting surface cracks in aluminum castings and forgings.

The technique for locating cracks in materials when only one side of the material is accessible.

Perform penetrant, chemical etching, and magnetic particle inspections.—Level 2:

The general procedure for performing magnetic particle inspection.

Demagnetize steel parts after magnetic particle inspection.

Clean parts in preparation for penetrant inspection.

The visual indications of a subsurface flaw or fracture during magnetic particle inspection.

Locate cracks and blowholes in welded assemblies.

The procedure for using dye penetrants.

Distinguish between heat-treated and non-heat-treated aluminum alloys when the identification marks are not on the material.

Perform basic heat-treating processes.— Level 2:

The types of aluminum alloys considered to be heat treatable.

Anneal copper tubing.

The steps in heat treatment of aluminum alloys.

The effects of various forms of heat treatment.

The effect of incorrect heat treatment on the corrosion-resistant properties of aluminum alloy.

Identify the degree of temper for aluminum alloy products from code designators.

The effect of heating a metal slightly above its critical temperature, and then rapidly cooling it.

The effect of strain hardening on the tensile strength of aluminum alloy.

The relationship between tensile strength and metal hardness.

Anneal a welded steel part.

Identify and select aircraft hardware and materials.—Level 3:

Identify aluminum alloys from code designators.

Identify steel from code designators.

The identification markings of AN standard steel bolts.

Identify aircraft cable.

The characteristics of a material that affect its ability to be hammered, rolled, or pressed into various shapes.

The SAE system of identifying steel.

Determine wrought aluminum alloy composition and condition by referring to aluminum codes.

Install self-locking nuts.

Determine the correct length bolt to use.

Determine correct torque values for tightening aircraft nuts and bolts.

Determine rivet composition, condition, shape, and dimension by referring to rivet code.

Identify materials suitable for use for firewalls and exhaust shrouds.

Install castle nuts.

The strength characteristics of type "A" rivets.

The characteristic of aluminum alloy rivet material that causes some rivets to require several days to reach their ultimate strength.

Determine that materials used in aircraft maintenance and repair are of the proper type and conform to the appropriate standards.

The characteristics of aluminum-clad sheet aluminum alloy.

Inspect and check welds.—Level 3:

The characteristics of a good weld.

The types of stress that welded joints can withstand.

The effect of welding over a previously brazed or soldered joint.

Perform precision measurements.—Level 3:

Use a micrometer and a caliper to make precise measurements.

Measure a small hole using a micrometer and a hole gage.

Read and interpret a vernier micrometer scale.

Use a dial indicator, V-blocks, and a surface plate to check alignment of a shaft.

F. GROUND OPERATION AND SERVICING.

Start, ground operate, move, service, and secure aircraft.—Level 2:

The procedure for extinguishing fires in the engine induction system during starting.

Use hand signals to direct aircraft movement.

Select and use external auxiliary power units for engine starting.

Tie down and secure aircraft for outside storage.

Protect aircraft fuel system from contamination during fueling operations.

Connect and operate an external source of hydraulic power.

Start and operate an engine equipped with a float-type carburetor.

Check a reciprocating engine for liquid lock.

Operate hand and electrical priming systems during engine starting.

Start and operate an engine equipped with a pressure injection carburetor.

Start and operate an engine equipped with an internal supercharger.

Identify and select fuels.—Level 2:

The effect of ethylene dibromide added to aviation gasoline.

The identifying color of various grades of aviation gasoline.

The characteristic of a fuel that affects its tendency to "vapor lock."

The significance of the numbers used to designate various grades of aviation gasoline.

The relative advantages of gasoline and kerosene for use as fuel for turbine engines.

Determine the type of fuel to be used with a specified aircraft.

The factors affecting the antiknock characteristics of fuel.

G. CLEANING AND CORROSION CONTROL.

Identify and select cleaning materials.—Level 3:

The effect of caustic cleaning products on aluminum structures.

The characteristics and use of chemical cleaners.

Clean aluminum and steel engine parts.

The type cleaner for use on high-strength metals.

The methods for cleaning turbine engine compressor blades.

Perform aircraft cleaning and corrosion control.—Level 3:

Protect tires and other rubber products from the deteriorating effects of cleaning materials.

The cause and corrective procedures for fretting corrosion.

Identify and control intergranular corrosion of heat-treated aluminum alloy.

Protect structure against dissimilar-metal corrosion.

Prevent and remove rust.

The effect of oily, dirty surfaces on the operation of high-performance aircraft.

Protect interior surfaces of closed steel and aluminum tubing against corrosion.

The methods of protecting aluminum alloy parts against corrosion.

Clean and protect battery compartments and adjacent areas.

Remove corrosion products such as metal flakes, scale powder, and salt deposits from aluminum.

Clean corrosion-resistant parts by blast cleaning methods.

Use paints and similar organic coatings for corrosion protection purposes.

H. MATHEMATICS.

Extract roots and raise numbers to a given power.—Level 1:

The method of determining the square or cube of a number.

The procedure for determining square root.

Determine areas and volumes of various geometrical shapes.—Level 2:

Calculate the area of rectangles, squares, triangles, circles and trapezoids.

Determine the volume of rectangles, cubes, and cylinders.

Compute the surface area of an airfoil.

Determine cylinder displacement of a reciprocating engine.

Solve ratio, proportion, and percentage problems.—Level 3:

Determine the ratio of two numbers.

Find what percent one number is of another.

Determine the rate percent of a given number.

Calculate the compression ratio of an engine.

Convert decimal numbers to their fractional equivalent.

Perform algebraic operations involving addition, subtraction, multiplication, and division of positive and negative numbers.—Level 3:

Locate the main-wheel weighing point with reference to the datum.

Determine the distance between the tail or nose gear and the main-wheel weight point.

Calculate the c.g. relative to the datum.

The effects of adding or removing equipment on the empty weight of the aircraft.

I. MAINTENANCE FORMS AND RECORDS.

Write descriptions of aircraft condition and work performed.—Level 3:

Describe the repairs made to an aircraft structure.

State aircraft condition based upon inspection.

Complete required maintenance forms, records, and inspection reports.—Level 3:

Enter the required information in the permanent maintenance records when a minor repair has been performed.

Prepare and properly dispose of FAA Form 337.

The minimum information required to be entered in the maintenance records after maintenance or alteration of aircraft.

Make record entries to indicate compliance with Airworthiness Directives.

The definition of "time in service" with respect to maintenance records.

The record requirements for returning aircraft to service after 100-hour inspection.

The requirement for maintaining a permanent record of aircraft maintenance.

The definition of "repair" as related to aircraft maintenance.

The requirements for a permanent maintenance record.

J. BASIC PHYSICS.

Use the principles of simple machines; sound, fluid, and heat dynamics.—Level 2:

The relationship between temperature and heat.

The methods of heat transfer.

The forces acting upon a body in circular motion.

The relationship between the pressure and the rate-of-flow of a liquid through an orifice.

The relationship between the pressure, volume, and temperature of an air mass.

The relationship of work, force, and power.

The effect of air density on engine power output.

The relationship between air velocity and pressure on the upper surface of an airfoil.

The effect of atmospheric temperature and humidity on airfoil lift.

The principles of transmission of power in a hydraulic system.

The relationship of pressure, area, and force.

K. MAINTENANCE PUBLICATIONS.

Select and use FAA and manufacturer's aircraft maintenance specifications, data sheets, manuals, and publications, and related Federal Aviation Regulations.— Level 3:

Determine the suitability of a propeller for use with a particular engine-airplane combination.

Determine the minimum diameter of a propeller type and model when used with a particular engine.

Locate aircraft leveling and weighing information.

Determine engine/propeller speed ratios.

The instrument markings required on a specified type and model aircraft.

The purpose and applicability of Technical Standard Orders.

The purpose and applicability of Supplemental Type Certificates.

Identify the useful load and empty weight c.g. of an aircraft by reference to data.

Use FAA Specifications and Type Certificate Data Sheets.

The applicability and requirements for aircraft airworthiness certificates.

Determine the control surface movement limits of a specified aircraft.

Determine seat locations of an aircraft, using aircraft specifications.

Use aircraft listing to find information about aircraft of limited production.

The purpose and applicability of FAA Airworthiness Directives.

Use "Table of Limits" to determine condition of parts.

Read technical data.—Level 3:

Find specified information in technical reports and manuals.

L. MECHANIC PRIVILEGES AND LIMITATIONS.

Exercise mechanic privileges within the limitations prescribed by FAR 65.—Level 3:

The criteria for determining the classification (major, minor, or preventive maintenance) of airframe repairs and alterations.

The criteria for determining the classification (major, minor, or preventive maintenance) of powerplant repairs and alterations.

The criteria for determining the classification (major, minor, or preventive maintenance) of propeller repairs and alterations.

Return an aircraft to service after installation of an engine type other than that for which the aircraft was originally certificated.

The minimum age requirement for issuance of a mechanic certificate.

The privileges of a mechanic in relation to 100-hour and annual inspections.

The requirements for reporting change of address.

The duration or effective period of a mechanic certificate.

The requirements an applicant must meet for issuance of a mechanic certificate.

Determine the maintenance classification (major repair, minor repair, preventive maintenance) of landing gear tire removal, installation, and repair.

Determine the maintenance classification (major repair, minor repair, preventive maintenance) of servicing landing gear shock struts.

Determine the repair classification (major repair, minor repair, preventive maintenance) of repairs to steel tubing structures by welding.

Determine the repair classification (major repair, minor repair, preventive maintenance) of replacing the fabric on fabric-covered parts such as wings, fuselages, stabilizers, and control surfaces.

The recency-of-experience requirements for certificated mechanics.

The privileges of a mechanic regarding return to service of aircraft after major repairs.

Determine the maintenance classification (major repair, minor repair, preventive maintenance) of the replacement of aircraft components with new, rebuilt, or repaired components of similar design.

AIRFRAME STRUCTURES

A. WOOD STRUCTURES.

Service and repair wood structures.— Level 1:

The general requirements of scarf splice joints.

The repair procedure for elongated holes in wood spars.

The permissible wood substitutes for use in making repairs to wood structures.

The procedures for repairing wood rib capstrips.

The characteristics of glue used in aircraft construction and repair.

The procedure for sealing the inner surfaces of a wooden structure that is to be assembled by gluing.

The general characteristics of the wood commonly used in aircraft construction.

Identify wood defects.—Level 2:

Recognize acceptable and nonacceptable wood defects.

Inspect wood structures.—Level 2:

The effect of moisture content on wood size and strength.

The strength characteristics of wood structures.

The characteristics of plywood and laminated wood.

B. AIRCRAFT COVERING.

Select and apply fabric and fiberglass covering materials.—Level 1:

The factors to consider in selecting aircraft fabric.

The types of seams commonly used in aircraft fabric coverings.

The general requirements for making doped and lapped seams.

The meaning of the term "warp" as used in reference to aircraft textile products.

The precautions to observe when installing surface tape on control surfaces.

Inspect, test, and repair fabric and fiberglass.—Level 3:

Determine the condition of aircraft fabric.

Apply a doped-on patch to aircraft fabric.

Make a sewed repair to a fabric-covered surface.

The areas on a fabric-covered aircraft most susceptible to corrosion.

C. AIRCRAFT FINISHES.

Apply trim, letters, and touchup paint.— Level 1:

The requirements for registration markings.

The relative proportions of identification markings.

The use of color and ornamentation when applying registration marks.

Identify and select aircraft finishing materials.—Level 2:

The characteristics of butyrate and acetate dopes.

The types of thinners used with various types of paint and dope.

The characteristics of fabric rejuvenators.

The types of priming paints generally used on aircraft.

The type paint used to coat the insides of battery compartments.

Apply paint and dope.—Level 2:

The purpose of fungicidal dope in aircraft finishing.

The application of rejuvenator to repair an aged dope finish.

The products and methods used to dope-proof airframe structures.

The effect of atmospheric conditions on dope during its application.

Sand and rub aircraft finishes.

Apply primer to aluminum alloy parts.

Use and maintain a paint spray gun.

The purpose of brushing the first coat of dope instead of spraying.

Inspect finishes and identify defects.—Level 2:

The type of painting defect caused by moving the spray gun in an arc instead of a straight line.

The cause of runs and sags in aircraft finishes.

D.   SHEET METAL STRUCTURES.

Install special rivets and fasteners.—Level 2:

Determine correct rivet length and diameter.

Install a hi-shear rivet.

The precautions concerning rivet fit.

Install deicer boot fasteners.

Install blind-type rivets.

The stresses that a rivet is designed to resist.

Inspect bonded structures.—Level 2:

The reason for using metal sandwich panels in high-speed aircraft construction.

The use of the metallic "ring" test to inspect for delamination damage of bonded structures.

Evaluate the extent of damage to a bonded structure and determine the type repair needed.

Inspect and repair plastics, honeycomb, and laminated structures.—Level 2:

Distinguish between transparent plastic and plate-glass enclosures.

Protect plastics during handling and repair operations.

Remove scratches and surface crazing from plastic enclosures.

Drill shallow or medium depth holes in plastic materials.

The effect of moisture entrapped in honeycomb structures.

Use a router to remove damaged area from honeycomb panels.

Clean honeycomb panels prior to patching.

Inspect, check, service, and repair windows, doors, and interior furnishings.—Level 2:

Clean transparent plastic window and windshield materials.

Inspection procedures and airworthiness requirements for safety belts.

The characteristics of acrylic plastic enclosure materials.

Maintain safety belts.

Secure transparent plastic enclosures to the aircraft structure.

Protect transparent plastic enclosure materials during handling and storage.

The physical characteristics of transparent plastic enclosure materials.

Form and shape acrylic plastic.

Repair shallow surface scratches in transparent plastic enclosures.

Inspect and repair sheet-metal structures.—Level 3:

Select and use twist drills.

Select and use a hand file for soft metals.

Prepare dissimilar metals for assembly.

Determine the type, size, and number of rivets for use in structural repairs.

Repair sheet-metal flight control surfaces.

The loads acting upon a semimonocoque fuselage.

The construction characteristics of monocoque and semimonocoque structures.

The construction characteristics of cantilever wing structures.

The types of loads carried by wing spars.

Drill holes in stainless steel.

Define bearing failure as related to sheet-metal structures.

Define shear failure.

Repair a hole in a stressed-skin metal wing.

Repair a section of damaged skin using a single-lap sheet splice.

Construct a watertight joint.

Countersink a hole.

Perform the dimpling process.

Select the correct rivet to accomplish a repair using a specified material.

Repair or splice stringers on the lower surface of a stressed-skin metal wing.

Determine the correct rivet layout and spacing for a specified repair.

Use proper riveting techniques.

Stop drill cracks in sheet metal.

Repair a slightly oversize hole.

Repair structural units, such as spars, engine supports, etc., that have been built from sheet metal.

Repair shallow scratches in sheet metal.

Determine the condition of a stressed-skin metal structure that is known to have been critically loaded.

Use a reamer.

Install conventional rivets.—Level 3:

Prepare sheet metal for installation of flush rivets.

Identify and select rivets.

Determine the correct rivet length and diameter.

Select and use the correct rivet set for specified rivet head styles.

Select and use bucking bars.

Remove rivets.

Determine the condition of a driven rivet.

Determine the circumstances under which 2117 rivets may be used to replace 2017 and 2024 rivets.

Define rivet tipping.

Determine the correct number of rivets to be used in making a structural sheet-metal repair.

Handle and install rivets that require heat treatment prior to use.

Adjust and use an air-operated riveting gun.

The circumstances under which type "A" rivets may be used in aircraft.

The mechanical properties of heat-treated rivets.

Hand form, lay out, and bend sheet metal.—Level 3:

Make a joggle or offset bend.

Bend sheet metal that requires the use of a large radius.

Determine the neutral axis of a bend.

Define bend radius.

Determine the amount of material required to make a specified bend.

Bend sheet metal to a specified angle.

Lay out and bend a piece of sheet metal using a minimum radius for the type and thickness of material specified.

Lay out a bend in relationship to metal "grain" to minimize the possibility of cracking.

Determine the flat layout dimensions of a component part to be formed by bending.

Form metal by bumping.

E.   WELDING.

Weld magnesium and titanium.—Level 1:

The method of cleaning magnesium in preparation for welding.

The main function of a flux while welding magnesium.

The types of gases to use when gas-welding magnesium.

The use of butt joints when gas-welding magnesium.

Solder stainless steel.—Level 1:

The use of silver soldering as a method of bonding metals.

The preparation of stainless steel for soldering.

The methods of cleaning material after soldering.

Fabricate tubular structures.—Level 1:

The types of tubing splices.

The proper welding sequence to use when welding fuselage tubes.

The characteristics of a welded tubing joint.

The protection of the interior of tubular steel that is to be closed by welding.

The methods used to control distortion of steel tube structures during welding repairs.

The preparation of tube ends for welding.

Solder, braze, gas-, and arc-weld steel.—Level 2:

Use cleaning operations to prepare sheet steel for welding.

Adjust oxyacetylene welding torch to produce the type flame needed to weld a specified material.

Select and use filler rod.

The effect of excessive heat on metal.

Operate a portable welding set.

Select the correct size welding torch tip.

The precautions regarding welding over a previously brazed or soldered joint.

Solder a wire or cable to an electrical component.

Sweat-solder a lap joint.

Normalize a steel part after welding.

Identify steel parts considered to be repairable by welding.

The preheating required prior to welding.

Weld aluminum and stainless steel.—Level 2:

Use a filler rod when welding aluminum with oxyacetylene.

Use flux when welding aluminum.

The purpose and effect of using inert gas to shield the arc in certain types of welding.

F. ASSEMBLY AND RIGGING.

Rig rotary-wing aircraft.—Level 1:

The condition of flight that a properly rigged aircraft should maintain.

The relationship of thrust and drag of an aircraft during level unaccelerating flight.

The relationship of lift and weight of an aircraft during level unaccelerating flight.

The meaning of the term "angle of attack" of an airfoil.

The type of control movement used to induce forward flight in a helicopter.

The method of controlling vertical flight of a helicopter.

The movement of an aircraft about its axes during normal flight maneuvers.

The factors affecting stability of an aircraft about its axes.

The methods of maintaining directional control of a helicopter.

The cause and effect of rotor blade stall in helicopters operating at high speeds.

The cause of vertical vibration in a two-blade helicopter rotor system.

The preparations required prior to rigging.

The method of tracking helicopter main rotor blades.

Rig fixed-wing aircraft.—Level 2:

The condition of flight that a properly rigged aircraft should maintain.

The factors to consider when rigging vertical stabilizer of single-engine, propeller-driven aircraft.

The relationship of thrust and drag of an aircraft during level unaccelerating flight.

The effect of incorrect wing incidence angle.

The effect of dihedral on aircraft stability.

Use wing "wash-in" and "wash-out" to correct aircraft rigging.

The relationship of lift and weight of an aircraft during level unaccelerating flight.

The meaning of the term "angle of attack" of an airfoil.

The effect of flaps on aircraft landing speed and approach angle.

The meaning of the term "incidence angle" of an airfoil.

The movement of an aircraft about its axes during normal flight maneuvers.

The relationship between the center of pressure of a wing and its angle of attack.

The factors affecting stability of an aircraft about its axes.

The usual location of aircraft c.g. in relationship to center of lift.

The changes in lift and drag of the wings when an aircraft is rolled about its longitudinal axis.

The procedure for establishing wing angle of incidence prior to repairing wing attachment fittings.

Check alignment of structures.—Level 2:

Prepare fuselage for alignment check.

Check alignment of internally braced wing structure.

The significance and method of expressing reference positions.

Check alignment of assembled aircraft.

Assemble aircraft.—Level 3:

The methods of safetying aircraft screws, bolts, and nuts.

Assemble, adjust, and safety cable turnbuckles.

The correct method of inserting bolts in aircraft fittings.

Install and inspect swaged cable terminals and fittings.

Balance and rig movable surfaces.—Level 3:

The inspection requirements for cable-operated primary flight control systems.

Handle and make up control cables.

The corrosion protection requirements of control cables.

The effect of overtightening control cables.

The relationship between specified movements of the cockpit controls and the control surfaces.

The relationship between specified control movements during flight and the movement of the aircraft about its axes.

The movement of the controls, control surfaces, and the aircraft about its axes during normal flight maneuvers.

Balance control surfaces after repair.

The relationship between specified movements of the trim tab operating device and the trim tab.

Secure the cockpit flight controls in preparation for control surface rigging.

The effect of a worn pulley in a cable-operated control system.

The means used to reduce or prevent control surface flutter.

The purpose and operation of control surface locks.

The purpose and operation of differential controls.

The purpose and applicability of fairleads in a cable-operated control system.

Install and rig the cables in a flight control system.

Splice control cables using Nicopress sleeves.

The probable causes of control surface flutter.

The maintenance requirements of control surface trim tab systems.

The purpose of counterweights incorporated into the leading edges of some primary control surfaces.

The purpose and function of "spring tabs" and "servo tabs".

Measure control surface movement and adjust control stops.

The effect of temperature changes on control system cable tension.

Assemble, adjust, inspect, and safety push-pull tube-type flight control systems.

The types and characteristics of cables used in aircraft primary control systems.

Jack aircraft.—Level 3:

Determine maximum allowable jacking weight.

The use of correct capacity jacks.

Protect aircraft from damage during lifting and lowering operations.

Use ballast when jacking aircraft with engine removed.

The effects of wind when jacking aircraft.

G. AIRFRAME INSPECTION.

Perform airframe conformity and airworthiness inspections.—Level 3:

The maximum period of time an aircraft can be flown before an annual inspection is required.

Determine the condition of airframes, airframe systems, and components.

The primary purpose of inspection.

The maximum time an aircraft that carries passengers for hire or is used in flight instruction can be flown before being inspected.

Determine that an aircraft is in conformity with FAA Specifications.

Determine that applicable Airworthiness Directives have been complied with.

Conduct a thorough and detailed inspection of an aircraft.

AIRFRAME SYSTEMS AND COMPONENTS

A. AIRCRAFT LANDING GEAR SYSTEMS.

Inspect, check, service, and repair landing gear, retraction systems, shock struts, brakes, wheels, tires, and steering systems.
—Level 3:

Determine aircraft tire inflation pressures.

The factors affecting the retreading of aircraft tires.

Adjust landing gear toe-in.

Install and remove aircraft wheel and brake assemblies.

Install tubes and tires.

Protect aircraft tires from hydraulic fluids.

Service brake deboosters.

Service landing gear shock struts.

The effects of increasing temperature on "parked" brakes.

Determine the cause of an oleo strut bottoming during taxi operations.

The pressure source for actuating power brakes.

Select and install air valves in oleo shock struts.

Observe safety precautions when demounting tire and wheel assemblies.

Determine if a brake system requires bleeding; perform brake system bleeding.

Inspect and adjust multiple-disc brakes.

Install new linings in hydraulically operated single-disc brakes.

Determine the cause of spongy brake action.

Inspect and service aircraft tires and tubes.

Determine the reason for "dragging" brakes.

The method of equalizing braking pressure on both sides of the rotating disc of a single-disc brake.

Operate and check retractable landing gear.

Determine the cause of fading brakes.

Replace actuating cylinders.

Install brake blocks in an expander-tube brake assembly.

Inspect brake drums.

The purpose and function of metering pins in oleo shock struts.

Determine the cause of excessive brake pedal travel.

The operating principles of oleo shock struts during landing.

The storage requirements for aircraft tires and tubes.

The effect of a broken return spring in a brake master cylinder.

Determine the cause of grabbing brakes.

The purpose and operation of a debooster in a hydraulic power brake system.

Detect internal leakage in a brake master cylinder.

The operating principles of servo, expander-tube, multiple-disc, and single-disc aircraft brakes.

The purpose and operating principles of brake master cylinders.

B. HYDRAULIC AND PNEUMATIC POWER SYSTEMS.

Repair hydraulic and pneumatic power system components.—Level 2:

Install packing seals and rings on hydraulic components.

Determine the correct seal type to use with ester-base, petroleum-base, and vegetable-base fluids.

Remove and install hydraulic selector valves.

Remove and install a spool-type or balanced-type pressure regulator.

Determine the cause of excessive oil in an aircraft pneumatic power system.

The operating principles of a pneumatic power system multistage reciprocating compressor.

Identify hydraulic seals and packings.

Protect packing rings or seals against thread damage during installation.

Identify and select hydraulic fluids.— Level 3:

Determine the fluid type for use in a specified aircraft hydraulic system.

The method of measuring the viscosity of a liquid.

Identify ester-base, petroleum-base, and vegetable-base fluids.

Inspect, check, service, troubleshoot, and repair hydraulic and pneumatic power systems.—Level 3:

Determine the air pressure in a hydraulic accumulator.

The location and use of quick-disconnect fittings in hydraulic and pneumatic systems.

The mounting position of diaphragm and bladder-type hydraulic accumulators.

Service hydraulic reservoirs.

Determine the causes of incorrect system pressure.

Service porous paper and micronic filtering elements.

Adjust the pressure setting of the main system relief valve.

Purge air from a hydraulic system.

The term used to indicate force per unit area.

Identify the types of hydraulic power systems.

The purpose, location, and operation of a hydraulic fuse.

Protect a hydraulic system against contamination during a component replacement.

Inspect a hydraulic system for water and metal contamination.

Service a pneumatic system moisture separator.

The purpose, location, and operation of an orifice check valve in the wing flap actuating system.

The purpose, location, and operation of a wing flap overload valve.

The purpose, location, and operation of a hydraulic system pressure regulator.

The purpose, location, and operation of a sequence valve.

The purpose, location, and operation of a crossflow valve.

The purpose, location, and operation of a hydraulic system pressure accumulator.

The purpose, location, and operation of a shuttle valve.

The purpose, location, and operation of a check valve.

Install and remove engine-driven hydraulic pumps.

The indications of a worn or damaged hydraulic pump shaft.

The operating principles of hydraulic hand pumps.

The cause of hydraulic pump chatter during operation.

The operating principles of a constant-displacement hydraulic pump.

The operating principles of a variable-displacement hydraulic pump.

The purpose of the shear section on the shaft of an engine-driven hydraulic power pump.

The purpose and operation of a hydraulic actuating cylinder.

Determine the cause if a constant-pressure hydraulic system with no external leakage will not hold pressure when the power pump is not operating.

Determine the cause if an engine-driven power pump will not maintain system pressure during the actuation of a unit in the system.

The general features and operating principles of aircraft pneumatic power systems.

The purpose of pressurized reservoirs in some hydraulic systems.

The purpose and location of a standpipe in some hydraulic reservoirs.

The causes of too frequent cycling of a constant-pressure hydraulic system.

Operate and check a hydraulically operated flap system.

The operating mechanism of most hydraulic pressure gauges.

The indications of a low fluid supply during system operation.

C. CABIN ATMOSPHERE CONTROL SYSTEMS.

Repair heating, cooling, air-conditioning, pressurization, and oxygen system components.—Level 1:

The usual reasons a surface combustion heater fails to operate.

The effects of cracks or holes in an exhaust-type heat exchanger.

The usual sources of contamination of a freon system.

The method of protecting a freon system from contamination during replacement of a component.

Inspect, check, troubleshoot, service, and repair heating, cooling, air-conditioning, and pressurization systems.—Level 1:

The operating principles of a thermostatically controlled surface combustion heater.

The methods used to control cabin pressure of a pressurized aircraft.

The protective features included in the control circuits of surface combustion heaters.

The purpose and operation of check valves in the delivery air ducts of a pressurization system.

The basic principles of providing and controlling aircraft pressurization.

The inspection requirements of cabin heating systems that utilize an exhaust heat exchanger as a source of heated air.

The method of checking a combustion heater fuel system for leaks.

The function of the condenser in a freon cooling system.

The function of the evaporator in a freon cooling system.

The function of an expansion valve in a freon cooling system.

The location, in relationship to each other, of the units in a freon cooling system.

The method of determining the liquid level in a vapor-cycle cooling system.

The procedure for servicing a vapor-cycle air-conditioning system that has lost all its freon charge.

The basic operating principles of an air-cycle cooling system.

The function of a jet pump in a pressurization and air-conditioning system.

The function of a mixing valve in an air-conditioning system.

The function of the negative pressure-relief valve in a pressurization system.

The function of the outflow valve in a pressurization system.

The function and principles of operation of an automatic cabin rate-of-climb control system.

Inspect, check, troubleshoot, service and repair oxygen systems.—Level 2:

Check oxygen system for leakage.

Service oxygen system with breathing oxygen.

Inspect a breathing oxygen system for contamination.

D. AIRCRAFT INSTRUMENT SYSTEMS.

Install instruments.—Level 1:

The installation and connection of shock-mounted vacuum instruments to their power system.

The types of hardware used to install instruments.

The application of operation markings to the glass face of an aircraft instrument.

The protection of instruments during handling.

The installation practices necessary to prevent damaging an instrument.

The installation practices used in making hose or tubing connections to the instruments.

Inspect, check, service, troubleshoot, and repair heading, speed, altitude, time, attitude, temperature, pressure, and position indicating systems.—Level 1:

The procedure for "swinging" an aircraft magnetic compass.

The methods used to test a static air system for leakage.

The significance of various types of marks on the face of an instrument.

The operating principles of a thermo-couple temperature-indicating circuit.

The service requirements of instrument system filters.

The effect of a ruptured or disconnected static pressure line located inside a pressurized cabin.

E. COMMUNICATION AND NAVIGATION SYSTEMS.

Inspect, check, and service autopilot and approach control systems.—Level 1:

The operating principles of the sensing device used in an autopilot system.

The purpose and operation of the autopilot.

The purpose of a servomotor in an autopilot system.

The installation requirements for autopilot units.

The function of a position transmitter in an autopilot system.

Inspect, check, and service aircraft electronic communication and navigation systems.—Level 1:

The FCC regulations pertaining to the operation of two-way radio.

The principal conditions which must be considered in the installation of radio.

The protection of radio equipment from shock and vibration.

The methods of reducing engine noise in radio receivers.

Inspect and repair antenna and electronic equipment installations.—Level 2:

The preferred location and methods of mounting external antennas.

The procedure for returning an aircraft to service after a radio installation has been made in accordance with approved installation data.

The preferred location for the VOR localizer receiver antenna on a small aircraft.

F. AIRCRAFT FUEL SYSTEMS.

Check and service fuel dump systems.—Level 1:

The reasons for requiring fuel dump systems.

The methods used to control the operation of fuel dump chutes and valves.

The principal safety requirements for a fuel dump system.

The purpose of jettison pumps in fuel dump systems.

Perform fuel management, transfer, and defueling.—Level 1:

The precautions required when defueling an aircraft.

The tank-to-engine combinations possible with a crossfeed system.

The method of maintaing c.g. limits using fuel transfer technique.

The arrangement of fuel system controls, indicators, and warning lights.

Inspect, check, and repair pressure fueling systems.—Level 1:

The method of controlling fuel level during pressure fueling operations.

The methods used to operate fueling valves.

The protection of integral tanks against overpressure during pressure fueling operations.

The arrangement of fueling system controls, indicators, and warning lights.

The precautions required when fueling an aircraft.

The purpose and operation of pilot valves.

Repair aircraft fuel system components.—Level 2:

Repair and seal fuel tanks.

Pressure test fuel tanks.

Remove and clean fuel strainers.

The precautions to follow when routing fuel lines.

The method of regulating fuel system pressure.

Inspect and repair fluid quantity indicating systems.—Level 2:

The methods used to determine the level of fluid in a tank.

The purpose of remote-reading electrical gages.

Calibrate liquidometer-type fluid quantity indicating systems.

The effect of aircraft attitude on fluid level measuring devices.

Troubleshoot, service, and repair fluid pressure and temperature warning systems.—Level 2:

Determine and adjust the pressure or temperature at which warning systems operate.

Determine the cause of incorrect warning system indications and make corrections.

Test the operation of temperature and pressure warning systems.

Inspect, check, service, troubleshoot, and repair aircraft fuel systems.—Level 3:

The fuel system inspection requirements for aircraft operating in areas of high humidity or wide temperature changes.

The design and installation requirements for aircraft fuel tanks.

The maintenance requirements of fuel tank sumps.

The marking requirements for fuel filler openings.

The purpose of potassium dichromate in a fuel system.

The reason for using booster pumps with engine-driven pumps.

The purpose of baffle plates in fuel tanks.

The installation and operation requirements of fuel valves.

The venting requirements of interconnected fuel tanks.

G. AIRCRAFT ELECTRICAL SYSTEMS.

Repair aircraft electrical system components.—Level 2:

Use a "growler" in generator and motor armature inspection and repair.

Check the condition of shunt and compound generator field circuits.

Locate and use overhaul information for electrical equipment.

The procedures for correcting generator brush arcing.

Dress or turn the commutator surface of a motor or generator armature.

Flash a generator field.

The effect of changes in speed and load on generator output.

The function of a commutator in a direct current electric motor.

Seat new or replacement generator brushes.

The effect of incorrect generator brush spring tension on generator operation.

The methods of reducing armature reaction in aircraft generators.

The operating principles of carbon-pile voltage regulators.

The design factors that determine the number of cycles-per-revolution an alternating current generator will produce.

Determine the speed (r.p.m.) of an electric motor.

The basic principles of generators.

The factors that affect the torque produced by an electric motor.

The methods used to protect armature shafts from overloads.

The speed and load characteristics of series-, compound-, and shunt-wound motors.

The means employed to control current and voltage output of aircraft generators.

The methods used to control output frequency and voltage of alternating current generators.

The general operating characteristics of vibrator-type current and voltage regulators.

The purpose and operation of reverse-current cutout relays in generator control circuits.

The basic internal electrical circuits of series-, compound-, and shunt-wound generators.

The operating principles of magnetic clutches and brakes commonly used with electric motors.

The purpose and operation of reversible electric motors.

Install, check, and service airframe electrical wiring, controls, switches, indicators, and protective devices.—Level 3:

The purpose, applicability, and operation of electrical fuses and circuit breakers.

The types and operation of electrical switches.

Install and wire electrical switches.

Splice wiring in aircraft electrical systems.

The characteristics of high-tension and low-tension electrical wiring.

The purpose, applicability, and use of electrical wiring terminal strips.

The criteria for selecting aluminum and copper electrical cables.

Replace terminals on aircraft aluminum and copper electrical cables.

Determine the current-carrying capacity of an electrical circuit.

The installation and maintenance of open wiring electrical systems.

Install electrical wiring in conduits.

The method of protecting electrically operated emergency systems from accidental actuation.

The strength requirements for electrical cable terminals.

Select and install electrical bonding jumpers.

The installation requirements for electrical junction boxes.

The characteristics of single-wire electrical systems.

The special requirements an electrical bonding jumper must meet if it is required to carry a ground load for a unit of electrical equipment.

The purpose of shielding electrical wiring and equipment.

The use of quick-disconnect electrical plugs and sockets.

The purpose of static wicks or dischargers.

The American Wire Gage (A.W.G.) system of designating electrical wire sizes.

Inspect, check, troubleshoot, service, and repair alternating current and direct current electrical systems.—Level 3:

The results of short or open circuits in a generator control circuit.

The effect of sticking points in a reverse-current cutout relay.

The cause and effect of solenoid switch chatter.

The installation and circuit requirements for anticollision light systems.

The installation and circuit requirements for position lights.

The method of providing direct current for battery charging on aircraft that operate only alternating current generators.

The common methods of controlling output current and voltage of compound direct current generators.

The operating principles and characteristics of inverters.

Determine the output frequency of an alternating current generator.

The operating principles and characteristics of rectifiers.

The method of providing alternating current in aircraft that operate only direct current generators.

The electrical device usually used to convert alternating current to a lower or higher voltage without a change in frequency.

The operating principles and characteristics of electrical induction coils.

The operating principles and characteristics of transformers.

The advantages of using alternating current in aircraft.

H. POSITION AND WARNING SYSTEMS.

Inspect, check, and service speed- and takeoff-warning systems, electrical brake controls, and antiskid systems.—Level 1:

The general requirements for installing skid detectors.

The operating principles of hydraulic brake antiskid systems.

Inspect, check, troubleshoot, service, and repair landing gear position indicating and warning systems.—Level 3:

Determine the cause of a gear unsafe warning signal.

The effect of various electrical faults in the operation of the landing gear warning system.

I. ICE AND RAIN CONTROL SYSTEMS.

Inspect, check, troubleshoot, service, and repair airframe ice and rain control systems.—Level 2:

Install deicer boots.

The operating principles of anti-icing systems that utilize heated air in the leading edges of airfoils and intake ducts.

The operating principles of electrically operated anti-icing systems.

Protect deicer boots from deterioration.

J. FIRE PROTECTION SYSTEMS.

Inspect, check, and service smoke and carbon monoxide detection systems.—Level 1:

The operating principles of smoke and carbon monoxide detection systems.

Inspect, check, service, troubleshoot, and repair aircraft fire detection and extinguishing systems.—Level 3:

The type of fire-extinguishing agent most suitable for use with electrical fires.

The fire-extinguishing agent normally used with built-in aircraft fire-extinguishing systems.

1. GENERAL SUBJECTS

The material in this section of the manual covers the requirements for the General section of the FAA written tests. This particular part of the test is required for both airframe and powerplant ratings, but the applicant is *not* required to take it if he can show that he has previously passed it. Proof of passing may be in the form of a mechanic certificate with the powerplant rating or an Airman Written Examination Report for the powerplant rating that shows a passing grade on the General section.

All text material in this section is pertinent to maintenance, repair, and alteration of airframes and their related systems and components. The applicant for the airframe rating, therefore, is urged to study this section of the manual even though he may have already passed the General section of the FAA powerplant test.

A. BASIC ELECTRICITY

The airframe mechanic applicant must understand basic electricity and its application to aircraft in general. Pertinent phases of electricity are covered in this section. Additional material on electricity is included in Section 3, Airframe Systems and Components.

In general, aircraft use both *alternating current (AC)* and *direct current (DC)*. However, AC has become more predominant in large aircraft because of its advantages: easier installation, higher efficiency, greater adaptability to aircraft, and fewer maintenance problems. AC can be efficiently changed to DC (where needed) by the use of rectifier-transformers.

Alternating current may be described as current which periodically changes direction and continuously changes in magnitude. The current commences at zero and builds up to maximum in one direction, then falls back to zero, builds up to maximum in the opposite direction, and returns to zero (zero to peak to zero to peak to zero). Similarly, the voltage builds up to maximum in one direction, drops to zero, rises to a maximum in the opposite direction, and then returns to zero. Effective value determines the amount of power derived from alternating current.

Frequency (F). The number of cycles (zero to peak to zero to peak to

37

zero) occurring per second is known as the frequency of the current and is measured in cycles per second (CPS). Typical domestic electrical current operates on a frequency of 60 CPS. AC current in aircraft may operate on 400 CPS, 800 CPS or some other specified cycle. Alternations in AC move a half-cycle; thus there would be 120 alternations in any 60-cycle current.

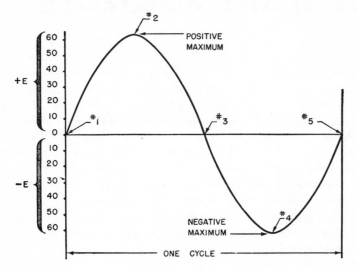

Fig. 1. Graph of AC voltage.

The *phase* of an AC or a voltage is the angular distance it has moved from zero to peak in a positive direction. The phase angle is the difference in degrees of actual rotation between two alternating currents or between a voltage and a current.

A common occurrence in AC systems is for the current to lag or lead the voltage.

When a capacitor (condenser) is in series in an AC circuit, it seems that the alternating current is passing through the capacitor. Actually, electricity is stored first on one side of the capacitor and then on the other side, permitting the alternating current to flow back and forth in the circuit without passing through the capacitor. The effect of a capacitor in an AC circuit is to cause the current to *lead* the voltage. If there was no resistance in an AC circuit, the current would lead the voltage by as much as 90°. The effect of *capacitance in AC circuits* is most pronounced as the frequency increases. Frequencies in electrical circuits have been increased to fantastic levels. Thus many special types of electrical equipment have been designed in order to reduce the effect of capacitance in electrical circuits.

The effect of capacitance in an AC circuit is the same in some aspects to that of electrical resistance, and is called *capacitive reactance*. Since it opposes the flow of current it is measured in ohms.

The effect of *inductance in AC circuits* is opposite to that of capacitance. Capacitance causes current to lead voltage and inductance causes the current to lag.

The end effect of inductance in an AC circuit is known as *inductive reactance* and is measured in ohms since it impedes (slows down) the flow of current in the circuit. The inductive reactance in a circuit is proportional to the inductance and the frequency of the alternating current. When inductance is increased, the induced electromotive force (EMF) which opposes the applied EMF is increased, hence the current flow is reduced. When the frequency of the current is increased, the rate of current change in the inductance coil is also increased. Thus, the induced opposing EMF is higher and the current flow is again reduced. It can now be noted that the effects of capacitance and inductance are opposite since inductive reactance increases as the frequency increases and the capacitive reactance decreases as the frequency increases.

In alternating current circuits it is necessary to consider capacitive reactance and inductive reactance before the net current in a circuit can be determined. The combination of reactance, capacitive reactance, and inductive reactance is called *impedance*.

Capacitance and inductance are known to have opposite effects in an alternating current circuit. The effect of one of these can be used to cancel the effect of the other. Remember that inductive reactance increases in proportion to the frequency. Capacitive reactance is inversely proportional to the frequency. Therefore, a frequency can be found that will balance any capacitance with any inductance—capacitive reactance of any capacitor may be equal to inductive reactance of any inductance coil at the proper frequency. At the time the capacitive reactance in a current is equal to the inductive reactance, the circuit can be described as *resonant*.

In addition to the current, voltage, and resistance of a circuit, the *power* must also be considered. Power is the rate of doing work and is generally measured electrically in watts (W). A *watt is the power* consumed in a circuit under a pressure of one volt. One horsepower is equal to 746 watts and one kilowatt (KW) is equal to 1,000 watts. Wattage of a circuit is easily found by multiplying the current (amperes) flowing through it by the voltage applied to the circuit. *Example:* an electrical motor using 70 amperes at a pressure of 24 volts develops 1,680 watts (70×24) or 1.68 kilowatts (1,680 ÷ 1,000), or 2.252 horsepower (1,680 ÷ 746).

ELECTRICAL MEASURING DEVICES, DIELECTRICS, AND CABLE CHART. During the inspection, maintenance, and operation of aircraft electrical equipment, the mechanic will often have to measure voltage, current (amperes) and resistance. Therefore, the aircraft mechanic should understand the use of the following measuring devices.

The voltmeter is used to measure the potential difference, or voltage, between two electrical points or stations. The voltmeter incorporates a galvonometer movement connected in series with a high-resistance unit. The purpose of the resistance is to limit the current flow through the meter movement. Most voltmeters give a full scale deflection with less than 0.01 of an ampere flowing through the movement.

Always connect voltmeters in parallel with the electrical unit to be tested; that is, across the unit or across the points between which the potential is to be measured. To obtain the correct direction of pointer deflection, connect the plus (+) terminal of the voltmeter to the positive side of the circuit or unit. *Do not connect* the voltmeter to a source of voltage which exceeds the voltmeter scale.

The ammeter is used to measure electrical current flowing in a circuit. Always connect ammeters in series with the load. To obtain the correct direction of pointer deflection, connect the plus (+) terminal of the ammeter to the positive side of the circuit. Never connect the ammeter across the terminals of a battery or generator as this will cause the ammeter to burn out immediately. An ammeter uses essentially the same movement as the voltmeter. However, instead of a current limiting resistor in series with the movement as in the voltmeter, the ammeter uses a *shunt* (a current by-pass) in parallel with the movement. The resistance of the shunt and movement are so proportional that most of the current flows through the shunt and only a small fraction through the movement, or just enough to give full-scale deflection when the total current is equal to the maximum range of the ammeter. Therefore, if a movement is used in a 300 ampere ammeter which requires 0.01 ampere to give a full-scale deflection, only 0.01 ampere will flow through the meter when 300 amperes is measured. The rest of the 300 amperes or 299.9 amperes will flow through the shunt. Generally, the shunt part of the ammeter is installed in a junction box and connected in either the positive or negative generator lead. The movement is mounted in a case in the instrument panel.

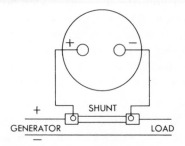

Fig. 2. External wiring diagram.

The loadmeter is merely an ammeter, but instead of indicating amperage of the electrical system it indicates percent of rated output being utilized. The loadmeter instrument card may read from .1 to 1.25. If the loadmeter for a generator were to read 1.00, the generator would be operating at 100% of its rated load. If the reading is more than 1.00, this would indicate that the particular generator is overloaded.

An ohmmeter measures resistance in *ohms*. It is composed of a meter movement, a network of resistors, and a small dry cell battery as a source of voltage. Since the voltage supplied to the ohmmeter is constant, the current through its meter movement, and hence the deflection of the pointer, depends on the resistance of the unit being tested. The meter, therefore, indicates the resistance of the unit measured. It is calibrated to read this resistance directly. Never connect an ohmmeter into a live circuit (one that has voltage applied) as this will burn it out.

A thermocouple thermometer is an excellent unit for measuring temperatures such as piston engine cylinder head temperature (CHT), turbojet engine exhaust gas temperature (EGT), etc. One of the important advantages of the thermocouple is that it needs no outside source of power and is self-sustained.

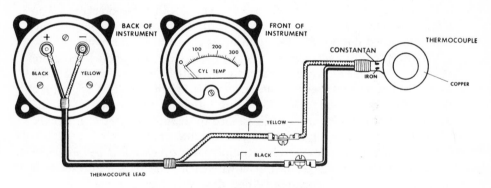

Fig. 3. Thermocouple thermometer.

The thermocouple thermometer consists of a thermocouple, connecting leads, and an indicating meter or instrument. The thermocouple itself is made of copper and incorporates two dissimilar metals joined together—generally constantan and iron. The thermocouple thermometer utilized to measure the CHT of a piston-propeller engine consists of a copper gasket that replaces the regular spark plug gasket. Attached to the copper gasket is the constantan and iron. The cylinder head temperature is conducted to the copper gasket, hence to the constantan and iron where unlike small electrical impulses are registered and sent through the connecting lead wires to the meter on the instrument panel. The meter receives the electrical impulses and converts the electrical impulses to a temperature reading. The meter is actually a delicate voltmeter calibrated in *degrees* rather than *volts.*

The principles involved in a *galvanometer* are employed in most voltmeters and ammeters used in electrical circuits. This instrument is basically utilized to detect the presence of an electric current in a circuit and to determine the current's direction and relative strength.

Electrical prefixes. There are several prefixes utilized to indicate a precise electrical value. Among these are micro, mega, kilo, and milli. Micro indicates one one-millionth, mega indicates one million times or $1 \times 1,000,000$, kilo indicates 1,000 times, and milli indicates one one-thousandth. Electrical terms or values that utilize these prefixes are micro-farads, mega-volts, kilo-watt, and milli-amperes.

Dielectrics, in general, are materials that will not conduct an electrical current. The specific inductive capacity, or the dielectric constant, of various insulating materials is the ability of these materials to have electric lines of force created or produced in them compared to the lines of force that would be produced in air. The following gives some values of specific inductive capacities of several insulating dielectrics relative to air: a vacuum 0.94; oil 2.2 to 4.7; rubber 2.0 to 3.5; mica 2.5 to 6.7; porcelain 5.7 to 6.8; water 81.0. Water has an exceptionally high dielectric constant but cannot be used due to the ionization of its electrical conductivity.

Dielectric substances contain few, if any, free electrons, which therefore resists the passage of an electric current. A dielectric can thus be used as an insulating material for cables, etc., or for the medium separating the electrodes of a condenser. Dielectric fatigue is the property of some dielectrics losing their power of resistance to disruption after a difference of potential has been applied for a considerable time.

Electrolytic corrosion results from the *electro-chemical action* when dissimilar metals are placed in contact with each other. This is also referred to as *dissimilar metal corrosion.* When dissimilar metals come in contact with each other a suitable insulation should be used to prevent this type of corrosion.

Shielding is a complete enclosure of the parts of an electrical system that is inter-connected and properly grounded. Metal housing is commonly used for this purpose and when adequate shielding is properly installed it reduces, to a great extent, electrical corona, electrical leakage, and radio interference (static interruption).

Standard *electrical cable charts* are utilized to determine the length, size, ampere load, and uses to which a given cable can be utilized. The chart is relatively easy to use if certain basic rules are followed. It is suggested that you read the following cable chart rules (one by one) and apply each

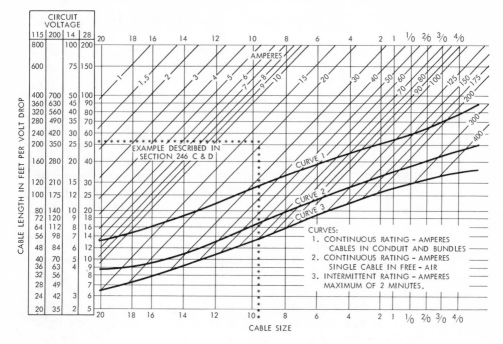

Fig. 4. Electric cable chart.

rule to the accompanying illustration. The rules end with a typical cable chart problem.

1. This chart is for copper electrical cable.

2. Electrical cable sizes are at the top and bottom of the chart and given in sizes from 20 to 4/0, 20 being the smallest and 4/0 the largest.

3. Cable lengths appear to the left of the chart with the shortest length at the bottom and longest length at the top.

4. Ampere value lines run diagonally upward from the left toward the right side of the chart. Ampere values vary from one ampere (upper left) to 400 amperes (upper right).

5. The "curve lines" 1, 2, and 3, run diagonally across the chart from lower left to middle right. These curve lines indicate the use to which the electrical cable can be installed and the corresponding number is explained at the lower right of the chart.

6. If a copper cable is to be replaced by aluminum cable, it is necessary to use an aluminum cable *two numbers* (not two sizes) smaller than the copper cable number. Thus, the aluminum cable that is used will be larger (in diameter) than the copper cable. This is necessary since aluminum cable has more resistance (per given diameter) than

copper cable. Do not, however, use an aluminum cable smaller than size #6.

7. The cable chart is based on a one-volt drop (see note at left of chart). Hence, if the problem (or circuit) calls for a volt drop of more than one, the length of the cable must be shortened accordingly before going into the chart to solve the problem. *Example:* if the length of the cable is to be 100 feet and the volt drop of the circuit is a two volt drop, divide 100 by 2 which equals 50 feet. Then go into the chart at 50 feet to solve the particular problem; if the length of the cable is to be 100 feet and the volt drop of the circuit is four, divide 100 by 4 and go into the chart at 25 feet; if the length of the cable is to be 100 feet and there is a .5 volt drop, the reverse procedure must be used. One hundred divided by .5 is 200; thus you would go into the chart at 200 feet to solve the problem.

8. If the original requirement calls for #8 aluminum cable, you must go into the chart at #10 size (copper) or two numbers higher to arrive at a correct conclusion.

9. After finding the correct intersection of a given cable length and ampere value, if the intersection is to the right of a given size cable, go to the next larger size cable (next smaller number) to the right for the proper cable size. Otherwise, if you go to the left for the cable size, the cable will be too small for the intended electrical load and possibly overheat.

10. Remember, the smallest aluminum cable to be used in aircraft is cable size #6.

Typical problem: The required cable length is 75 feet, volt drop is 3, and ampere load is 30 amperes. What is the required size of copper cable? Go into the chart at 25 feet (75 divided by 3), then to the right to 30 amperes. The point of intersection is to the right of size #12. Thus you will have to use copper size #10 or aluminum cable #6. The intersection is also below the curve 1 line; thus the cable cannot be used for curve 1 purposes (continuous rating—amperes, cables in conduit and bundles). This cable can be used for curve 2 and 3 purposes. It is suggested that for practice you try the following cable chart problems. Correct answers follow the problems.

1. What size copper cable will be required for a length of 70 feet, a one-volt drop, and an ampere load of 100 amperes?

2. Can this cable be used for a continuous rating in conduits and bundles?

3. The length of a required copper cable is 50 feet, a volt drop of 2, and the size of the cable is #12. How many amperes will this cable allow?

4. The length of a required copper cable is 30 feet, the volt drop is

3, and the cable size is #16. The maximum amperes this cable can carry is?

5. The cable in the preceding question can be used for what curve purposes?
6. A copper cable length is 100 feet, volt drop is .5 (½), and the cable size is #10. The cable will carry how many amperes maximum?
7. You must substitute a #6 copper cable for aluminum cable. You would use aluminum cable #?
8. You plan to use #2 aluminum cable. To obtain the proper cable information you would refer to which copper cable size?

Correct answers: (1) 1/0; (2) yes; (3) 22 amperes; (4) 20½ amperes; (5) #2 and #3; (6) 4½ amperes; (7) #4; (8) #4.

Note: Additional electrical data will be found in Section 3, Airframe Systems and Components.

AIRCRAFT STORAGE BATTERIES. A basic source of electrical energy in aircraft is the storage battery. It stores electricity by means of a chemical reaction. The *lead-acid storage battery* is made with a given number of cells. Each cell is composed of an acid-proof "jar" that holds a number of lead plates. There are two types of plates, one containing spongy lead and the other a chemical known as lead dioxide. The plates are prevented from touching by means of porous insulating separators. The spongy lead is negative and the lead dioxide plate is positive. There are always an uneven number of negative plates in a cell. All similar plates are connected to a common terminal. The plates and separators are immersed in a solution of sulphuric acid and water which is termed the electrolyte.

Storage batteries of the lead-acid type are prepared for use by allowing a low voltage of direct current (DC) to flow into the battery through the positive terminal. This is the basic process of charging a battery. The current flowing into the battery causes chemical changes in the materials of the plates. The charging electrical energy is thus converted to chemical energy. When the charging is completed, the battery is ready for connecting into a circuit. The conditions are then reversed when drawing electricity during the use of the battery. The chemical energy is converted to electrical energy and there is thus a flow of current through the circuit when needed. This action causes discharging of the battery.

When a cell or cells are being discharged, hydrogen and oxygen are being liberated. *Excessive gassing* occurs when a cell or cells are overcharged, charged, or discharged at too high a rate. These gasses are explosive. Therefore, when servicing, charging, handling, or installing storage batteries no smoking should be allowed. Caution must be exercised to prevent sparks and other phenomena that may prove hazardous. Battery acid is a poison that is highly destructive to clothing, and very harmful to the skin if not washed off immediately. If battery acid comes in contact with the skin, apply a mild antiseptic after washing with water.

The voltage of each battery cell is two volts (actually 2.2 volts when in good charge). Thus a 12-volt battery will have 6 cells and a 24-volt battery will have 12 cells.

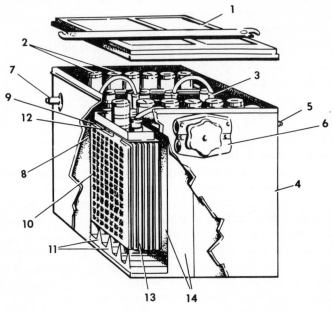

1. Metal cover	8. Cell container
2. Filler cap and vent plug	9. Positive plate group strap
3. Cell connectors	10. Plate
4. Metal container	11. Plate supports
5. Vent	12. Negative plate group strap
6. Quick-disconnect receptacle and plug	13. Separators
7. Vent	14. Cells

Fig. 5. Typical aircraft storage battery.

The rated capacity of a battery depends on the number, area, and thickness of the plates in each cell. Storage batteries are rated in ampere-hours. This indicates the number of amperes the battery will deliver over a period of hours. A 45-ampere-hour battery will deliver 9 amperes for 5 hours, a 60 ampere-hour battery will deliver 1 ampere for 60 hours, etc. In other words, any combination of amperes and hours that total the ampere-hour rating will indicate the output of the particular battery. This, of course, indicates a static battery or one that is not under charge at the time the ampere-hour draw is made.

The condition of charge is measured by a hydrometer. The specific gravity of the acid (electrolyte) is ascertained and from this reading it can easily be determined whether the particular battery is in good charge, medium charge, or poor charge and ready for the battery charging line. The reading of a new fully charged battery will be 1.300. The electrolyte at this value is 1.3 times the weight of water as the specific gravity of water is 1.000. A fully discharged battery has a specific gravity of 1.150. A battery should be recharged when the reading of one cell is 1.240 or lower.

Due to the extreme heavy drain of battery current when starting an aircraft engine, especially the larger types, the battery should not be used if possible. A battery cart or other auxilliary ground power unit should be used.

A storage battery loses water (due to evaporation) when in use, and so it's important that clean (distilled) water be added to the electrolyte to a proper level in the battery cells.

During extremely cold or below freezing weather, storage batteries should be kept in a warm place. A *discharged* battery will freeze at plus 5°F. However, a fully charged battery will not freeze until the temperature drops below minus 80° to minus 90°F. When not in use, batteries tend to lose their charge. Thus, they should be checked frequently and recharged when necessary.

Adequate ventilation must be provided for storage batteries regardless of their location in the aircraft. This reduces the fire hazard and insures that the harmful battery fumes will not infiltrate the aircraft interior.

In the *nickel-cadmium* type of storage battery the positive plates consist of a number of perforated steel tubes that are welded together. Each of the tubes is filled with nickel peroxide. The negative plate is constructed of many small perforated steel pockets; each pocket is filled with pure powdered iron.

Similar to the lead-acid battery, alternate positive and negative plates, separated by rubber insulation, are stacked together. All positive plates are connected together and all negative plates are connected together. The electrolyte is a solution of potassium hydroxide in water. The perforations in these tubes and pockets of the plates permit the electrolyte to come in contact with the active ingredients.

Housed in a nickel-plated steel container are the assembled plates. The steel container is utilized because there is no acid to attack it. In fact, the potassium hydroxide of the electrolyte helps to protect the steel container from corrosion. The posts, both positive and negative, extend from the top of the container through insulated bushings. A vent is incorporated through which not only gases can escape but distilled water can be added as the battery requires.

Each cell can produce an electromotive force (EMF) of about 1.9 volts when fully charged, and this falls to 0.9 volt when the cell is discharged.

The advantages of the nickel-cadmium battery as compared to the lead-acid battery are: (1) it is stronger, (2) electrolyte undergoes very little change, (3) less maintenance, (4) it can remain in a low-charge state without damage, (5) the freezing point is well below any temperature likely to be encountered, and (6) it does not produce fumes that are harmful or corrosive.

B. AIRCRAFT DRAWINGS

Aircraft drawings are based on the same principles as other technical drawings. However, there are some drawings that may require special techniques, symbols, and applications. The aircraft engineer and draftsman generally work together in preparation of a given set of drawings even though there may be a wide gap in the knowledge that each possesses.

Conventional representation in aircraft drawings must conform to those known and used by the trade to insure an over-all understanding and consistency for those concerned with the project. In some instances, aircraft mechanics may have to make drawings.

There are, in general, three major types of drawings. They are (1) major and sub-assembly drawings, (2) detail or working drawings, and (3) illustrated or expanded drawings.

The basic tools of the aircraft draftsman are generally well known. They are: (1) the drawing board, (2) drawing paper, (3) drawing set (instruments), (4) tracing paper, (5) T square, (6) triangles, (7) drafting tape, (8) protractors, and (9) a good source of light.

Drawing lines are somewhat standardized and each has a particular meaning. Refer to the accompanying illustration for drawing line familiarization.

White or buff drawing paper may be used. However, if blue (or black and white) prints are desired, the drawing should be made on transparent paper, or the original drawing will have to be traced later on transparent paper. Drawing the original projects on tracing or transparent paper can save time and expense.

The drawing set (drawing instruments) generally consists of ink ruling pens, compasses, dividers, etc. The average set is a valuable aid in drawing many intricate patterns and lines.

It is important that the aircraft mechanic be able to properly visualize an object as shown on a drawing or blueprint. The accompanying illustration is a good example of drawing visualization.

BORDER LINES: (VERY HEAVY)
HEAVIEST LINE ON DRAWING.

OUTLINE OF PART: (HEAVY)
THE PART OUTLINE IS THE OUTSTANDING FEATURE OF THE DRAWING.

HIDDEN LINES: (HEAVY)
THIS LINE IS MADE UP OF SHORT DASHES AND INDICATES A LINE OR
LINES HIDDEN FROM VIEW BY SOME PORTION OF THE PART.

CROSS SECTION LINES: (HEAVY)
THESE LINES SIGNIFY MATERIAL THAT HAS BEEN CUT BY A PLANE.
LIGHT DIAGONAL LINES ARE SO SPACED AS TO GIVE A SHADED EFFECT.

BREAK LINES: (HEAVY)
ON SHORT BREAKS A FREEHAND LINE IS USED. ON LONG BREAKS A RULED
LINE AND FREEHAND ZIG-ZAG LINES AT SPACED INTERVALS.

ADJACENT PARTS: (HEAVY)
BROKEN LINE, MADE UP OF ONE LONG AND TWO SHORT DASHES.

CUTTING PLANE: (MEDIUM)
A BROKEN LINE MADE UP OF ONE LONG AND TWO SHORT DASHES.

CENTER LINES: (MEDIUM)
MADE UP OF LONG AND SHORT DASH LINES.

DIMENSION & EXTENSION LINES:(LIGHT)
UNBROKEN EXCEPT AT DIMENSION.

LINES OF MOTION OR ALTERNATE POSITION:
BROKEN LINE MADE UP OF LONG DASHES.   (LIGHT)

Fig. 6. Blueprint lines.

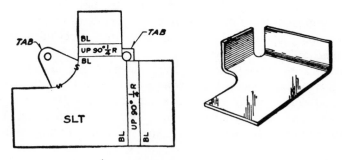

Fig. 7. Template with two tabs.

To achieve good lettering on drawings, these facts must be considered: (1) the instrument with which letters are to be made, (2) form or shape of the letter, and (3) techniques employed to execute good letters.

The information generally present in a drawing or blueprint box is as follows:

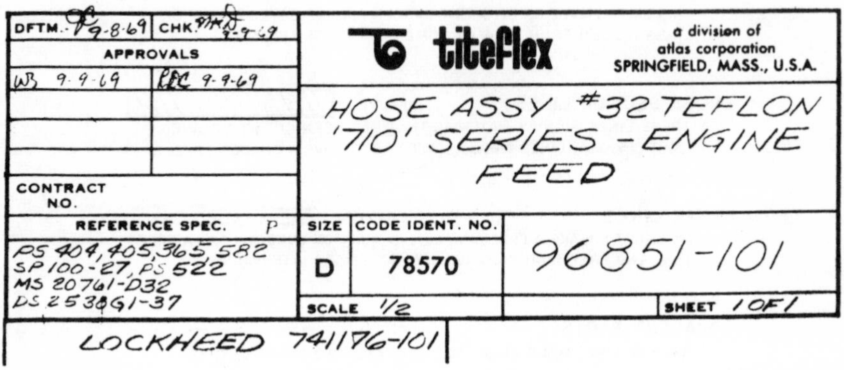

DFTM.- *9-8-69*	CHK.*9-7-69*		**ᏫᎾ titeflex**		a division of atlas corporation SPRINGFIELD, MASS., U.S.A.
APPROVALS					
WB 9-9-69	*REC* 9-9-69		*HOSE ASSY #32 TEFLON*		
			'710' SERIES – ENGINE		
			FEED		
CONTRACT NO.					
REFERENCE SPEC. P		SIZE	CODE IDENT. NO.		
PS 404,405,365,582 *SP 100-27, PS 522* *MS 20761-D32* *DS 25306/-37*		D	78570	*96851-101*	
		SCALE *1/2*		SHEET *1 OF 1*	

LOCKHEED 741176-101

Fig. 8. Blueprint box.

Drawing and blueprint principles should be known by the aircraft mechanic. It is not only essential to read drawings and blueprints, but he may have to make drawings himself when it is necessary to execute a change or modification not previously specified. An example of this would be the presentation of data, sketches, drawings, etc., on some new or different modification of an aircraft that must be approved by the FAA Engineering Section.

In instances where a number of similar parts are needed, the mechanic will work from a template. *Templates* are patterns used in transferring the required dimensions to a number of pieces of material or stock. The basic purpose of working from a pattern (template) is to achieve a higher degree of uniformity in the work being accomplished and to save time.

Other drawing accessories. In addition to the basic drawing tools previously mentioned, there are a number of other items the mechanic may use when doing layout work, etc. They are scribes, steel rules, measuring tapes, punches, and French curves.

Layout work. There are basic steps the mechanic should take when making a layout of wood or metal from a drawing or blueprint. He should find a good working surface such as a flat bench or table. When working with large sheets of material it is important to avoid bending. A helper should be utilized when placing large sheets on the working surface. The bottom of the material should be protected by placing heavy paper or felt between the material and the working surface.

There are several *layout fluids* available for layout work. It is important that the pattern placed on the material by drawn lines stands out clearly while the mechanic is cutting along these lines. Flat white paint, bluing fluid, and zinc chromate are considered to be good layout fluids.

Lines drawn with metal scribes are satisfactory providing the scribed lines on the material stand out prominently. This is sometimes a problem, especially when the material to be scribed is of a neutral color.

Layout planning. Mechanics should closely examine the blueprint or drawing from which a material layout is to be transferred and plan his work accordingly. Where a part or unit required is small, it may be possible to make it from a piece of satisfactory scrap metal. If satisfactory scrap is not available, it is advantageous to cut the small part from a corner of a large sheet rather than from an inner section. This saves a larger portion of the remaining sheet for future jobs.

The layout on the material should be completed with the least amount of damage to the material such as scratches, dents, etc.

C. WEIGHT AND BALANCE

The aircraft mechanic must know and understand the principles of weight and balance of the airplane. Load distribution is important to safe flight, and if not properly carried out may contribute to hazardous instability.

Weight and balance computation is not a difficult subject nor does it require extensive study. Once the basic principles and computations are learned, weight and balance becomes an interesting subject and a challenge.

The weight and balance problems in the FAA examination are general in nature and do not require the use of specified aircraft weight and balance computers or graphs, as used by airline personnel. In view of the above, this section does not require them either.

The following definitions are common to all methods of weight and balance control and should be studied:

The maximum authorized weight of the aircraft and its contents is the *gross weight* and is listed in the particular aircraft specifications.

Empty weight includes all operating equipment that has a fixed location and is actually installed in the aircraft.

The *useful load* is found by subtracting the empty weight from the maximum gross weight of the aircraft. It consists of the crew, maximum oil, fuel, passengers, baggage, cargo, etc., unless otherwise noted. The useful load consists of both the pay and non-pay load.

Aircraft weight check consists of checking the sum of the weights of all items of useful load against the authorized useful load (maximum weight less empty weight) of the aircraft.

Datum is an imaginary vertical plane or line from which all horizontal measurements are taken for balance purposes with the aircraft *in level flight position.* The datum may be at the leading edge of the wing, nose of the aircraft, or in a position in front of the nose. If not specified, the datum may be located in any convenient position on the aircraft.

Arm (or moment arm) is the horizontal distance, in inches, from the datum to the center of gravity of an item. A plus (+) arm indicates the item is located aft of the datum. A minus (−) arm indicates the item is located forward of the datum.

The moment of an item about the datum is obtained by multiplying the weight of the item by its horizontal distance from the datum.

Center of gravity is generally known as the balance point of an aircraft. It is an imaginary point about which the nose-heavy and tail-heavy moments are exactly equal in magnitude.

Empty weight center of gravity. The center of gravity of an aircraft in its empty weight condition.

Empty weight center of gravity range is determined so that when the empty weight center of gravity falls within this range the specification operating center of gravity limits will not be exceeded under standard specification loading arrangements.

Operating center of gravity range is the distance between the forward and rearward center of gravity limits shown in the aircraft specifications.

Mean Aerodynamic Chord (MAC) is the mean chord of the wing. For weight and balance purposes it is used to locate the center of gravity range of the aircraft. MAC data will be found in the aircraft specifications, flight manual, or aircraft weight and balance records.

Weighing point. When required to locate the center of gravity by weighing, it is first necessary to obtain horizontal measurements between the points on the scales at which the airplane's weight is concentrated. In usual weighing practice, a vertical line passing through the centerline of the landing gear axles will locate the point on the scales at which the weight is concentrated.

METO: maximum except takeoff horsepower.

Minimum fuel for weight and balance purposes. To determine the weight of fuel in pounds (fuel weighs 6 lbs. per gallon), multiply METO horsepower by .55 and divide the sum by 2.

Full oil is the quantity of oil shown in the aircraft specifications as *oil capacity.*

Tare. The weight of chocks, blocks, etc., used when weighing aircraft; included generally in the scale readings. Tare must be deducted from the scale reading to obtain the actual aircraft weight.

Unit weights. Fuel—6 pounds per U.S. gallon; oil—7½ pounds per U.S.

gallon; water—8 pounds per U.S. gallon; crew and passengers—170 pounds per person.

Algebraic signs. In computing weight and balance, care must be taken to insure retention of the proper algebraic sign (+ or -). Always visualize the aircraft with the nose to the left.

Determination of algebraic signs (+ or -). Any weight item added to the aircraft either side of the datum is a plus (+) weight. Any weight item removed is a minus (-) weight. Visualizing the nose of the aircraft to the left, any arm to the left (forward) of the datum is a minus (-). Any arm to the right (rearward) of the datum is a plus (+).

Weight and balance extreme conditions. The maximum forward and rearward center of gravity position for the aircraft.

A complete *list of the equipment* included in the certificated empty weight is found in either the approved airplane operating manual or the weight and balance data.

Equipment changes. The owner(s) of an aircraft is responsible for a continuous record of each aircraft, listing all changes affecting the weight, center of gravity location, and equipment, in order that a computed weight and center of gravity location may be established at any time.

Installation of ballast. Ballast is sometimes permanently installed for center of gravity balance purposes resulting from the installation or removal of equipment items and should not be used to correct a nose-up or nose-down tendency of the aircraft.

The *loading schedule* should be kept with the aircraft and usually forms a part of the airplane flight manual.

Weighing an aircraft for the purpose of determining the empty weight center of gravity is an important function and should be undertaken only by experienced personnel. The general procedure is as follows, but it may vary depending on the type of aircraft and conditions at the time of weighing:

(1) The weighing should take place inside a closed hangar to prevent error in scale readings due to wind or other related weather. Place the aircraft in level attitude and check that the scales are correct, then level, set at zero, and note the tare prior to placing aircraft on the scales.

(2) Ascertain that the fuel tanks are empty. Oil tanks may either contain the number of gallons noted on the filler cap or they may be drained. The weight and balance reports must indicate whether the weights include full oil or oil drained.

(3) Equipment and special items such as tools, anchors, ropes, etc., must be in place if they are to be carried as standard aircraft equipment.

(4) Examine, on completion of the above steps, the aircraft and weighing equipment to be sure everything is in order.

COMPUTING THE EMPTY WEIGHT CG. In computing the empty weight center of gravity of small aircraft, the procedure of weighing is the same as for large aircraft as indicated in the preceding steps. However, in computing the location of the empty weight center of gravity the placement of the datum determines the method of computations.

The datum is a reference line in a vertical plane from which all horizontal measurements are taken for the purpose of establishing balance with the airplane in level flight attitude. Any convenient place on the airplane may be used as a datum position. However, the airplane manual generally notes the exact position to be used. The datum on light aircraft is generally located at the leading edge of the wing and on large air transport type aircraft the datum will be found at the nose or at a convenient position ahead of the nose.

On light aircraft with the datum at the leading edge, moment arms will be plus or minus depending on whether the location is behind or ahead of the datum. Arms behind the datum will be plus (+) and arms ahead of the datum will be minus (−).

After the airplane has been weighed, the formula used to find the empty weight center of gravity is determined by whether the aircraft is of the nose-wheel or tail-wheel type and whether the datum is located forward or rearward of the main wheels.

The following is a typical problem explanation to establish the empty weight and the empty weight center of gravity of a nose-wheel type of airplane.

The airplane has been placed on the scales (one for each wheel) and the scale readings have been placed in a typical schedule arrangement with the tare noted. Subtracting the tare from the scale readings gives the net weight at each of the three wheels. This results in a total net weight for the airplane (1,673 pounds) and is given the designation W. The front wheel net weight is 454 pounds and is given the designation F. The distance from centerline to centerline of the main and nose wheels is designated L and the distance from the centerline of the main wheels to the leading edge is designated D. Summarizing:

D—distance centerline of main wheels to datum = 34″.
F—net weight of the nose wheel = 454 pounds.
L—distance from centerline of main wheels to centerline of nose wheels
    = 67.8″.
W—net empty weight of the airplane = 1,673 pounds.
With the above known facts, the following formula is needed:

$$\text{Empty weight center of gravity (EWCG)} = D - \left(\frac{F \times L}{W}\right)$$

$$\text{EWCG} = D(34'') - \left(\frac{F(454 \text{ lbs.}) \times L(67.8'')}{W(1,673 \text{ lbs.})}\right)$$

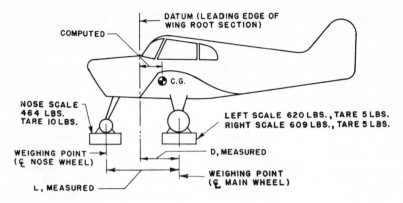

DATUM (LEADING EDGE OF WING ROOT SECTION)

COMPUTED

C.G.

NOSE SCALE
464 LBS.
TARE 10 LBS.

LEFT SCALE 620 LBS., TARE 5 LBS.
RIGHT SCALE 609 LBS., TARE 5 LBS.

WEIGHING POINT
(℄ NOSE WHEEL)

D, MEASURED

WEIGHING POINT
(℄ MAIN WHEEL)

L, MEASURED

TO FIND: EMPTY WEIGHT AND EMPTY WEIGHT CENTER OF GRAVITY

Datum is the leading edge of the wing (from aircraft specification)
(D) Actual measured horizontal distance from the main wheel weight point (℄ main wheel) to the Datum--- 34.0"
(L) Actual measured horizontal distance from the front wheel weighing point (℄ front wheel) to the main wheel weighing point--------------------------------- 67.8"

SOLVING: EMPTY WEIGHT

Weighing Point	Scale Reading #	Tare #	Net Weight
Right	609	5	604
Left	620	5	615
Front	4 64	10	454
Empty Weight (W)			1673

SOLVING: EMPTY WEIGHT CENTER OF GRAVITY

Formula: $C.G. = D - \dfrac{F \times L}{W} = 34" - \dfrac{454 \times 67.8}{1673} = 34" - 18.3" = 15.7"$

Fig. 9. Empty weight and empty weight center-of-gravity—nose-wheel aircraft.

EWCG = 34″–18.3″ = + 15.7″ (plus value due to being aft of datum).

EWCG = + 15.7″ (distance from leading edge datum to position of EWCG).

It is important to understand that the part of the formula $\dfrac{F \times L}{W}$ actually gives the distance from the main wheel centerline to the EWCG. Thus 34″ (distance from centerline of main wheels to the datum at the wing leading edge) minus 18.3″ gives the EWCG distance from the datum at the wing leading edge of 15.7″.

Having found the EWCG of this particular airplane we can proceed with the second step in weight and balance procedures by loading this airplane (using the original empty weight data) and finding the CG in the loaded condition. After establishing the exact weights of items to be placed at given arms (distance) from the datum, a typical loading schedule is made up as follows:

Item	Weight (lbs.)	X	Arm (ins.)	=	Moment (inch lbs.)
Aircraft empty	1,673.0		15.7		26,266
Oil	17.0		-24.0		-408
Fuel (40 gals.)	240.0		22.0		5,280
Pilot	170.0		15.0		2,550
Passengers (2)	340.0		47.0		15,980
Baggage	50.0		72.0		3,600
	2,490.0 (TW)				53,268 (TM)

Fig. 10.

To find the new center of gravity, the following standard formula is used:

$$CG = \frac{\text{total moments (TM)}}{\text{total weights (TW)}}$$

$$CG = \frac{53,268}{2,490} = 21.39$$

Thus the new CG is 21.39″ aft of the datum at the leading edge. *Note:* plus (+) signs are omitted, but minus (–) signs are used and computed.

Since the original loading data of this aircraft permits a rearward CG position 21.90″, the above loading places the CG within allowable limits. *Note:* The total moment is always found by adding the plus moments and then subtracting any minus moment values such as the "–408."

The maximum gross weight permitted on this particular aircraft is 2,503 pounds. The total weight (gross) as loaded is 2,490 pounds. Thus the gross weight as loaded is within limits.

Having established the above facts, the mechanic and/or pilot in charge of this aircraft would clear it for flight from a weight and balance standpoint.

Once the EWCG is established and found to be within the particular aircraft's limitations, weights may be added and/or removed, a new loading schedule made up, and a new total weight and total moment established. Then proceed as before to find the new CG using the formula: CG = TM ÷ TW. Always check the results against the particular aircraft's gross allowable weight and the center of gravity forward and rearward limits.

In many cases the CG position is given in percent of Mean Aerodynamic Chord (%MAC). The MAC is known as the distance from the leading edge of the wing to the trailing edge and is measured in inches. To find the %MAC you must know the distance in inches from leading edge to trailing edge of the wing, and the CG in inches from the leading edge. *Example:* The MAC is 100 inches and the CG is 25 inches aft of the leading edge. Obviously, the CG is ¼ aft of the MAC (100 ÷ 25) or 25% of the MAC (25% MAC). The %MAC is generally found by dividing the MAC into the CG distance. *Example:* 25 ÷ 100 = .25; remove the decimal and the value becomes 25%.

In the airplane whose weight and balance was previously computed, let's assume the MAC to be 78″. Knowing the new CG to be 21.39″ aft of the leading edge, the %MAC can be found by dividing 21.39″ by 78″. The result is .2741, or 27.41% by moving the decimal two places to the right. The %MAC would thus be 27.41, and is generally found to the nearest two decimal points.

BAR-TYPE WEIGHT AND BALANCE PROBLEM. The bar-type of weight and balance problem illustrates, in a simplified manner, the principles of aircraft weight and balance procedures. The bar typifies an airplane since the datum is placed well to the left of the center of the bar. This is similar to the airplane because the center of gravity is generally located forward or just aft of the leading edge of the wing. It is well to stress here that all bar-type weight and balance computations are done with the nose of the airplane to the left, and the wing and engine also to the left of the bar's center where the bulk of the airplane weight is concentrated.

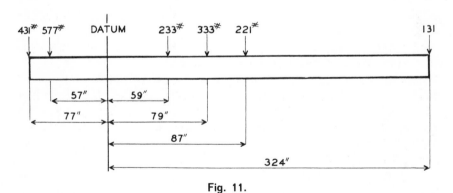

Fig. 11.

In solving for the center of gravity of the above bar problem we must find the total of the weights and moments created by each weight at its particular location from the datum. Once the total weight and moments are known, the formula $CG = \dfrac{\text{total moment}}{\text{total weight}}$ can be used.

The following solution to the problem is such that all figures and computations are organized in such a manner as to eliminate errors. Common errors are generally caused by carelessness in basic arithmetic and disorganized computations.

Item	Weight (lbs.) X	Arm (ins.) =	Moment (inch lbs.)
1	431	−77	−33,187
2	577	−57	−32,889
3	233	59	13,747
4	333	79	26,307
5	221	87	19,227
6	131	324	42,444
	1,926 (TW)		35,649 (TM)

Fig. 12.

The total net moment is the sum of the plus and minus values. Plus values are not preceded by a plus (+) sign, but minus values are *always* preceded by a minus (−) sign.

Applying the formula $CG = \dfrac{\text{total moment}}{\text{total weight}}$ we find that the center of gravity of the bar $= \dfrac{\text{(TM) } 35,649}{\text{(TW) } 1,926} = 18.5''$ from the datum. Since the total of the plus moments is greater than the total of the minus moments the center of gravity is aft of the datum.

In the bar problem just completed, it is obvious that any change of weights and/or moment arms can be made for any specified condition of loading. Regardless of loading changes, one has only to use the above method of computation to arrive at a new total weight and total moment and then use the formula $CG = TM \div TW$ to find the distance of the new CG from the datum.

Finding new EW and EWCG after engine change. This type of weight and balance problem is quite common and can be easily solved by using the same basic principles as explained previously.

The airplane has an EW of 1,321 lbs. and an EWCG of 38.15″ with its original engine installed that weighs 265 lbs. and located at station −30.00″. You have replaced the original engine with one that weighs 285 lbs. at the same station. After completion of the engine change (assuming no other weight changes), what is the new EW and EWCG? (Remember, in solving problems the plus sign is omitted.)

Item	Weight (lbs.) × Arm (ins.)		Moment (inch lbs.)
Original aircraft empty	1,321	38.15	50,396
Removal of engine	−265	−30.00	7,950
Installation of new engine	285	−30.00	−8,550
	Total weight 1,341 (TW)	Total moment	49,796 (TM)

The new EW is 1,341 lbs. and the new EWCG is 37.13″ (EWCG $= \dfrac{TM}{TW} = \dfrac{49,796}{1,341} = 37.13″$)

D. FLUID LINES AND FITTINGS

Aircraft require both rigid and flexible fluid lines and fittings. Their basic purpose is to permit the flow of fluids such as fuel, oil, hydraulic, etc., to essential parts and units of the aircraft. The lines (tubing) and associated fittings must be of high quality to insure against engine and other component failures.

The materials used for metal tubing can be aluminum, aluminum alloys, copper, steel, etc. Flexible lines can be made of natural rubber, synthetic rubber, etc., and used where there will be relative motion between the ends of metal tubing or vibration prone areas of the aircraft.

Aircraft systems that require the use of rigid or flexible lines or tubing are quite numerous. A few examples are: fuel systems, lubrication systems, oxygen systems, hydraulic systems, heating systems, etc. Tubing can also be used for the routing of electrical wiring. The tubing, in this instance, gives protection to the wiring and also acts as a shielding to reduce static waves that cause radio interference (static disturbance).

Several factors are to be considered in the selection of tubing, flexible hose, and fittings for the various aircraft installations. These factors are strength, minimum weight, and resistance to corrosion and deterioration. Lines of any type must be able to resist and withstand maximum operating temperatures, pressures, and mechanical stresses. When selecting tubing, the manufacturer's recommendation should be followed.

Fitting materials. The material used for tubing fittings is determined by the material utilized in the particular tubing. Aluminum alloy fittings must be used with aluminum alloy tubing, steel fittings with steel tubing, and brass or bronze fittings with copper tubing. The basic purpose for matching like materials that come in contact with each other is to prevent dissimilar metal corrosion.

Tubing maintenance. When doing maintenance work on lines or tubing, pay particular attention to the alignment of the tubing and the tubing fittings. Be sure you are using the correct radii of tubing bends, exercise caution so that the tubing exterior is not unduly marred or scratched, and that careless handling is prevented. Check that proper clamps and other holding devices are utilized when attaching tubing or lines to the aircraft structure. Maintain, service, and adjust aircraft lines, tubing, and systems in accordance with manufacturer's maintenance manuals and pertinent component maintenance manuals.

All-metal flexible hose. A vibration absorbing hose that is pressure-tight is made from flexible inner metallic tubing and covered with layers of braided metal. This type of hose retains its original flexibility and does not become stiff when under pressure.

Various types of hose are made from natural or synthetic rubber. They are used for hydraulic, fuel, ventilation, oil, and vacuum systems. This type of hose may be used for oxygen breathing lines and other component purposes. The great development in the field of *rubber-based hose* has added to its strength, durability, resistance to deterioration, and great flexibility. These are the basic reasons why this type of hose may be used for many purposes. They are rated for low pressure, medium pressure, medium-high pressure, and high pressure installations depending on the type of aircraft system in which they are to be used. Flexible hose is most valuable when used in parts of a system where there is considerable vibration or relative motion between the sections to be connected together.

Beaded tubing. The correct arrangement of utilizing and installing flexible hose connections is to have a bead near the end of the tubing and fittings where the connection is to be made between the tubing and the fitting. The bead not only helps to provide a seal between the hose and the tubing but also resists any tendency of the hose connection being pulled or blown off due to system pressures. When hose clamps are used it is important that they be tightened properly. The correct tightness can be obtained by turning the clamp finger-tight and then an additional one-half to one and one-half turns with conventional pliers.

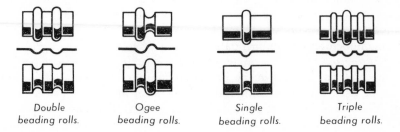

Double beading rolls.   Ogee beading rolls.   Single beading rolls.   Triple beading rolls.

Fig. 13. Beaded tubing.

Some *aircraft fuel hose* is made with a synthetic (aromatic) fuel resistant tube. This type of hose generally has a covering reinforced with two braided inserts. Aromatic resistant fuel hose is identified by a broken red line and a solid white line along its entire length.

Hydraulic system hose is made for various working pressures and consists of synthetic tubes. The tubes are reinforced with cotton or wire braids and impregnated with synthetic rubber.

Oxygen systems hose. In general, hose utilized for oxygen breathing systems is similar in construction to hydraulic system hose as described above.

Tubes and pipes may be connected to other tubes or pipes or to the various units of an installation by approved and/or standard type *fittings.* Fittings are fabricated in many shapes and sizes and each is designed to fulfill certain specific requirements. Tube fittings are identified under standard categories such as AN (Army and Navy), MS (Military Standards), etc.

Repairs to tubing. In many instances metal tubing or piping will have to be replaced when found to be damaged beyond acceptable tolerances. To replace a section of metal tubing, first select a section to be used that has the same material, wall thickness, and outside diameter (OD). Be sure the new section is straight and round. Small kinks, dents, etc., may be removed by striking the tube lightly with a wooden paddle while rolling it back and forth on a felt surface.

Dents can also be removed by pulling a "bullet" (or oval shaped piece of steel of the correct diameter) attached to a cable through the tube. The bullet or piece of steel will press the dent outward. The tube, in the area of the dent, will thus be returned to its original contour.

When *cutting the ends of tubing* be sure the cut is clean and square (90° to the centerline of the tube). After cutting remove the inside burrs with a reamer, knife, or scraper. The outside edge of the cut should be cleaned carefully with a file but not rounded off too much. Copper or aluminum tubing may be cut with a small tube cutter. Straight, short sections of tubing are to be avoided in the repair or replacement of metal tubing or lines. When bending tubing, it is best to allow for contraction, expansion, and to permit the tubing to absorb vibration.

Tube bending patterns. It is best to use the section of tubing to be replaced as a pattern for the new tubing. If this is not possible, use a piece of welding rod or stiff wire as a suitable pattern. Correct bends maintain the tube's circular shape and presents a smooth appearance throughout the bend. Though proper specifications for bends should be used, a general rule may be followed when necessary. The minimum bend radius should be at least five times the tube diameter. When bending tubing, close the end of the tube, whenever possible, with a plug and fill the tube

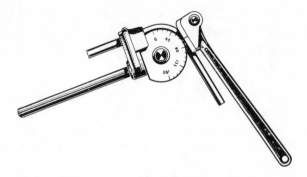

Fig. 14. Making a bend with a small tube bender.

with a substance such as sand. This will give the tube an all-around support and prevent deformation during the bending operation. This bending procedure is appropriate when making field repairs. However, approved bending tools should be utilized whenever possible.

Flares in metal tubing. Flaring the ends of tubing serve a dual purpose and prevent the end of the tube from slipping out of the fitting and also acts as a seal. *Flaring tools must be used* as it is almost impossible to make a satisfactory tube flare by hand; the least imperfection may create a leak at the fitting area. Flares are generally made at a 45° angle. However, hydraulic tube flares are generally made at a 37° angle.

Fig. 15. A well formed double flare and a double-flaring tool.

When *routing fluid lines and electrical cable* along the same run in an aircraft, the fluid lines should always be installed under the electrical cables. With the electrical power cables installed above the fluid lines there is no possibility of fluid leaks causing an electrical fire hazard.

Aircraft fitting identification. AN (Army-Navy) fittings and other standard type fittings are numerous but each has a specific purpose. AN fittings are identified by a number following the AN designation. This identifies the *design* of the fitting and not the *size.* The size is determined by a "dash number" following the design number. *Example:* AN 911-6D is a nipple (911), ⅜" pipe thread (6), and made of aluminum alloy (D). Another example: AN 777-5 is a universal elbow (777), 5/16" outside diameter (OD) (5), and made of steel (absence of code letter). No attempt will be made here to cover *all* types of fittings. The aircraft mechanic should have as an important "tool" a copy of a typical "Aircraft Mechanic's Handbook" of which there are several types available covering the same general information. Not only do these handbooks give illustrations and identification of aircraft fittings but many other standard type items normally used in shop repair procedures.

Fitting material code. Aluminum alloy (Code D), steel (absence of a code letter), brass (Code B), and aluminum bronze (Code Z for AN819 sleeve).

Visual identification is provided by the color of the material for brass or bronze and by a dye in the case of aluminum alloy or steel fittings. Aluminum alloy fittings are colored *blue* and steel fittings are colored *black.* This makes it possible to determine the material of a fitting at a glance.

E. MATERIALS AND PROCESSES

All materials used in the repair and maintenance of aircraft must always be as good or better than the original material. Where the quality of a material may be in question or where a repair or modification is not of a usual nature, it may be necessary to submit data, including engineering, to the FAA Administrator for approval.

NONDESTRUCTIVE TESTING METHODS. There are a number of approved testing methods available. Depending on the particular testing method, one or more materials can be tested by the same method, but some methods may be used for only one type of material.

The following is not intended to be a complete treatise on material testing but is sufficient to acquaint the aircraft mechanic with standard methods and procedures.

Magnaflux. Also known as *magnetic particle inspection,* this method is utilized to inspect magnetic materials such as iron and steel. It is an excellent means to determine the presence of cracks, etc., that may not otherwise be visible to the naked eye. The item (such as an engine crankshaft) to be inspected is placed in the magnaflux tester in such a manner that a strong magnetism can be induced into the item. A liquid (iron oxide) is sprayed over the item. Iron particles in the liquid are thus attracted to any surface crack, etc., and will show as a dark line or dark area which can then be easily seen by the naked eye. From this inspection it can be determined if the imperfection is of a major or minor proportion. The disposition of the particular item can then be determined.

After a magnaflux inspection, the item should be carefully demagnetized, cleaned, and the inspection completed by coating the item with a suitable preservative.

X-Ray or radiograph. This type of inspection can be used on either magnetic or non-magnetic materials to detect subsurface voids such as open cracks, blowholes, etc.

Flourescent penetrant (Zyglo) inspections can be made on metal, plastics, etc., and it is highly efficient in the detection of cracks and other flaws.

Ultrasonic flaw detection can be used to inspect all types of materials. This type of inspection can detect small cracks, checks, and voids that may not otherwise be detected by X-ray. An ultrasonic test instrument requires access *to only one surface* of the material to be inspected, and can be used with either straight line or angle beam testing techniques. The instrument electronically generates ultrasonic vibrations and sends them in a pulsed beam through the part to be tested. Any discontinuity within the part, or the opposite end, will reflect the vibration back to the instrument. The instrument measures the elapsed time between the initial pulse and the return of all reflections and indicates such time lapse on a cathode ray indicator or paper recorder. Ultrasonic inspection requires a skilled operator who is familiar with the equipment being used as well as the required inspection method of testing the many different parts.

Magnifying glass inspection of welding. Careful examination of all joints with a medium-power magnifying glass (at least a 10-power), after first removing all scale, is considered an acceptable method of inspection for repaired structures.

Several *dye penetrant* type inspection kits are available which will reveal the presence of surface cracks or defects and subsurface flaws which extend to the surface of the part being inspected. These penetrant type inspection methods are considered acceptable provided the part being inspected has been thoroughly cleaned, that all areas are readily accessible for viewing, and the manufacturer's recommendations as to method of application are closely followed.

BASIC HEAT-TREATING PROCESSES. Generally speaking, heat treating is any acceptable method for the controlled heating and cooling of metals. Heat-treatment develops a desired metal hardness or softness, ductility, grain structure, and tensile strength. Among heat-treating processes are annealing, normalizing, tempering, and hardening. Important to heat-treatment of metals is the metal quenching (cooling) in oil, water, brine, or air.

Ferrous metals such as iron and steel and some *non-ferrous metals* such as copper and aluminum can be heat-treated satisfactorily. The method of heat-treatment for the various metals is dependent on the kind of metal and the desired results.

Basic steps in heat-treating. There are definite general steps in the heat-treatment of metals. They are: (1) heating the particular metal to a temperature within or above the metal's critical temperature. This must be done under conditions that are correctly controlled; (2) holding (or soaking) is necessary to keep the metal at a given temperature for a given period of time. Under these conditions the metal will become saturated completely with heat and the anticipated changes in the grain structure will take place; (3) returning the metal to room temperature by the process of quenching the metal in oil, water, air, etc.

Annealing. This is a form of heat-treatment of metals that consists of heating and cooling procedures for the purpose of removing stresses, gases, inducing softness, altering ductility, toughness, or other physical properties. Annealing also produces a refinement of metal grain structure.

The means of annealing is accomplished by gradually heating the material to a temperature above the critical temperature, holding it at this given temperature for a specified length of time (based on when the grain structure has been refined), and cooling it by the method prescribed for the particular metal being annealed.

Nitriding is a method of holding special alloy steels at temperatures below the critical temperature in ammonia (anhydrous ammonia). This permits nitrogen to be absorbed into the surface of the metal as iron nitride and results in a grain hardness.

Carburizing. In this heat-treating method the metal is held at an elevated temperature while it is in contact with gaseous, liquid, or solid material that is rich in carbon content. It is necessary to allow sufficient time for the surface metal to soak up (or absorb) adequate carbon to become a high-carbon steel.

Cyaniding is a very fast means of producing surface hardness on iron-clad alloy that normally has a low content of carbon. In this method, the steel can be immersed in a molten bath of salt (cyanide). Another means of cyaniding is to apply powdered cyanide to the surface of the heated steel. The temperature of the steel, in this process, should range from 1,300 to 1,600°F. The temperature used will depend on the type of steel, depth of the hardening desired, type of cyanide compound, and the length of time the steel is exposed to the cyanide.

Normalizing is the process by which iron base metals are heated above their critical temperature. This is done to obtain better solubility of the carbon in the iron. The heating process is followed by cooling the metal in still air.

Hardening. In this process, the metal is heated slightly above its critical temperature and then cooled rapidly by quenching it in water, brine, or oil. This method of heat-treatment produces a fine-grain structure, excellent hardness, the metal's maximum tensile strength, and a very minimum ductibility. The metal, in this particular condition, is generally too brittle for many uses. However, this heat-treatment is the first step in producing high-strength steel.

Heat-treatment of aluminum alloy parts. All structural aluminum alloy parts are to be heat-treated in accordance with instructions issued by the manufacturer of the particular part or material. If the heat-treatment produces warping, straighten the parts immediately after quenching. Heat-treat riveted parts before riveting to prevent warping and corrosion. When riveted assemblies are heated in a salt bath, the salt cannot be entirely washed out of the crevices, thus causing corrosion.

Quench aluminum alloy material from the solution heat-treated temperature as rapidly as possible or with a minimum delay after removal from the furnace. Quenching in cold water is preferred although less drastic chilling such as hot or boiling water is sometimes employed for bulk sections, such as forgings, to minimize quenching stresses. The transfer of 2017 alloys from the heat-treatment medium to the quenching tank must be accomplished as quickly as possible. An elapsed time of 10 to 15 seconds will, in some cases, result in noticeably impaired corrosion resistance.

Reheating of 2017 and 2024 alloys at temperatures above that of boiling water after heat-treatment and the baking of primers at temperatures above that of water after heat-treatment will not be considered acceptable without subsequent complete and correct treatment. Such practices tend to impair the original heat-treatment of the material.

IDENTIFY AND SELECT AIRCRAFT HARDWARE AND MATERIALS. Aluminum as the basic material used in the construction of aircraft is due to its desirable physical and chemical properties which are either inherent or can be introduced into the aluminum by mixing with other metals which become alloy. Aluminum is easily formed, can be held to close tolerances, is corrosion resistant, and has a very high strength-weight ratio.

Aluminum alloys may be obtained in cast or wrought form. Castings are produced by pouring or forcing molten aluminum alloy into metal or sand molds or dies. Aluminum in wrought form is obtained in three different ways. They are: (1) rolling slabs of hot aluminum through rolling mills which produce sheet, plate, and bar stock; (2) extruding hot aluminum through dies to form "T" sections, channels, angles, etc.; and (3) forging or hammering a heated billet of aluminum alloy between a female and male type of die to form a particular desired part.

Aluminum alloys are identified by designation of the various alloys. *Example:* 2024-T3. The first digit identifies the alloy type. In this example, the "2" indicates an aluminum-copper alloy. The second digit identifies the alloy modification. The digit "0" used here will indicate an original alloy and the digits "1" through "9" will indicate further modifications. The third and fourth digits ("24" here) identifies the specific aluminum alloy. In this example, the aluminum alloy consists of 3.8 to 4.9% copper, 1.2 to 1.8% magnesium, 0.1% chromium, and the remainder aluminum. The "T3" is the temper designation. The "T" indicates a heat-treated alloy. The numeral following the letter "T" shows the type of heat-treatment. Any variation of actual time, specific temperatures, etc., will be indicated by a second numeral.

Wrought alloys that depend on strain hardening for their degree of temper and strength are the 1000, 3000, and 5000 series alloys. All *non-heat-treatable aluminum alloys* may be fusion welded or resistant welded.

Some characteristics of working with non-heat-treatable alloys are: (1) they cannot be heat-treated for greater strength; (2) additional strength is obtained by cold working; (3) they have low tensile strength but improved corrosion resistant properties over other aluminum alloys; (4) if they become too hard during cold working processes, soften the part by stress "relieving" (heating to a specific temperature); and (5) these alloys have a higher degree of workability and, where ultimate strength is not an important factor, they are used in preference to the heat-treatable alloys.

Heat-treatable alloys. Where high strength is required, the heat treatable 2000, 4000, 6000, and 7000 series alloys are used in aircraft construction.

These alloys (in the annealed "0" condition) may be worked with nearly the same degree of success for severe forming as the non-heat-treatable alloys. If additional strength is required after forming, the part may be heat-treated to give it the desired strength.

It is important to remember that a heat-treated part is more resistant to corrosion than the same part in the annealed condition.

Heat-treatable alloys should not be gas welded with the exception of 6061 designated alloy. This is because welding destroys the heat treatment and corrosion resistant properties of the part.

Alclad or clad sheets. This metal derives its name from the application of a thin coating of pure aluminum covering the external surfaces. This aluminum coating protects the sheet against corrosion. Mechanics and other personnel must exercise care in handling clad sheets to prevent surface scratching. If the protective coating is marred in any manner, the center alloy will be exposed to corrosive action at that point. If corrosion occurs due to marring, scratching, or other reasons, the corrosion may be "arrested"

Fig. 16. Clad aluminum and aluminum alloy.

or temporarily stopped by an application of bicarbonate of soda. To prevent further corrosion, a coating of zinc chromate (or other approved corrosion inhibitors) should be applied to the affected areas.

Precautions in handling aluminum sheets. (1) Always use a recommended bend radius when bending aluminum sheets; (2) use a proper protective layer between adjacent clad sheets. However, do not use a brown sulphate paper. When wet, this type of paper will cause corrosion; (3) do not use a scribe to mark aluminum sheets—use a soft lead pencil or other non-scratching marking item; (4) use extra care not to scratch or mar the surface of aluminum sheets while moving them from one location to another; (5) always bend aluminum sheets so that the bend line runs perpendicular to the metal grain (90° to the grain or directly across the grain). Metal grain can easily be identified as it is the direction of the sheet "rolling" during fabrication; and (6) always store aluminum sheets in a dry (low humidity) area since moisture is the greatest cause of metal corrosion in general.

Identification of steels (steel classification). Steels are classified according to established standards. In order to easily identify the various steels on blueprints, drawings, etc., a simple numbering system is utilized. A few classes of steels and their numbering systems follows:

1. Carbon steel   4. Molybdenum steel
2. Nickel steel   5. Chromium steel
3. Nickel-chromium steel 6. Chromium-vanadium steel

The basic identification rules are: (1) the first number indicates the class of steel; (2) the second number indicates the percentage of the dominating alloy; and (3) the last two numbers indicate the carbon content in hundredths of 1%.

According to these rules, 2330 steel would be decoded as follows: the "2" indicates the class is nickel steel; the "3" indicates 3% nickel steel; and the "30" indicates the carbon range to be 0.30%.

1040 steel would be decoded as follows: carbon steel, 0% or no carbon alloy, and .40% carbon range.

In aircraft construction, common steels used are 1020, 1025, and 1040 alloys. These are the lower strength steels. The higher strength steels used in high-stress areas of aircraft must be tough and able to withstand vibration. These steels are of the 4130 and 2330 alloys.

Plain carbon steels. These are divided into three grades: (1) low carbon (0.10 to 0.25%); (2) medium carbon (0.25 to 0.50%); and (3) high carbon (0.55 to 0.95%).

Low carbon steels are used in the manufacture of nails, wire, structural steels, machine parts, screws, etc. *Medium carbon* steels are used in the manufacture of forged, cast, or mechanical parts which require heat treatment to develop their highest strength properties. *High carbon* steels are used in the manufacture of tools, springs, high tensile wire, shear blades, etc. This grade of steel is *hard* and of high tensile strength.

Alloy steels are chromium, nickel, molybdenum, tungsten, and vanadium.

Titanium and its alloys are noted for their outstanding strength-to-weight ratio, high temperature properties, corrosion resistance, and a fatigue strength generally unsurpassed by other metals.

In general, metals and materials used in aircraft are of a higher quality than the same type of materials used in other manufactured items. Aircraft mechanics should be knowledgeable on all materials including aluminum, steels, magnesium, bronze, brass, plastics, wood and glass fiber, beryllium, titanium, glass cloth, aircraft fabrics, lithium, and paper. Paper seems an unusual material for aircraft, but it can be used as a core material in a sandwich of two-sheet metal panels to form floorboards or shelves. The floorboards absorb considerable amounts of high-frequency noise. These panels also provide a strong, rigid support for aircraft structural furnishings or cargo.

INSPECT AND CHECK WELDS. The importance of good welding cannot be overstressed because of the strength and security each weld must develop.

Fusion welding is used extensively in aircraft manufacture and repair. It is a process in which enough heat is applied to melt the edges or surfaces of metal. In turn, this permits the molten parts to flow together, leaving a single solid piece of metal after adequate cooling has taken place. In order to build up the welding seam to a thickness greater than the base metal, additional metal is added to the weld by the use of welding rods. The type of welding rod (material ingredients) is dependent on the material to be welded and the size of the welding rod is dependent on the thickness of the material to be welded.

The *strength of a good weld* is dependent on the skill and knowledge of the person doing the welding. He has two important steps to take. They

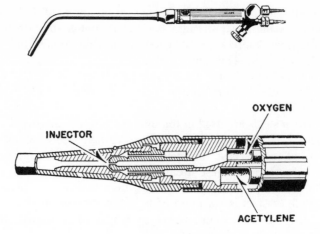

Fig. 17. Welding torch.

are: (1) correctly prepare the metal in advance of the welding process; and (2) insure adequate knowledge of the equipment in order to use the correct welding flame for the particular job.

The weld must be formed correctly to provide strength and to resist fatigue failure at any point along the joint. Welds, if properly made, will develop excellent strength qualities. If improperly made, welds may develop only a nominal percentage of the strength they should develop.

Welding temperatures. During welding procedures, temperatures must be rigidly controlled as some metals will have very little or no strength when they have been exposed or raised to extremely high temperatures. This can occur if the welder is careless or lacks the knowledge to develop correct welding flames for the particular metal being welded.

In order to *reduce distortion* and residual stresses that adversely affect a finished weld, the welder should adhere to the following rules: (1) distribute the heat very evenly; (2) keep in mind to apply the smallest necessary amount of heat to the weld; and (3) use appropriate jigs to hold the metal in place.

Characteristics of a good weld. The welder should carefully inspect a finished weld and note the following: (1) the weld should be free of blowholes, porosity, and globules; (2) there should be no sign of pitting, burning, cracking, or distortion; (3) the weld seam should be of a uniform thickness and smooth; (4) no oxides should be formed on the base metal more than one-half inch from the weld; and (5) the weld should taper off smoothly into the base metal.

Basic weld testing. There are a number of reliable tests for welds. However, the mechanic should be able to apply some very simple tests to help judge his ability. The bend test is simple and reliable. After welding, the metal is allowed to cool slowly. The welder then grasps it with a pair of pliers. He then clamps it in a vice so that the weld is parallel to the top of the vice jaws and just above the top of the vice. Striking the top part of the metal with a hammer, the metal should gradually bend to an angle of at least 90° without cracking. If the weld breaks off due to the hammering, obviously the weld has not been made correctly and is thus unsatisfactory.

Visual inspection of completed welds can be made to determine its worth. The welder should look for smoothness of the weld bead, amount of reinforcement, cleanliness of the weld, contour of the weld, and be certain there are no blowholes present.

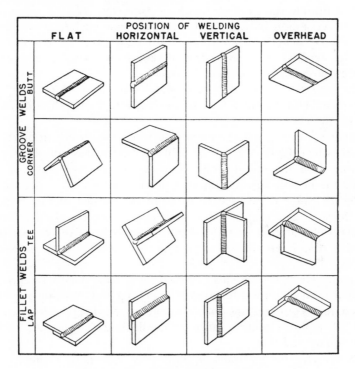

Fig. 18. Four basic welding positions.

PERFORM PRECISION MEASUREMENTS. There are many precision measuring tools and devices available to the aircraft mechanic. How many he should know how to use accurately depends on his general work assignments. However, he should make an endeavor to learn how to use as many as possible. The following information describes the more common everyday type of precision measuring instruments that all mechanics should be familiar with:

The vernier scale. This type of measuring scale is an auxilliary scale for estimating fractions of a scale division when the reading to the nearest whole division on the main scale of an instrument is not sufficiently accurate for the job to be accomplished.

Machinist's steel rule is one of the most frequently used measuring devices for the aircraft mechanic. It is accurately graduated in 1/8, 1/16, 1/32, and 1/64 of an inch.

Folding steel rule, flexible steel rule, or a steel tape. There are many measurement needs beyond the capacity of the machinist's steel rule. The folding steel rule is particularly handy for measuring curved objects. The rule can be shaped or bent to fit almost any contour. It has a small hook on the zero end so that measurements can be made quick and easily.

Most flexible rules are six feet long. However, the steel tapes can be 50 to 100 feet or more in length and are used for making long measurements.

These types of rules and tapes must be kept clean and dry. Dust, sand, etc., tend to wear away the markings and when combined with moisture causes rust to form on the steel.

The *depth rule* has a narrow blade which slides through a slotted locking arrangement. This type of rule is used to measure the depth of holes, slots, keyways, and other recesses.

The *hook rule* is a convenient rule when measuring from an edge, especially when the edge is out of sight. Normally, it is not good practice to measure from the end of a rule. However, it is permitted with the hook rule and the flexible rule as indicated previously.

Outside calipers are used to measure the outside diameter of rods, shafts, bolts, rivets. They are also used to measure the width, thickness, and length of materials that are within the capacity of the caliper jaws.

Inside calipers are utilized to take measurements of slots, grooves, and openings.

Squares for angles may also be used for making measurements. However, their most important function is laying out and checking angles. The steel square, try square, and double square are all designed to check right angles. All types of squares must be handled carefully so that they will remain square and true.

Micrometers for precision measurements. The most common is the "one-inch" type designed to make precision measurements not over one inch thickness or diameter. Large micrometers are available for jobs beyond the capacity of the one-inch type.

The *spindle* of the "mike" is threaded through a sleeve and fastened to the thimble. The standard spindle has 40 threads per inch so that it moves 1/40 of one inch for each revolution of the thimble. The 1/40 of an inch is the same as 25/1000 of an inch or 0.025 inch. The thimble scale is marked off into 25 spaces with each space representing 1/1000 of an inch or 0.001 inch. This makes it possible to read a measurement

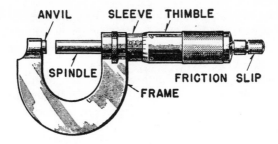

Fig. 19. 1-inch outside micrometer.

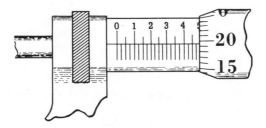

Fig. 20. The micrometer scale here indicates 0.4945 inch.

to the nearest ten-thousandths of an inch. Mechanics can become experts at reading micrometers after some diligent practice.

The *vernier scale* may be added to the better micrometers so that the user will not have to "read between the lines" of the micrometer scales. This scale added to the micrometer permits the user to measure accurately to the nearest ten-thousandths of an inch. Micrometers with a vernier scale are often referred to as a *ten-thousandths mike.*

Vernier calipers are basically a ruler with which precision measurements can be made since they have their own vernier scale.

The mechanic and measuring instruments. Making precision measurements is a must for the aircraft mechanic and can be learned easily by familiarization with the various instruments and/or tools. Diligent practice is a great help to the mechanic in this phase of his work.

F. GROUND OPERATION AND SERVICING OF AIRCRAFT

Ground operation and servicing of aircraft covers a very wide scope of activities. The procedures and activities involved are dependent, to a great extent, on the category and type of aircraft involved. However, the major purpose remains the same for any aircraft. This involves extreme caution when moving aircraft on the ground either by tow or under its own power, being on the alert for unexpected emergencies, and servicing the aircraft with the correct fuel, oil, etc.

The following is general in nature and is not intended as an overall guide. The person responsible for a particular aircraft should use the following information in conjunction with instructions outlined in the particular aircraft's operating and maintenance manuals.

PREFLIGHT INSPECTIONS. Prior to flight an aircraft is given a preflight inspection. This is basically the responsibility of the pilot in command but should be understood thoroughly by the aircraft mechanic. On large aircraft the pilot in command is also the person responsible for the preflight inspection but the flight engineer is generally given the duty of actually making the inspection or walk-around prior to flight.

A preflight inspection should be made by always starting at a certain point of the aircraft and proceeding in a thorough and orderly manner until all pertinent items have been checked and/or inspected. A suggested preflight pattern follows:

1. *Flight deck or cockpit.* Check to be sure that all electrical and ignition switches are in OFF position. Check the landing gear control for DOWN. Be sure that all controls are in a "shutdown" attitude.

2. *Propellers.* Check the propellers for nicks, cracks, safetying, etc. Use caution when in the vicinity of any propeller.

3. *Engine.* Check engine and visible accessories for security and safetying of nuts, connections, etc. Check fuel, oil, hydraulic, and other lines for security and possible leaks. Inspect exhaust system for security and the absence of cracks and holes. Any fault in the exhaust system may create a fire or carbon monoxide hazard.

4. Check for ample *fuel, oil,* etc. Where possible, check fluid quantity levels visually. Do not depend on gages. Drain and/or check fuel system sufficiently to eliminate any possible contamination such as condensation (water), etc. Check security of liquid supply tank caps and that necessary vents are open.

5. *Landing gear.* Check tires for cuts, cracks, adequate tread, and proper inflation. Check struts and fittings for safetying and any evidence of cracks, bends, or excessive wear. Check brake assemblies visually and be sure there are no leaks of brake fluid or damp oily spots that might indicate a leakage of fluid.

6. *Wing, fuselage, and tail surfaces.* Check all external surfaces (covering), whether metal or fabric, for holes, wrinkles, corrosion, etc. Inspect ailerons, rudder, elevator, etc., for security and freedom of movement. Be sure that all surfaces are free from mud, snow, ice, and frost.

7. *Pitot-static system.* Be sure that the static vents are open and that

the pitot tube is unobstructed. Remove any protective device that may have been placed over the pitot tube for protection against dust, dirt, etc.

8. *Controls.* Check controls in cockpit (flight deck) for proper movement. Set stabilizer or elevator trim tab for proper takeoff position.

9. *Weight and balance.* Check the anticipated loading to be sure it is within allowable gross weight and that the center of gravity will not exceed forward and aft limitations.

The preceding pre-flight data is general in nature and reflects the more important items to be checked. The person responsible for the pre-flight inspection should refer to the particular airplane's manual for specific information on pre-flight procedures.

BASIC ENGINE STARTING PROCEDURES. Engine starting procedures will vary depending on the type and size of the engine. There are, however, several fundamental procedures that should be observed.

1. Be sure the immediate area of the engine is clear of persons and ground equipment, that the area is free of gravel, rocks, etc., that may be stirred up causing injury to the propeller, be ingested in the air inlets, or cause some property damage. The wheels should be held secure by adequate parking brakes and/or securely placed blocks or chocks in front of the wheels.

2. The precise starting procedures will differ considerably among reciprocating, turbojet, and turboprop engines. The person starting the particular engine must be adequately familiar with the engine or follow precisely the starting procedures outlined in the engine operating instructions.

3. For small general aviation type aircraft, the person starting the engine should call "clear" as a warning to anyone in the area that engine starting procedures are to commence. If there is no response the starter may then be engaged. In large aircraft operation, handlers and ground crewmen are very alert to the hazards of engine starting where reciprocating, turboprop, or jet turbine engines are commonly used. In turn, these personnel will not permit unauthorized persons to be within the hazard area.

4. When the engine starts, advance the throttle or fuel control unit to obtain the recommended engine RPM, check the engine instruments immediately for normal warmup indications. If the oil pressure gage does not indicate proper engine warmup pressure within 30 seconds, shut the engine down and check or have checked the cause for the low or no oil pressure indication.

5. An engine equipped with a float type carburetor (no idle cutoff) should be started with the mixture control in full-rich position. If the engine carburetor is equipped with an idle cutoff, the engine should be started with the mixture control in the idle-cutoff position. Immediately after starting, move the mixture control to full-rich and complete the engine

warmup. Engine warmup periods must be consistent with the specific engine instructions due to generally inadequate cooling. In other words, limit ground operation to a minimum of time.

When large engines equipped with pressure carburetors, fuel injection, etc., are to be started use a check list for the particular engine. This also applies to turboprop and jet turbine engines. Do not rely solely on memory.

6. Aircraft engines, in general, must reach normal operating temperatures before they will run smoothly and dependably. Oil temperature, oil pressure, and CHT (cylinder head temperature) must be carefully monitored. In the warmup of turboprop and jet turbine engines, EGT (exhaust gas temperature), TIT (turbine inlet temperature), oil pressure, and oil temperature are a few of several indications that must be carefully monitored.

Preflight inspections, engine start, and engine warmup are often a part of the aircraft mechanics responsibilities. Use a check list, particularly when working with large, complicated aircraft and aircraft engines.

ENGINE FIRE PRECAUTIONS. Engine fire precautions must be taken during engine starting. In general aviation, it is not feasible to have ground personnel stand by with fire extinguishing equipment. This practice is done, however, in large commercial operations.

The mechanic (or pilot) that is starting an engine must understand how an engine induction fire might occur and the steps to be taken to prevent or to alleviate the fire.

An engine induction fire during starting operations is an extreme hazard. Generally, it is the result of excess fuel accumulation in the induction system due to overpriming or starting with lean mixtures. A subsequent backfire (when the intake valve is off its seat or open) will cause the flame element to flow back through the induction system and expose the carburetor area to be set afire.

The best preventive step in keeping induction fires to a minimum or absolute prevention is to *know* the particular engine's starting procedures. Study the particular engine's operating manual (use a check list) and be particularly alert to the precise priming instructions and mixture control placement.

A proper method of handling an induction fire, should it occur, is to open the throttle wide. This will draw the fire into the engine cylinders and dissipate the fire out through the exhaust system. A fire extinguisher should be easily available should the fire get out of control.

HAND SIGNALS FOR GROUND OPERATION. When aircraft are being moved on the ground, especially large aircraft in congested areas, it is essential that ground personnel be available to direct the pilot (or qualified mechanic) by use of proper hand signals. During hours of darkness, ground personnel should use suitable flashlights or other approved accessories.

GROUND POWER FOR STARTING ENGINES. In order to reduce the "drain" on the aircraft batteries, there are several types of external auxiliary power units that are utilized for engine starting and other ground servicing. They may be in the form of a battery cart, rectifier-transformer units (APU's), etc. In general, these ground units must be able to serve the aircraft ground receptables with direct current (DC).

Caution must be exercised during the use of ground power units. Some precautions are: to prevent sparking due to the potential fire hazard; keep clear of all aircraft components; know precisely how to connect and disconnect; and be sure all ground units are clear of aircraft prior to any aircraft movement.

OUTDOOR STORAGE OF AIRCRAFT. When aircraft are tied down or secured outdoors it is absolutely necessary that all tiedown appliances such as cables, anchors, chains, etc., be well secured and of adequate strength. Ropes should not be used for tiedown of aircraft due to their deterioration and loss of strength when exposed continually to heat, moisture, etc.

Many aircraft have been severely damaged or completely destroyed due to careless tiedown practices, and/or unexpected high winds.

Airplanes should be tied down "into the wind" when possible and the control surfaces should be protected from unnecessary "flapping" by the temporary installation of gust locks. Wheels should be secured with properly designed chocks that will secure the tires and create as much ground friction as possible should the airplane rock back and forth. Suitable blocks or angle iron can be used for securing the tires when chocks are not available. When ground secured cables or other ground secured tiedown supports are not available, heavy blocks of concrete or metal may be used as a temporary substitute. All blocks or chocks, regardless of their type, should have an eyelet or other suitable holding device to which a rope or chain can be attached. This is an important safety feature that permits ground personnel to remove them from a safe distance. This is especially a good practice when the engine and/or engine-propeller is operating at the time the blocks or chocks are removed to permit movement of the aircraft.

Prior to moving or taxiing an aircraft on the ground be sure that all blocks, chocks, chains, etc., are clear of the aircraft and that any temporary gust locks have been removed from the control surfaces.

FUEL CONTAMINATION. A major probable cause of erratic engine operation or total engine failure can be traced to fuel contamination in the form of dirt, dust, condensation, etc.

A few precautions can easily be taken to prevent fuel contamination. They are: (1) keep the airplane, especially around the fuel tank filler openings, free of dust, dirt, and other foreign matter; (2) when fueling under gusty wind conditions, precautions must be taken to prevent foreign matter from entering the fuel filler openings: (3) if it is necessary to fuel an aircraft from fuel drums or other like containers, filter the fuel through a clean chamois skin; (4) fill the fuel tanks after the last flight of the day to prevent condensation (moisture) from forming on the internal surfaces of the tank; and (5) drain the gascolator (fuel filter) and/or tank drains prior to flight to insure that any moisture or foreign matter that may have entered the fuel or fuel system will be eliminated.

POWERPLANT LIQUID LOCK (HYDRAULIC LOCK). Oil accumulation in the engine cylinders, especially the lower cylinders of a radial engine, can create excessively high compression during engine starting procedures with possible engine structural damage. Liquid lock presence should be determined prior to turning the engine over for starting purposes. The propeller is turned by hand and if it stops or becomes difficult to turn, liquid lock should be suspected. To eliminate this condition, remove the spark plugs of the affected cylinders to permit the accumulated oil to drain from the cylinder combustion chambers. When assured that any accumulation has completely drained, replace the spark plugs, wiring, etc., and then commence engine starting procedures.

PRIMER OPERATION. It is preferable to prime the engine after the engine is being turned over by the starter. Priming before engine rotation may be ineffective, in many cases, because it may cause fuel to dilute with the oil on the walls of the cylinders being primed. After the engine is turning over with the starter, hand prime with three to six strokes of the primer pump handle (check the particular engine's operation manual for precise instructions).

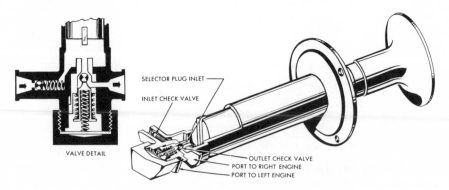

SELECTOR PLUG INLET

INLET CHECK VALVE

VALVE DETAIL

OUTLET CHECK VALVE
PORT TO RIGHT ENGINE
PORT TO LEFT ENGINE

Fig. 21. Manual primer, twin-engine aircraft.

The primer may be of the electrically operated type. In this case you would hold the primer button (or control) to ON for a given number of seconds according to the engine operating manual.

When using a hand operated primer, be sure that the primer handle is closed and secured (locked) after completing the priming operation. If not properly secured, excess fuel will continue to flow from the primer to the primed cylinders.

Use engine primers diligently, whether hand operated or electrical, to insure sufficient priming fuel to enter the primed cylinders and to prevent excess fuel from "flooding" the engine cylinders with subsequent backfiring or difficult starting.

The number of cylinders actually primed will depend on the particular engine. However, the nine-cylinder radial single-row engine may be used as an example. In this instance, the top five cylinders (No. 1, 2, 3, 8, and 9) actually receive primer fuel.

SUPERCHARGED ENGINES. When an engine incorporates an internal two-speed (low blower and high blower) supercharger, be sure the supercharger control is placed in "low blower" setting prior to starting the particular engine.

On any engine that incorporates any type of supercharger, starting instructions must be followed correctly to prevent over-boosting the engine with possible subsequent structural engine damage.

IDENTIFICATION AND SELECTION OF FUELS. Aviation gasoline (fuel) is classified by octane ratings *(anti-knock value),* and *performance number power ratings.* Correct fuel ratings for a particular engine are to be found in the engine manufacturer's specifications, the operating limitations records, the FAA Approved Operating Manual, particular aircraft's owners manual, and/or this information is in the form of a placard. In general, the type of fuel used will be noted at or near the fuel filler openings.

The use of fuel that has a higher octane than that required for a particular engine *does not* improve engine performance and may, in fact, be harmful. Higher than recommended octane fuel will create higher temperatures and cylinder pressure than that required creating excessive engine operating stresses.

The use of aviation fuel of a lower octane than that recommended by the engine manufacturer is definitely harmful under all operating conditions. Lower than recommended octane ratings can cause excessive cylinder temperatures, fouled spark plugs, burned and/or stuck valves, high oil consumption, loss of power, and a combination of preignition and/or detonation. Excessive engine wear will be the end result of operating engines on too low octane fuels.

Preignition. An engine operating condition where excessive cylinder head temperatures will create "hot spots" inside the combustion chamber. This in turn can ignite the fuel too early and disrupt smooth engine operation.

Detonation. A condition whereby a portion of the fuel ignites and burns properly. The unburned fuel is compressed to such a high state of compression that it self-ignites in the form of an "explosion." The result is engine vibration, excessive cylinder head temperatures, and if permitted to occur over an extended period of time, structural damage to the engine will result. The best deterrent in the prevention of detonation is use of correct fuel and control of cylinder head temperatures.

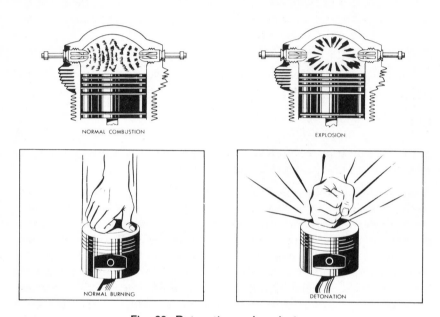

NORMAL COMBUSTION   EXPLOSION

NORMAL BURNING   DETONATION

Fig. 22. Detonation and explosion.

THE ACCEPTABLE TYPE OF FUEL. The type of fuel that may be determined "perfect" for a particular engine is one that vaporizes well at low temperatures for easy engine starting. It must not be susceptible to "vapor lock" and it must have an anti-knock or octane rating high enough to prevent detonation over a varied heat range of operation especially when used in high compression engines. Acceptable fuel must be easily available and in large quantities.

Aromatic fuels. Some high octane fuels are of the aromatic or high solvency power type. These fuels should not be used in aircraft engines that are unmodified for aromatic fuels. In addition, when aromatic fuels are used, a special type flexible hose must be utilized since these fuels are highly deteriorating to natural rubber flexible hose and the hose connections.

Fuels for turbine (jet) engines. The basic fuel requirement for turbine engines is that the fuel (basically kerosene) have a high BTU (British Thermal Unit) value. It must be suitable for proper burning (combustion) in the jet burner "cans," be relatively inexpensive, and be easily available in very large quantities.

Fuel colors. Dye is added to aviation fuel to make it easily identifiable by all concerned with ground servicing and flying of each particular aircraft. Standard aviation fuel colors are: (1) 80/87 octane—red; (2) 91/98 octane—blue; (3) 100/130 performance rating—green; (4) 115/145 performance rating—purple; and jet fuel (JP4, etc.)—colorless. *Note:* The first number indicates the fuel value at lean mixtures and the second number indicates the fuel value at rich mixture settings.

VAPOR LOCK OF FUELS. Vapor lock is a condition that occurs when liquid fuel vaporizes in the fuel system and creates sufficient vapor pressure to impede or completely stop fuel flow. Obviously, this condition can cause engine malfunction or complete engine stoppage.

Causes of vapor lock. There are several conditions that contribute to the problems of vapor lock, all of which can easily be controlled. The more important causes are: (1) servicing the aircraft with incorrect fuel that has a low vapor point; (2) routing fuel lines too close to sources of heat such as the exhaust system; (3) looping fuel lines in such a manner that fuel must abruptly flow "uphill"; and (4) fuel in outboard tanks some distance from the engine-driven fuel pumps that does not have enough "boost" from the electrical centrifugal boost pumps that are installed in the fuel tanks on large aircraft.

Control of vapor lock. To control this condition, adhere to these basic rules: (1) use proper fuel for the particular engine; (2) route fuel lines away from sources of heat; (3) if fuel lines *must* be routed near sources of heat, the lines should be "lagged" (wrapped with suitable heat resistant material) or heat shields installed in vulnerable areas; (4) use correct fuel line length to prevent taking up excess length with loops; and (5) if necessary to bend or loop a fuel line, keep the bend or loop from having an abrupt upward attitude because a lag in fuel flow can occur at this point.

ANTI-KNOCK FUEL CHARACTERISTICS. The anti-knock values of aviation fuel have been improved through the years. This improvement along with improved engineering and construction of aircraft engines assists in the production and operation of highly efficient aircraft powerplants.

Fuel improvements are the results of better refining, addition of fuel chemical ingredients such as tetraethyl leads, and freedom from contamination and impurities.

G. CLEANING AND CORROSION CONTROL

Corrosion of metals is caused by their exposure to acids, fumes, water, moist air (especially moist *salt* air), etc. During aircraft inspections, particular attention must be given to detecting the presence of metal corrosion, especially in areas not normally visible such as corners and crevices.

In general, corrosion can be materially reduced and, in many instances, prevented by the use of the better grades of base metals, adding special metal ingredients such as nickel and chromium, coating metal surfaces with paint, corrosion inhibiters, tin, cadmium, or by an electro-chemical process known as anodizing.

Commercially pure aluminum is a white lustrous metal that is fairly malleable and ductile. It is non-corrosive and resists the oxidizing action of the atmosphere. The non-corrosive properties of pure aluminum are lost, to a certain extent, in the process of producing the various aluminum alloys. This loss is offset most commonly by the application of a protective coating of pure aluminum. The resulting material is called "clad" and marketed under the trade names of Alclad and Pureclad.

In order that the *clad aluminum* product will have maximum desired physical and strength properties, the thickness of the surface coating is kept at a minimum but sufficient to afford adequate surface protection to the inner alloy. Clad sheet aluminum alloy stock, as used in aircraft work, has a pure aluminum surface coating approximately 5% of the core (or alloy) thickness.

CLEANING AND POLISHING OF AIRCRAFT SURFACES. To maintain a good appearance and, more important, to prevent surface corrosion, the cleaning and polishing of aircraft that have aluminum surfaces must be done with proper equipment, approved cleaners and polishes, and polishers or buffers that "wear" the surfaces a very minimum amount. Improper cleaning and/or polishing of surfaces can result in the removal of the thin protective coating of pure aluminum and expose the alloy or core to corrosion and deterioration.

CLEANING OF ENGINES AND COMPONENTS. Proper cleaning of engine parts and components is essential because a thorough inspection can be accomplished only if they are well cleaned.

Engine parts and components may be cleaned by one or several different operations. Parts can be cleaned (or washed) in gasoline or kerosene if the removal of oil or grease is the only problem. A vapor type of degreaser is also satisfactory to use.

When there are carbon deposits to be removed from the part, it is generally necessary to immerse the part or parts in a suitable cleaning solution. This will soften the carbon deposits and make them relatively

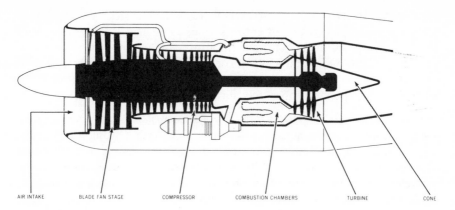

AIR INTAKE   BLADE FAN STAGE   COMPRESSOR   COMBUSTION CHAMBERS   TURBINE   CONE

Fig. 23. Turbofan airflow.

easy to remove. Hard carbon deposits are best removed by "grit" blasting or scraping with soft aluminum scrapers. Some approved types of grit used in grit blasting are crushed wheat, crushed prune seed, crushed peach seed or weed seeds. Crushed walnut shells or walnut shell grit is very satisfactory in the cleaning of turbine engine compressor blades.

Sand blasting and shot blasting of engine parts is not recommended in most instances because the sand or shot blasting tends to leave a rough finish on the part/s. This condition will hide cracks and also some of the sand or shot blast may become imbedded in the part material. The imbedded blast could get into the oil stream and cause bearing or other failure during engine operation.

After grit blasting, the parts should be washed with a suitable solvent to remove loose oil or sludge. The parts then should be sprayed with a proper solvent to remove all traces of grit and then dried with compressed air. If a steel part that has been blasted is not to be inspected immediately, it should be covered with a thin coat of light oil after cleaning. This will prevent rust or some other deterioration after the cleaning process.

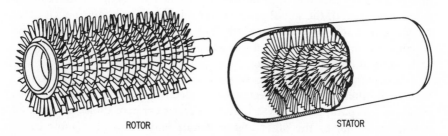

ROTOR   STATOR

Fig. 24. Rotor and stator elements of a typical axial-flow compressor.

ALUMINUM ALLOY CORROSION. Corrosion attacks on aluminum alloy may take place over an entire surface or it may localize and result in deep pits in the metal. This condition may be accentuated from a lack of homogeneous metal characteristics or improper heat treatment. Corrosion attacks are promoted by contact of metals with materials that absorb water such as wood, sponge rubber, felt, dirt, surface films, etc.

BASIC TYPES OF CORROSION. Fretting corrosion occurs when relative motion of small amplitude takes place between two close fitting components. The rubbing contact destroys the protective film on the metal surface and also removes small particles of "virgin" material from the surfaces. These particles act as an abrasive and prevent the formation of any protective oxide film. In addition, this exposes fresh active metal to the atmosphere which further compounds the problem.

If the contact areas are small and sharp, deep grooves resembling Brinell markings or pressure indentations may be worn into the rubbing surfaces.

Most aluminum alloy structural surfaces are electrically anodized in chromic acid tanks prior to fabrication. This gives a good protection against fretting corrosion since it creates a thin film of aluminum oxides to form on the exterior surface of the aluminum. This film is resistant to corrosion and forms a good paint base. Other processes that are somewhat satisfactory but do not form as good a corrosive protection as anodizing are: (1) alkaline cleaning followed by a chromic acid dip; (2) phosphoric acid cleaner; and (3) alkaline dichromate cleaner.

A selective attack along the grain boundaries of metal alloys is known as *intergranular corrosion.* This type of corrosion results from a lack of uniformity in the alloy structure. It is particularly characteristic of precipitation hardened alloys of aluminum and some stainless steels. Intergranular corrosion is difficult to detect in its original stage. Ultrasonic and eddy current inspections are utilized in determining its presence in metals. When this type of corrosion is present and well advanced, the metal may blister or delaminate. This condition is referred to as "exfoliation." In many instances the only solution is to replace the affected parts or sections.

The general cause of intergranular corrosion is due to heat-treated aluminum alloy that has been improperly quenched. As the metal boundaries are destroyed, the metal grains lose their inherent bond to each other resulting in a loss of metal strength. This, of course, is a dangerous type of corrosion because of the seriousness of the damage and the difficulty in locating the areas of damage.

Dissimilar metal corrosion. When two dissimilar metals are in contact with each other and are connected by an electrolyte (such as a continuous liquid or gas path, salt spray, exhaust gases, or condensation), accelerated corrosion of one or the other of the metals may occur. The most easily oxidized surface becomes the "anode" and corrodes. The less active metal of the two becomes the "cathode" of the galvanic cell. The degree of the attack depends on the relative activity of the two surfaces. The greater the difference in activity, the more severe the attack to create dissimilar corrosion.

With few exceptions, whenever metals from two different groups are to be in contact with each other, some special protection is required to assure that dissimilar metal corrosion does not occur.

Direct surface attack. The most common type of general surface corrosion results from the direct reaction of a metal's surface with oxygen in the atmosphere. Unless properly protected, steel will rust and aluminum and magnesium will form corrosion products. The corrosion attack can be accelerated by the metal's exposure to salt spray, salt bearing air, industrial gases, or by engine exhaust gases that come in direct or indirect contact with the particular metal.

Stress corrosion. This type of corrosion results from the combined effect of static tensile stresses applied to a surface over a period of time under corrosive conditions. In general, cracking susceptibility increases with stress, particularly at stresses approaching the yield point of the metal.

CLEANING AND PROTECTION OF MATERIALS. All *exterior surfaces* of aircraft should be kept clean and free of dirt, grime, and other detrimental contamination. After thorough cleaning of the surfaces, they should be polished with approved materials using care not to use any abrasive materials in the process. This not only lessens the possibility of corrosion, rust, etc., but will also add to the performance and speed of the aircraft, especially so in the high performance type.

Use only approved cleaners and polishes in order to maintain the thin protective coating of the metals.

It is important that *tube interiors* be protected against corrosion. A small amount of water, for instance, trapped inside a metal tube can corrode entirely through the tube walls in a relatively short period of time. A good precaution against internal tube corrosion is to coat the tube interior by flushing with hot linseed oil, paralkatone, or other suitable corrosion inhibitor.

Storage battery compartments. The cleaning of battery compartments can be accomplished by the use of a solution of bicarbonate of soda for lead-acid batteries and a boric acid solution for nickel-cadmium batteries. After proper cleaning and adequate drying, battery compartments should be protected from further corrosion by coating the area with acid paint, asphalt base paint, or a bituminous base paint.

Paint or other coatings. In general, aluminum surface corrosion can be "arrested" or temporarily halted by the use of bicarbonate of soda solution. After the surface is cleaned and dried, a coating of zinc chromate can be applied to prevent further corrosion. The surfaces can then be covered with aircraft quality paint or other similar organic coating for further surface protection. Creating an attractive exterior appearance should also be considered during the last stages of surface finishing.

H. BASIC MATHEMATICS

The aircraft mechanic must be able to solve basic mathematical problems not only for the requirements of the examination but also because he will run up against these same problems quite often in aircraft maintenance. This section of the book is not intended to be a treatise on mathematics but simply provides you with a few examples of the kind of mathematics a mechanic should be able to handle.

Addition: accomplished by lining up figures or digits that are to be added together for a total sum:

(1) 25	(2) 251	(3) 2,510	(4) 2.57	(5) .7274	(6) 257,901
37	377	7,640	3.17	.8210	368,201
41	499	8,920	9.20	.7977	457,301

Answers: (1) 103; (2) 1,127; (3) 19,070; (4) 14.94; (5) 2.3461; (6) 1,083,403.

Subtraction: accomplished by lining up figures or digits where one is to be deducted from the other:

(1) 47	(2) 1,792	(3) 1.2750	(4) .8750	(5) 23,517	(6) 1,075,627
−23	−1,476	−0.0270	−.6275	−17,502	− 995,729

Answers: (1) 24; (2) 316; (3) 1.2480; (4) .2475; (5) 6,015; (6) 79, 898

Division: accomplished by finding the number of times a given number will go into a given total:
(1) 472 ÷ 2 (2) 1.2940 ÷ 5 (3) 10,740,000 ÷ 10 (4) .7805 ÷ .0156

Answers: (1) 236; (2) .2588; (3) 1,074,000; (4) 50.0

Multiplication: accomplished by finding the total when one number is multiplied by another number:
(1) 49 × 5 (2) 7,294 × 9 (3) .784 × .25 (4) 5.724 × 1.7 (5) .727 × .44

Answers: (1) 245; (2) 65,646; (3) .19600; (4) 9.7308; (5) .31988

Mathematical problems using plus (+) and minus (−) signs:
(1) +29 and −22 (2) −7 × −8 (3) +97 × −17 (4) The sum of +75,− 87,+62, and −17 (5) −.075 × +.675 (6) −97 × +23 (7) −49 × −7 (8) +89 × −91

Answers: (1) +7; (2) +56; (3) −1,649; (4) +33; (5) −.050625; (6) − 2,231; (7) +343; (8) −8,099

Squaring and cubing numbers:
(1) The square of 4 (4²) (2) The square of 13 (13²) (3) The cube of 3 (3³) (4) The cube of 17 (17³)(5) The square of 91 (91²) (6) The cube of 67 (67³)

Answers: (1) 16; (2) 169; (3) 27; (4) 4,913; (5) 8,281; (6) 300,763

Converting fractions to decimals:
(1) Convert 13/64 to a decimal equivalent (2) Convert ¾ to a decimal equivalent (3) Convert 11/16 to a decimal equivalent (4) Convert ¼ to a decimal equivalent

Answers: (1) .203125; (2) .75; (3) .6875; (4) .25

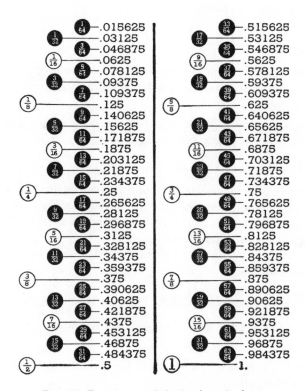

Fig. 25. Fractions and decimal equivalents.

Decimals can be easily converted to fractions by dividing the decimal value by the decimal equivalent of 1/64 inch (.0156). Example: The decimal .2031 is equivalent to 13/64 (.2031 divided by .0156).
(1) Convert .7500 to a fraction (2) Convert .6875 to a fraction (3) Convert .2500 to a fraction (4) Convert .9375 to a fraction

Answers: (1) 48/64 or 24/32 or 12/16 or 6/8 or ¾; (2) 44/64 or 22/32 or 11/16; (3) 16/64 or 8/32 or 4/16 or 2/8 or ¼; (4) 60/64 or 30/32 or 15/16

To convert fractions to percentage values, divide the whole into the part, carry the answer to four decimals (.0000), and then move the decimal over two places and add a percentage sign (%). The first step is the same as converting fractions to decimals. Example: 4/5 is equal to what percent? Dividing 5 into 4 results in .8000. Moving the decimal to the right two digits results in 80.00%.

(1) Convert ⅞ to a percent (2) Convert ¾ to a percent (3) Convert 2/5 to a percent (4) Convert 11/16 to a percent (5) Convert 15/16 to a percent (6) Convert 2/9 to a percent

Answers: (1) 87.50%; (2) 75.00%; (3) 40.00%; (4) 68.75%; (5) 93.75%; (6) 22.22%

An easy solution to multiplying fractions is to multiply the respective values together and conclude with a fractional answer.
Example: ¼ × ⅛ equals 1/32.

(1) Multiply ⅜ × ½ (2) Multiply ⅞ × ⅝ (3) Multiply 1/64 × ½ (4) Multiply ⅛ × ½ (5) Multiply 9/16 × ⅞ (6) Multiply ¾ × ⅛

Answers: (1) 3/16; (2) 35/64; (3) 1/128; (4) 1/16; (5) 63/128; (6) 3/32

The area of a rectangle can be found by multiplying the length times the width using common values for each, such as inches, feet, etc.
(1) A rectangular wing has a span of 30 feet and an average chord (width) of 6 feet. The area in square feet is?
(2) A rectangle has a length of 144 inches and a width of 24 inches. The area in square inches is?
(3) An airplane wing has a span of 110 feet (wing tip to wing tip) and an average chord (width) of 10 feet. The area in square feet is?

Answers: (1) 180 square feet; (2) 3,456 square inches; (3) 1,100 square feet

The area of a square can be found by multiplying the length of any two of its sides together.
(1) A platform is 10 feet square. The area in square feet is?
(2) A storage box is 56 inches square. The area of the bottom of the box in square inches is?
(3) A hangar is 37 feet square. How many square feet of floor space will be available to aircraft stored in this hangar (assume the 37 feet is inside measurement)?

Answers: (1) 100 square feet; (2) 3,136 square inches; (3) 1,369 square feet

The volume of a container (square or rectangular shape) can be found by multiplying the length times the width times the height: L×W×H.
(1) A rectangular container is 5'0" long, 3'0" wide, and 7'0" high. The volume of this container (cubic feet) is?
(2) A square container is 12" long, 12" wide, and 12" high. The volume of this container (cubic inches) is?

Answers: (1) 105 cubic feet; (2) 1,728 cubic inches

Reciprocating engine piston displacement (PD), engine displacement (ED), combustion chamber volume (CCV), and compression ratio (CR) can be found by formulas. The solutions for a typical 18-cylinder, air-cooled, twin-row radial engine are given below.
   Given:
Cylinder bore (diameter) is 6 inches.
Piston stroke (distance of travel from top to bottom position) is 6 inches.
Total volume in each cylinder (piston displacement plus the combustion chamber volume) is 198.50 cubic inches.

1. Find the piston displacement in cubic inches.
   Solution:
Formula to use is PD = π (pi) × radius^2 × stroke
                    PD = $3.1416 \times 3^2 \times 6$
                    PD = $3.1416 \times 9 \times 6$
                    PD = 169.65 cubic inches

2. Find total engine displacement (for 18 cylinders) in cubic inches.
   Solution:
Formula to use is TED = PD (one cylinder) × number of cylinders
                    TED = 169.65 cubic inches × 18
                    TED = 3,053.70 cubic inches

3. Find the combustion chamber volume (CCV) in cubic inches.
   Solution:
Formula to use is CCV = Total volume (TV) minus piston displacement (PD)
                    CCV = 198.50−169.65
                    CCV = 28.85 cubic inches

4. Find the compression ratio of this engine.
   Solution:
Formula to use is $CR = \dfrac{CCV + PD}{CCV}$

$$CR = \dfrac{28.85 + 169.65}{28.85}$$

$$CR = \frac{198.50}{28.85}$$

$$CR = 6.88 \text{ to } 1$$

Given:

9 cylinder, air-cooled, single row, radial engine.

Cylinder bore is 5 inches.

Piston stroke is 5 inches.

Total volume in each cylinder is 118.65 cubic inches.

(1) Find the piston displacement in cubic inches.

(2) Find the total engine displacement (for 9 cylinders) in cubic inches.

(3) Find the combustion chamber volume in cubic inches.

(4) Find the compression ratio of this engine.

Answers: (1) 98.20 cubic inches; (2) 883.80 cubic inches; (3) 20.45 cubic inches; (4) 5.80 to 1

I. MAINTENANCE FORMS AND RECORDS

Prescribed rules and regulations govern the maintenance, preventive maintenance, rebuilding, and alteration of any aircraft holding a U.S. Airworthiness Certificate and the airframe, engine, propeller, or appliance of such an aircraft.

Fig. 26. FAA Form 8320-2.

The information in this section is general in nature, but inside the rear cover of this book are up-to-date regulations that pertain specifically to the aircraft mechanic, his responsibilities, and the paper work with which he must be familiar.

Repairs must be entered in the aircraft permanent records (airframe logbook, engine logbook, or other FAA approved maintenance documents).

Fig. 27. Engine and aircraft logs.

The minimum information entered in these records must include a description of the work done and certification that all pertinent Airworthiness Directives (AD's) have been complied with and that all work has been accomplished using approved methods and materials. The person completing the work and returning the aircraft to service must indicate the date, his name, and FAA certificate number. An approved FAA repair shop (repair station), before returning the aircraft to service, must enter the description of the repair (or other work) in the aircraft permanent records and indicate the date, repair station number, and signature of an authorized representative of the repair station.

Form 337, "Major Repair and Alteration Form," must be completed after major repair and/or alteration to an aircraft. The information in Form 337 includes the aircraft description, name and address of the aircraft owner, name, address, and certificate number of the person or repair station responsible for executing the form, an indication of the type of major repair or alteration, an inspection certification (if needed) and a description or sketches of the work completed. A line should be drawn across the form at the completion of the described work and a notation "Nothing Further" indicated along this line.

Two copies of Form 337 must be executed by the aircraft mechanic and/or repair station. The original copy is submitted to the aircraft owner

DEPARTMENT OF TRANSPORTATION FEDERAL AVIATION ADMINISTRATION **MAJOR REPAIR AND ALTERATION** (Airframe, Powerplant, Propeller, or Appliance)	*Form Approved* *Budget Bureau No. 04-R060.1*
	FOR FAA USE ONLY OFFICE IDENTIFICATION

INSTRUCTIONS: Print or type all entries.   See FAR 43.9, FAR 43 Appendix B, and AC 43.9-1 (or subsequent revision thereof) for instructions and disposition of this form.

1. AIRCRAFT	MAKE		MODEL	
	SERIAL NO.		NATIONALITY AND REGISTRATION MARK	
2. OWNER	NAME (As shown on registration certificate)		ADDRESS (As shown on registration certificate)	

3. FOR FAA USE ONLY

4. UNIT IDENTIFICATION				5. TYPE	
UNIT	MAKE	MODEL	SERIAL NO.	REPAIR	ALTER-ATION
AIRFRAME	◆◆◆◆◆◆◆◆◆◆◆ (As described in item 1 above) ◆◆◆◆◆◆◆◆◆◆◆				
POWERPLANT					
PROPELLER					
APPLIANCE	TYPE				
	MANUFACTURER				

6. CONFORMITY STATEMENT

A. AGENCY'S NAME AND ADDRESS	B. KIND OF AGENCY	C. CERTIFICATE NO.
	U.S. CERTIFICATED MECHANIC	
	FOREIGN CERTIFICATED MECHANIC	
	CERTIFICATED REPAIR STATION	
	MANUFACTURER	

D. I certify that the repair and/or alteration made to the unit(s) identified in item 4 above and described on the reverse or attachments hereto have been made in accordance with the requirements of Part 43 of the U.S. Federal Aviation Regulations and that the information furnished herein is true and correct to the best of my knowledge.

DATE	SIGNATURE OF AUTHORIZED INDIVIDUAL

7. APPROVAL FOR RETURN TO SERVICE

Pursuant to the authority given persons specified below, the unit identified in item 4 was inspected in the manner prescribed by the Administrator of the Federal Aviation Administration and is ☐ APPROVED ☐ REJECTED

BY	FAA FLT. STANDARDS INSPECTOR	MANUFACTURER	INSPECTION AUTHORIZATION	OTHER (Specify)
	FAA DESIGNEE	REPAIR STATION	CANADIAN DEPARTMENT OF TRANSPORT INSPECTOR OF AIRCRAFT	
DATE OF APPROVAL OR REJECTION	CERTIFICATE OR DESIGNATION NO.	SIGNATURE OF AUTHORIZED INDIVIDUAL		

FAA Form 337   (8320)

Fig. 28. FAA Form 337.

and the second copy is forwarded to the FAA. It is optional as to whether the aircraft mechanic or the repair station retain a copy of the completed Form 337.

Time in service, with respect to maintenance records, means the time from the moment an aircraft leaves the surface of the earth until it touches the surface of the earth at the next point of landing.

In order not to confuse *flight time* with time in service, flight time means the time from the moment the aircraft first moves under its own power for the purpose of flight until the moment it comes to 'rest at the next point of landing. This is generally referred to as *block-to-block time.*

Maintenance of aircraft is the responsibility of the registered owner or operator of the particular aircraft. The owner or operator is responsible for maintaining an aircraft in an airworthy condition and being sure that all required inspections of the aircraft are completed within their required time limits. The owner or operator is responsible for the keeping of permanent and accurate maintenance records in an orderly manner. In addition, the owner or operator must ensure that maintenance personnel make appropriate entries in the aircraft and engine log books according to their proper classification. The owner or operator must also ensure that appropriate entries indicate that all regulations, Airworthiness Directives, etc., have been complied with and that the aircraft has been "released for service."

Annual Inspection. No person may operate an aircraft (of U.S. Registry) unless, within the preceding 12 calendar months, the aircraft has had (1) an annual inspection in accordance with current regulations and has been approved for return to service by an authorized person, or (2) an inspection for the issuance of an airworthiness certificate has been made and the aircraft certified to be placed in service by an authorized person.

100-hour inspection. No person may carry any person (other than a crewmember) for hire or give flight instruction for hire unless, within the preceding 100 hours of time in service, the aircraft has been inspected in accordance with current regulations and approved for return to service by an authorized person.

The 100-hour limitation may be exceeded by *not more than 10 hours* if necessary to reach a point at which the inspection can be made. The excess time, however, must be included in computing the next 100 hours of time in service. When the owner or operator of an aircraft complies with the *progressive inspection* requirements of the regulations, the 100-hour inspection and the annual inspection requirements shall not apply to the aircraft. *Note:* The annual and 100-hour inspections do not apply to an aircraft that carries a special flight permit, a current experimental flight permit, or other certificates of exemption.

Returning aircraft to service. The person approving or disapproving the return of an aircraft, airframe, aircraft engine, propeller, or appliance after an annual, 100-hour, or progressive inspection shall make an entry in the permanent maintenance records of that equipment, containing the following information:

(1) The type of inspection (for progressive inspections, a brief description of the extent of the inspection).

(2) The date of the inspection and the aircraft time in service.

(3) The signature (and, if a certificated mechanic, the certificate number) of the person approving or disapproving for return to service, the aircraft, airframe, aircraft engine, propeller, or appliance.

(4) For annual or 100-hour inspections, if the aircraft is approved for return to service, the following or a similarly worded statement must be entered: "I certify that this aircraft has been inspected in accordance with (insert type) inspection and was determined to be in airworthy condition." *Note:* If the aircraft is not approved for return to service, a list of discrepancies and unairworthy items is provided for the aircraft owner or lessee and this fact is also noted in the maintenance records of the aircraft. After all discrepancies and unairworthy items have been properly taken care of, the aircraft can be returned to service with appropriate notations in the aircraft maintenance records.

1. REGISTRATION NO. N–	DEPARTMENT OF TRANSPORTATION FEDERAL AVIATION ADMINISTRATION MALFUNCTION OR DEFECT REPORT			Form Approved Budget Bureau No. 04–R0003	FOR FAA USE ONLY CONTROL NO.
	A. MAKE	B. MODEL	C. SERIAL NO.	7. COMMENTS *(Describe the malfunction or defect and the circumstances under which it occurred. State probable cause and recommendations to prevent recurrence.)*	
2. AIRCRAFT					
3. POWERPLANT					
4. PROPELLER					
5. APPLIANCE/ COMPONENTS	D. NAME OF A/C AND SERIAL NUMBER				
6. SPECIFIC PART THAT CAUSED TROUBLE					
A. NAME OF PART	B. PART NO.		C. PART/DEFECT LOCATION		
FAA USE D. ATA CODE	E. TOTAL TIME	F. TIME SINCE OVERHAUL	G. CONDITION FOUND (worn, bent, broken, etc.)		
					CONTINUE ON REVERSE
FAA FORM 8330–2	SUPERSEDES FAA FORM 1226 AND FAA FORM 8330–1				ENTER ONLY READILY AVAILABLE DATA

Fig. 29. FAA Form 8330-2.

J. BASIC PHYSICS

An aircraft is a machine—one of the most magnificent and complicated machines designed by man. It is subject to the same mechanical laws that govern every other physical object, and the mechanic must be familiar with these laws and their application to aircraft.

Basic elements. An aircraft has weight, mass, and a center of gravity, and it exerts force and force is exerted upon the aircraft. The powerplants (whether reciprocating, turbojet, or turboprop types) do work. The aircraft travels through the air and is subject to velocities and accelerations.

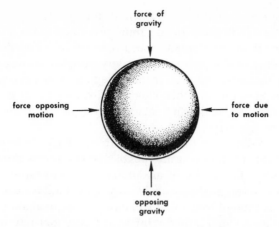

Fig. 30. Equal forces acting on a body.

A body has mass and it has weight. Mass is the amount of matter which the object contains. Weight is a measure of the force of gravity exerted between an object and the earth. Weight (or the effects of gravity) may change, depending on how far the object is from the earth. However, mass does not change.

Gravity is a constant force acting toward the earth's center. The heavier the object the greater will be the force of gravity upon the object.

Force is an action that produces or tends to produce a change in motion or velocity of an object. It requires force to propel an aircraft through the air and, at the same time, the air exerts force on the aircraft.

A combination of forces acting at a given point can be replaced by a *single* force that will produce the same effect. This single force is known as the *resultant* force. The four major forces acting on an aircraft in flight are *lift, drag, thrust, and gravity (weight)*.

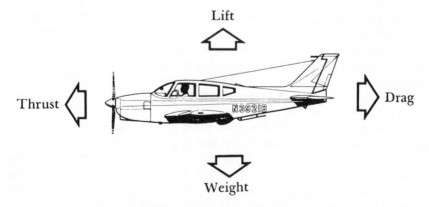

Fig. 31. Forces acting on an aircraft.

Equilibrium is a condition in which all forces acting on an object are in balance. The relationship of the four forces acting on an airplane in level flight and flying at a steady speed (or unaccelerated flight) are: (a) thrust equals drag, and (b) lift equals gravity (weight). During acceleration of the aircraft thrust will be greater than drag and when decelerating drag will be greater than thrust. When climbing (ascending) lift will be greater than weight (gravity) and when descending (gliding or diving) weight (gravity) will be greater than lift.

Stability. To obtain best performance from a flight standpoint, aircraft must be stable. There are two basic types of stability incorporated in aircraft: (1) *static stability*—the ability of an aircraft, when disturbed in flight, to tend to return to its original attitude by way of oscillations; and (2) *dynamic stability*—the creation of forces that will dampen the oscillations and return the aircraft to its original attitude. Inherent forces that return the aircraft to its original attitude (both static and dynamic stability) are created *without* the pilot applying pressure to the controls in the cockpit. Considered to be inherent aircraft stabilities, these can be somewhat nullified if the aircraft has not been properly rigged by the persons responsible for the maintenance or if the pilot has not properly trimmed the aircraft in flight.

Work. In a general sense, work is performed when a force acts (or moves an object) through a given distance. A common unit of work is the *foot-pound* and indicates that a force, measured in pounds, has moved an object through a distance measured in feet. Work can easily be found, in this instance, by multiplying the force applied to an object times the distance the object is moved, or, Work = Force × Distance.

Energy: The ability to accomplish work. Example: a 30-pound rock on the edge of a 10-foot cliff has *potential* energy because it requires 30 pounds lifted through 10 feet (or 300 foot-pounds) of energy to place the rock up on the 10-foot ledge. Therefore, the rock has a potential energy of 300 foot-pounds that will be converted to work should it fall 10 feet from the ledge.

Power: The *rate of doing work*. The basic unit of work is the *horsepower*. One horsepower is equivalent to the work required to create 33,000 foot-pounds in one minute of time (33,000 FT. LBS./MIN.) or 550 foot-pounds in one second (550 FT. LBS./SEC.). Power can be determined by dividing the amount of work done by the time it takes to accomplish the work.

Electrical power is measured by the unit *watts*. One horsepower, in this case, is equal to 746 watts.

Velocity. A given velocity is obtained by an object having moved a given distance in a given amount of time, or knowing the distance travelled and dividing the distance by the time that has elapsed (V = D ÷ T). Velocity is more commonly referred to as speed. Thus (S = D ÷ T). Velocity (or speed) is measured in several units such as statute miles per hour (MPH), nautical miles per hour (knots), feet per second (FPS), etc.

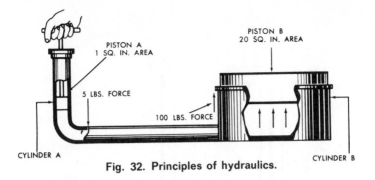

Fig. 32. Principles of hydraulics.

Hydraulic principles. The following hydraulic information is basic. Aircraft hydraulic systems are covered in a later section of this book.

For all practical engineering purposes, a fluid is non-compressible and can transfer work in graduated degrees. A small piston, having a given amount of pressure applied to it, will cause a larger piston to develop a large force that can be applied to the operation of landing gears, flaps, etc. The hydraulic system is highly adaptable to the aircraft since relatively small and light-weight piston-cylinder units (actuators) can create large amounts of force.

The principle of hydraulics pertains to the creation of force utilizing any type of fluid such as oil, water, hydraulic fluid, etc. Thus, do not confuse hydraulic pressures as only being created by hydraulic fluids.

Many units and/or components of an aircraft utilize hydraulic principles in their operation. Among these are retractable landing gears, flaps, wheel brakes, automatic pilot installations, etc.

A basic hydraulic system may consist of a one-square-inch piston displacing fluid against a 10-square-inch piston. The one-square-inch piston has a 10-pound force acting on it, expressed as a pressure of 10 pounds per square inch. The 10-square-inch piston will therefore have a pressure of 10 PSI acting on it but a total force of 100 pounds (10 PSI × 10 square inches). Thus, a hydraulic system can convert a small force into a large force and create a mechanical advantage.

Hydraulic pressure is derived from pumps (gear type in small aircraft and piston type in large aircraft). These pumps are driven directly by the engine or by some other means such as electric motors.

Temperature. The measurement and knowledge of relative temperatures is of utmost importance in the operation of aircraft. Temperature may be generally described as the relative hotness of a substance or it is an indication whether a substance can be considered cold, warm, or hot.

Temperature has a great effect on aircraft and engine performance and is a major factor in the development of operating and performance charts for each particular aircraft-engine combination.

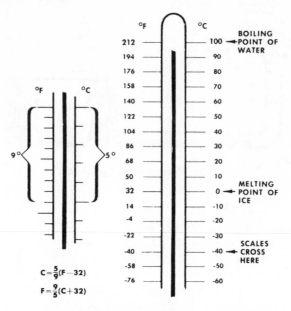

Fig. 33. Fahrenheit-centigrade temperature scales.

Standard sea level temperatures are 59°F. or 15°C. Temperature in the atmosphere decreases with altitude at a stable or standard rate. The average temperature lapse rate per 1,000 feet of altitude is 3.5°F. or 2°C.

Heat. In general, heat is a physical form of energy that can be generated by combustion, chemical action, friction, etc. The generation of heat energy is the basic principle by which all combustion engines derive their power.

When fuels are vaporized, mixed with air, and the fuel-air mixture ignited, a tremendous amount of heat energy is released for the purpose of creating engine power output regardless of the type of combustion engine involved.

Heat can be transferred by radiation, conduction, and convection. These methods of heat transfer are essential to the atmosphere and are applied to the operation of aircraft and aircraft engines.

Radiation is a means of transferring heat by waves, conduction is a means of transferring heat through solids, and convection a means of transferring heat vertically through the atmosphere.

The relationship between temperature, pressure, and volume of an air mass is important to aircraft operation. A decrease in volume of a given air mass will create an increase in pressure and vice-versa. In a fixed volume container, such as an air bottle, if the pressure is increased from 10 PSI to 20 PSI the volume remains the same but the pressure will be doubled. It is important to remember that pressure measurements are indicated in pounds per square inch and not pounds only. The total output of a unit in pounds is described as force. When a force acts through a given distance the resultant is work.

Matter: (1) solid, (2) liquid, and (3) gas. All substances come under one of these three forms or a combination thereof. Examples: (1) an engine crankshaft would be a solid; (2) fuel is a liquid; and (3) carbon monoxide is a gas.

The atmosphere is an envelope of air surrounding the earth's surface. The continual changes of temperature, pressure, and moisture (relative humidity) in the atmosphere have a pronounced effect on both aircraft and engine performance.

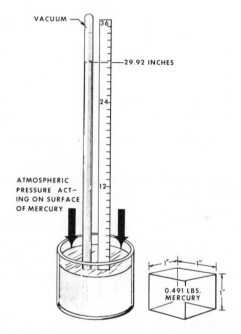

VACUUM

36

29.92 INCHES

24

ATMOSPHERIC
PRESSURE ACT-
ING ON SURFACE
OF MERCURY

12

1"   1"

0.491 LBS.
MERCURY

Fig. 34. Measurement of atmospheric pressure.

Temperature, pressure, and moisture effect on performance. Temperature decreases with altitude, pressure decreases with altitude. This normal atmospheric phenomenon creates adverse flight conditions that are overcome by engineering, ingenuity, production, and maintenance of aircraft. Heating, oxygen, and cooling systems must be incorporated to offset these atmospheric conditions. Their incorporation must be done with the least amount of added weight and the greatest amount of efficiency.

Moisture in the atmosphere affects air density to a great extent. High moisture content (high relative humidity) of the air decreases its density and low moisture content (low relative humidity) increases air density. Water vapor is considered the lightest component of air; thus, the greater concentration of water vapor the lighter (or less dense) the air. A good way to observe this phenomenon is to note that clouds (moist air) float above dry air (or air that has less moisture content than the cloud layers).

Airfoils. The term "airfoil" identifies any object that receives a useful reaction from the air through which it moves. Some examples of airfoils are wings, control surfaces, trim tabs, flaps, etc.

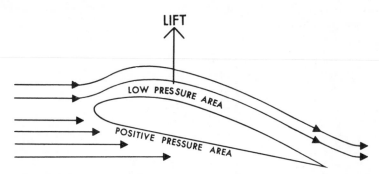

Fig. 35. Difference in pressure between upper and lower wing surfaces produces lift.

The airplane wing (a major airfoil) is so designed that when it is moved through the air a low pressure is created along the upper surface (upper camber or upper curvature). At the same time a higher pressure is created along the lower surface, resulting in a "lifting" action. As a rule, the upper surface of the wing has a greater curvature than the lower surface. As the air divides at the leading edge of the wing, the portion of air moving over the upper surface must do so at a greater velocity. This greater velocity creates a lower pressure resulting in wing lift. Obviously, the more dense the air the greater will be the lift reaction.

More lift is obtained under conditions of low temperature, high pressure, and low humidity (dry air). Lift is decreased under the opposite conditions of high temperature, low pressure, and high humidity (moist air).

General aircraft performance. In general, aircraft of the conventional type equipped with reciprocating engines have better performance and higher engine power output in the lower altitudes. When this type of aircraft ascends into lighter or less dense air its performance is decreased and engine power decreases appreciably. This results in a relatively low *service ceiling,* which is defined in the U.S. as that altitude at which an airplane climbs less than 100 feet per minute. It is assumed that at such a low rate of climb the airplane has lost the ability to be properly controlled and/or maneuvered; therefore, it is not considered to be "serviceable" or safe.

Absolute ceiling of an aircraft is the altitude at which the aircraft will no longer climb at all, and any attempt to make it climb higher will result in a stall.

To increase power in reciprocating engines, several means have been incorporated: fuel injection, blowers, superchargers (both internal and external), compounded by the use of power recovery turbines'(PRT's), low tension ignition, etc.

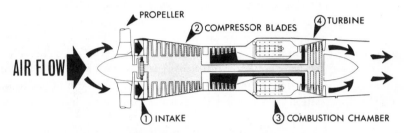

Fig. 36. Turboprop operation.

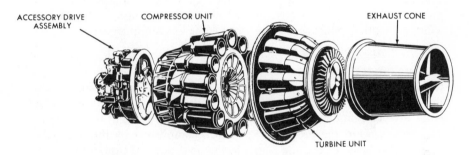

Fig. 37. Major assemblies, centrifugal-flow type.

The combined turbine engine and propeller development—the turboprop—represented a big advance over the piston-engine-propeller combination, but the pure jet or turbojet engine was an even more significant breakthrough as far as higher speeds and altitudes of operation were concerned.

K. MAINTENANCE PUBLICATIONS AND OTHER RELATED DATA

The aircraft mechanic must be familiar with the manufacturer's aircraft maintenance specifications, data sheets, manuals, and pertinent publications.

There will be no attempt here to describe each particular type of publication and its purpose since the mechanic's need in general will depend on the type, class, and category of aircraft that he is working on or the extent of the maintenance that he becomes involved with.

The aircraft mechanic will find many publications available to him—either free or at a nominal charge.

All Federal Aviation Administration offices will cooperate 100% with maintenance personnel by furnishing lists of suitable publications and, in many instances, will have maintenance forms available in limited quantity.

GENERAL MAINTENANCE INFORMATION. An example of the type of information a powerplant mechanic might have to find by reference would be the suitability of a given propeller for use with a particular engine-airplane combination or the minimum diameter of propeller types and models required for incorporation with a particular engine. This information would be found in the manufacturer's publications. Another example would be the location of an aircarft's leveling point and datum in order to determine the empty weight and the empty weight center of gravity. This information would be found by reference to the aircraft manufacturer's specifications for the particular aircraft. In this instance, the manufacturer's specifications will designate a given location on the aircraft from which proper leveling can be ascertained. It may be the bottom sill of a given window or other point convenient to use for leveling purposes.

ENGINE-PROPELLER SPEED RATIOS. Engine-propeller speed ratios are found in the manufacturer's specifications and can be readily converted to RPM by the mechanic. Examples: In a typical large reciprocating engine, the manufacturer's specifications indicate the engine-propeller speed ratio is 2:1 (two revolutions of the crankshaft to one revolution of the propeller). This reduction of the propeller RPM is accomplished by the installation of planetary gears (reduction gears) in the nose or front case of the engine. The reduction is necessary to control the propeller tip speeds of large propellers to keep the tip speeds lower than the speed of sound. An engine-propeller ratio of 16:9 indicates the crankshaft will rotate 16 times to 9 rotations of the propeller. At 2,400 RPM, with a 2:1 ratio, the engine will rotate at 2,400 RPM and the propeller at 1,200 RPM. At 3,000 RPM, with a 16:9 ratio, the engine will rotate at 3,000 RPM and the propeller at 1,687.5 RPM (9/16 of 3,000).

INSTRUMENT MARKINGS. Instrument markings of various colors are required on the face of aircraft instruments to indicate certain operation limitations. These limitations are established by the manufacturer of a particular type and model of aircraft.

Airspeed indicator markings. Standard color markings are painted on the glass face of an airspeed indicator, or colored decals may be used. (1) The green arc indicates the normal operating speeds of the aircraft; the lower segment of the green arc indicates stalling speed with power off, flaps up and gear up; the upper segment of the green arc indicates the normal maximum cruise speed. (2) The white arc indicates the flap operating speeds; the lower segment is the stalling speed, power off, flaps down and gear down, and the upper segment is the maximum allowable flap operating speed. (3) The yellow arc indicates the caution speed range. (4) The red line indicates the never-exceed speed of the aircraft.

Engine instrument markings. In general, two basic colors are used on engine instrument glass faces to indicate operating limitations: (1) green to indicate normal operating limitations, and (2) red to indicate maximum and/or minimum operating limitations.

Glass-case white marking. Markings are painted or placed on the glass face of instruments with decals. It is important that the glass, when installed, be properly lined up with the values on the actual face of the instrument. In order to insure the proper alignment, a white line is painted across the edge of the instrument case and on to the glass face the first time all color markings indicate correct limitations. Thus, if the glass is removed it can be replaced in its correct position.

TECHNICAL STANDARD ORDER (TSO'S). Technical Standard Orders are published and distributed to interested persons in the aircraft industry. They contain minimum performance and quality control standards for specified materials, parts, or appliances. The performance standards in each TSO are those that the Administrator finds necessary to ensure that the components concerned will operate satisfactorily and accomplish their intended purpose under specified conditions.

The TSO system lists the material, parts, and appliances for which the Administrator has received statements of conformance under the Technical Standard Order System.

No person may identify an article with a TSO marking unless he holds a TSO authorization and the article meets applicable TSO standards.

SUPPLEMENTAL TYPE CERTIFICATES. Supplemental type certificates are applicable when a major modification has been made to a given aircraft. A Supplemental Type Certificate will then be issued if all airworthiness requirements are met.

WEIGHT AND BALANCE DATA. All aircraft must have permanent maintenance records containing complete and current weight and balance information. This information is obtained from the latest required aircraft weighing procedure and the manufacturer's approved loading schedule.

When an aircraft has been modified or has had permanent items removed from or added to it, the original weight and balance data should be revised or the aircraft re-weighed.

All weight and balance information must be kept up to date and included as an integral part of the permanent records of the particular aircraft. This information must include the empty weight, empty weight center of gravity, useful load, maximum allowable gross weight, datum (reference or zero station) location, and the forward and rearward center of gravity limits. Precise information on allowable loading that would include occupants, fuel, oil, and baggage must be adjusted when modifications cause a change in the empty weight of the aircraft.

AIRCRAFT ENGINE AND PROPELLER SPECIFICATIONS, AND TYPE CERTIFICATE DATA SHEETS. The specifications as noted are of extreme importance to the mechanic, especially if he is in business for himself or works under the privileges of an Authorized Inspector (AI) rating. (An AI rating may be issued to an aircraft mechanic who has worked under the privilege of his Airframe and Powerplant Certificate for a period of at least three years, has been considered eligible by the FAA, and who has satisfactorily passed the required government examinations.)

The following publications are available from the Government Printing Office, both with monthly supplementary service for approximately one year: (1) Aircraft Specifications; and (2) Aircraft Engine and Propeller Specifications and Type Certificate Data Sheets.

The aircraft mechanic is not expected to remember all the varied specification data but he is expected to have available a reference library covering the airframe, engines, propellers, and appliances he is responsible for maintaining.

AIRWORTHINESS CERTIFICATES. Each aircraft registered in the U.S. that has met the standards of performance, maintenance, and required inspections is considered to be airworthy and is issued an Airworthiness Certificate. This certificate must be in the aircraft at all times while the aircraft is considered airworthy. It is renewed each year. All aircraft registered in the U.S. must have an annual inspection which is very complete and comprehensive. Any items found not to be airworthy must be repaired, modified, or replaced. After all discrepancies have been brought to an airworthy condition the aircraft is re-inspected, proper entries are made in the aircraft permanent records, and the aircraft is "returned to service." The person making the inspections and returning the aircraft to service must possess a current and valid Authorized Inspector Rating.

Aircraft on an approved Progressive Inspection System will be considered airworthy only if the initial inspection and all further routine and detailed inspections are conducted as prescribed in the Progressive Inspection schedule.

AIRWORTHINESS DIRECTIVES (AD's). Airworthiness Directives are issued for pertinent items, maintenance, or inspections that have to be dealt with in regards to certain designated aircraft, and may pertain to the airframe, engine, propeller, or an appliance. These Directives reflect the correction of a flaw or other weakness that may have occurred in a particular aircraft. The issuance of AD's may pertain to a certain make, model, type, and year of aircraft or they may be more far reaching in regards to the aircraft affected by their issuance.

Airworthiness Directives are issued for small aircraft (12,500 pounds or less, maximum certificated takeoff weight) and large aircraft (more than 12,500 pounds, maximum certificated takeoff weight) at regular intervals by the Administrator.

The initial AD summaries, available from the GPO, are (1) Summary of Airworthiness Directives for Small Aircraft, and (2) Summary of Airworthiness Directives for Large Aircraft. Each volume may be purchased separately on subscription.

Airworthiness Directives are mandatory and must be complied with in accordance with the time or date limitation incorporated in each directive.

ADVISORY CIRCULARS. The FAA issues Advisory Circulars to inform the aviation public of nonregulatory material of interest. Advisory Circulars are identified by number and subject corresponding to the numbering system used for Federal Aviation Regulations.

Many advisory circulars are very useful to the aircraft mechanic. Some are free of charge and others available at nominal charges. Advisory Circular Checklists are available on request from the FAA. From this checklist the mechanic can determine which AC's may be of interest to him.

L. MECHANIC PRIVILEGES AND LIMITATIONS

Aircraft mechanics must be familiar with the privileges and limitations imposed upon them as set forth in the Federal Aviation Regulations, Parts 65.15, 65.21, 65.81, 65.83, 65.85, and 65.89. Refer to the FAR booklet inside the back cover of this manual.

2. AIRFRAME STRUCTURES

This section covers the general types of airframe structures and the phases of maintenance and materials utilized in repair and alterations of structures.

A. WOOD STRUCTURES

Three forms of wood are commonly used in aircraft: solid wood, plywood, and laminated wood. All wood and plywood used in the repair of aircraft structures should be of aircraft quality.

When the moisture content of a piece of wood is lowered, its dimensional change is greatest in a tangential direction (across the fibers and parallel to the growth rings), somewhat less in a radial direction (across the fibers and perpendicular to the growth rings), and is negligible in a longitudinal direction (parallel to the fibers).

Shrinking effects may be minimized by (a) the use of bushings that are slightly short so that when the wood member shrinks, the bushings do not protrude and the fittings can be tightened firmly against the member and (b) gradual dropping off of plywood face plates either by feathering or by shaping.

GLUES AND GLUING. Satisfactory glue joints in aircraft develop the full strength of wood under all conditions of stress. The most important considerations in the gluing process are (1) properly prepared wood surfaces, (2) glue of good quality, properly prepared, and (3) good gluing technique.

It is recommended that no more than eight hours be permitted to elapse between final surfacing and gluing. The gluing surfaces should be machined smooth and true with planers, jointers, or special miter saws.

Wood surfaces for gluing should be free from oil, wax, varnish, shellac, lacquer, enamel, dope, sealers, paint, dust, dirt, old glue, crayon marks, and other extraneous materials.

Glues used in aircraft repairs fall into two general groups: casein glues, and resin glues.

Casein glues are probably more widely used than any of the resin glues in wood aircraft repair work. The forms, characteristics, and properties of water-resistant casein glues have remained substantially the same for many years except for the addition of preservatives. Casein glues for use in aircraft should contain suitable preservatives such as the chlorinated phenols and their sodium salts. This increases their resistance to organic deterioration under high humidity exposures. Most casein glues are sold in powder form ready to be mixed with water at ordinary room temperatures.

Resin glues (synthetic) for wood are outstanding in that they retain their strength and durability under moist conditions and even after exposure to water. The best-known and most commonly used synthetic resin glues are the phenol-formaldehyde, resorcinol-formaldehyde, and urea-formaldehyde types. Materials, such as walnut-shell flour or wood flour, are often added by the glue manufacturer to give better working characteristics and joint-forming properties.

Most glues are of the room temperature setting type; however, the suitable curing temperatures for the urea-formaldehyde type vary from 70° to 75°F. For resorcinal glues from 70°F. up. The strength of the joint cannot be depended upon if assembled and cured at temperatures below 70°F.

To make a satisfactory glue joint, glue should be spread evenly on both of the surfaces to be joined. It is recommended that a clean brush be used and care taken to see that all surfaces are covered. The spreading of glue on but one of the surfaces is not recommended.

Generally, the pressing time for casein and resin glue joints should be seven hours or more. Other types of glue require various times and temperatures for curing. Glue joints increase in strength mainly as a result of drying. Where it is convenient to do so, it is better to maintain pressure from one day to the next as this enables the joints to reach a higher proportion of their final strength before being disturbed.

Pressure should be applied to the joint before the glue becomes too thick to flow and is accomplished by means of clamps, presses, or other mechanical devices.

The amount of pressure required to produce strong joints in aircraft assembly operations may vary from 125 to 150 PSI for softwoods and 150 to 200 PSI for hardwoods.

On small joints such as found in wood ribs, the pressure is usually applied only by nailing the joint gussets in place after spreading.

SPLICING OF WOOD SPARS. A spar may be spliced at any point except under wing attachment fittings, landing gear fittings, engine-mount fittings, or lift and interplane strut fittings. These fittings should not overlap any part of the splice.

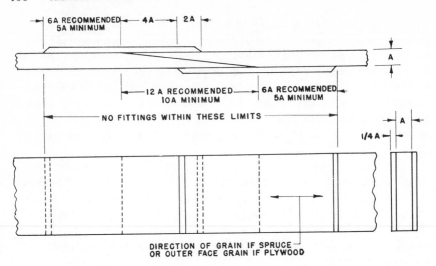

REINFORCEMENT PLATES TO BE SPRUCE OR PLYWOOD AND SHALL BE GLUED ONLY.

SOLID SPARS MAY BE REPLACED WITH LAMINATED ONES OR VICE VERSA,
PROVIDED THE MATERIAL IS OF THE SAME HIGH QUALITY.

Fig. 38. Method of splicing solid or laminated rectangular spars.

The splice reinforcing plate may overlap drag or antidrag wires or compression strut fittings if the reinforcing plates are on the front face of the front spar or on the rear face of the rear spar. In such cases it will be necessary to install slightly longer bolts.

In the splicing of box spar webs, solid wood should never be used to replace plywood webs as plywood is stronger in shear than solid wood of the same thickness. This is due to the variation in grain direction of the individual plies.

Solid spruce spars may be replaced with laminated ones or vice-versa, provided the material is of the same high quality. External plywood reinforcement should always be replaced with plywood as in the original structure.

In cases of elongated bolt holes in a spar or cracks in the vicinity of bolt holes, a new section of spar should be spliced in or the spar replaced entirely. Otherwise the method of repair must be specifically approved by a representative of the FAA.

TYPES OF WOOD PATCHES. Four types of patches are used in the repair of wood structures: (1) the surface or over-lay patch; (2) the splayed patch; (3) the oval plug patch; and (4) the scarf patch. Surface patches should not be used on skins over ⅛″ thick. Splayed patches should not be used on skins over 1/10″ thick. There are no skin thickness limitations for the use of scarf patches and plug patches.

Repairs to skins of single curvature can usually be formed from flat plywood, either by bending it dry or after soaking it in hot water.

B. AIRCRAFT FABRIC COVERING AND REPAIRS

All fabric, surface tape, reinforcing tape, machine thread, lacing cord, etc., used for recovering or repairing an aircraft structure should be of high-grade aircraft textile material of at least as good quality and equivalent strength as those originally used.

Reinforcing tape should be of similar quality to the fabric. Surface tape and finishing tape should have approximately the same properties as the fabric to be used. Lacing cord should have the strength of at least 80 pounds double or 40 pounds single strand. Machine thread should have a strength of at least five pounds single strand. Hand-sewing thread should have a strength of at least 14 pounds single strand.

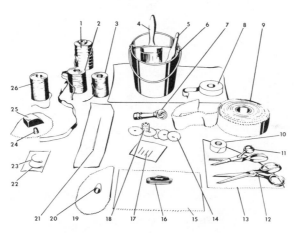

1. Reinforcing tape	15. Reinforcing patch, fitting
2. No. 8 cotton thread	16. Knife
3. Linen lacing cord	17. Sewing needles
4. Large 4-inch flat dope brush	18. Celluloid grommets
5. Small 3-inch flat dope brush	19. Reinforcing patch, fairing
6. Clear dope	20. Rubber finger guard
7. Roller	21. 12-inch rib-stitch needle
8. 1-inch adhesive tape	22. Curved rib-stitch needle, round
9. 2¼-inch pinked tape	23. Curved rib-stitch needle, tri-
10. Cellulose tape	angular
11. 8-inch pinking shears	24. Thimble
12. 8-inch scissors	25. Brown beeswax
13. Reinforcing patch, tip fairing	26. No. 8 cotton thread, waxed
14. Circular patches	

Fig. 39. Tools and materials used in fabric repairs.

In fabric covering practices, the fabric may be applied so that either the warp or fill threads are parallel to the line of flight. Either the envelope method or blanket method of covering is acceptable.

Flutter precautions must be taken when repairing control surfaces, especially on high-performance airplanes. This means that care should be exercised so that the repairs do not involve the addition of weight aft of the hinge line.

Dope-proofing is an important step in the covering of aircraft. Treat all parts of the structure which come in contact with doped fabric with a protective coating such as aluminum foil, dope-proof paint, or cellulose tape. Clad aluminum and stainless steel parts need not be dope-proofed.

Seams parallel to the line of flight are preferable although span-wise seams are acceptable. Sewed seams parallel to the line of flight (chordwise) should not be placed over a rib or be so placed that the lacing will be through or across such a seam.

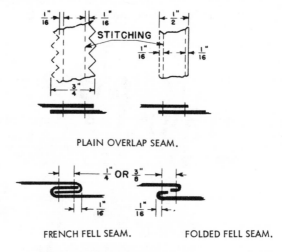

PLAIN OVERLAP SEAM.

FRENCH FELL SEAM.   FOLDED FELL SEAM.

Fig. 40. Typical seams.

Machine sewed seams should be of the folded fell or French fell types. Where selvage edges or pinked edges are joined, a plain lap seam is satisfactory.

Hand-sewing should be locked at intervals of six inches and the seams should be properly finished with a lock stitch and a knot. In hand-sewing there should be minimum of four stitches per inch.

A sewed spanwise seam on a metal or wood covered leading edge should be covered with pinked-edge surface tape at least four inches wide. A sewed spanwise seam at the trailing edge should be covered with pinked-edge surface tape at least three inches wide.

The double stitched lap joint should be covered with pinked edge surface tape at least four inches wide.

A lapped and doped spanwise seam on a metal or wood covered leading edge should be lapped at least four inches and covered with pinked edge surface tape at least eight inches wide. A lapped and doped spanwise seam at the trailing edge should be lapped at least four inches and covered with pinked edge surface tape at least three inches wide. It should be notched at intervals not exceeding six inches.

COVERING METHODS. The envelope method of covering is accomplished by sewing together widths of fabric cut to specified dimensions and machine-sewn to form an envelope. This can then be drawn over the frame to be covered.

The blanket method of covering is accomplished by sewing together widths of fabrics of sufficient lengths to form a blanket over the surfaces of the frame to be covered.

ANTI-TEAR STRIPS IN COVERING. On aircraft with never-exceed speeds in excess of 250 MPH, anti-tear strips are recommended under reinforcing tape on the upper surface of wings, and the bottom surface of that part of the wing in the slipstream. Where the anti-tear strip is used on both the top and bottom surfaces, pass it continuously up to and around the leading edges and back to the trailing edge.

REPAIR OF TEARS. Tears should be repaired by sewing the torn edges together using a baseball stitch and doping a piece of pinked-edge fabric over the tear.

Clean the surfaces to be covered by the patch by rubbing the surface with a rag dipped in dope and wiping it dry with a clean rag. You may also scrape the surface with a putty knife after it has been softened with fresh dope. Dope solvent or acetone may be used for the same purpose but care should be taken that it does not drop through on the inside of the opposite surface causing the dope to blister. A patch of sufficient size should be cut from airplane cloth to cover the tear and extend at least 1½″ beyond the tear in all directions. The edges of the patch should either be pinked similar to surface tape or frayed out about ¼″ on all edges.

SEWED PATCH REPAIR. When the damage is such that it will not permit sewing the edges together, a sewed-in repair patch may be used if the damage is not longer than 16″ in any direction. When the damaged area exceeds 16″ in any direction a new panel should be installed.

Apply surface tape over the seams with the second coat of dope. If the opening extends over or closer than one inch to a rib or other load member, the patch should be cut to extend three inches beyond the member.

After sewing has been completed, the patch should be laced to the rib over a new section of reinforcing tape using approved methods. The old rib lacing and reinforcing tape should not be removed.

UNSEWED REPAIRS. Unsewed (doped on) repairs can be made on all fabric-covered surfaces provided the never-exceed speed is not greater than 150 MPH. A doped patch repair may be used if the damage does not exceed 16″ in any direction. When damage exceeds 16″ in any direction, make the repair by doping in a new panel. Cut out the damaged section making a round or oval-shaped opening trimmed to a smooth contour. Clean the edges of the opening which are to be covered by the patch with grease solvent. Sand or wash off the dope from the area around the patch with dope thinner. Support the fabric from underneath while sanding.

Materials	Specification	Minimum tensile strength new (undoped)	Minimum tearing strength new (undoped)	Minimum tensile strength deteriorated (undoped)	Thread count per inch	Use and remarks
Airplane cloth mercerized cotton (Grade "A").	Society Automotive Engineers AMS 3806 (TSO–C15 references this spec.).	80 pounds per inch warp and fill.	5 pounds warp and fill.	56 pounds per inch.	80 minimum, 84 maximum warp and fill.	For use on all aircraft. Required on aircraft with wing loadings greater than 9 p.s.f. Required on aircraft with placarded never-exceed speed greater than 160 m.p.h.
Do_____	MIL–C–5646 ____	____ do ____	____ do ____	____ do ____	____ do ____	Alternate to AMS 3806.
Airplane cloth cellulose nitrate pre-doped.	MIL–C–5643 ____	____ do ____	____ do ____	____ do ____	____ do ____	Alternate to MIL–C–5646 or AMS 3806 (undoped). Finish with cellulose nitrate dope.
Airplane cloth cellulose acetate butyrate, predoped.	MIL–C–5642 ____	____ do ____	____ do ____	____ do ____	____ do ____	Alternate to MIL–C–5646 or AMS 3806 (undoped). Finish with cellulose acetate butyrate dope.
Airplane cloth mercerized cotton.	Society Automotive Engineers AMS 3804 (TSO–C14 references this spec.).	65 pounds per inch warp and fill.	4 pounds warp and fill.	46 pounds per inch.	80 minimum, 94 maximum warp and fill.	For use on aircraft with wing loadings of 9 p.s.f. or less, provided never-exceed speed is 160 m.p.h. or less.
Airplane cloth mercerized cotton.	Society Automotive Engineers AMS 3802.	50 pounds per inch warp and fill.	3 pounds warp and fill.	35 pounds per inch.	110 maximum warp and fill.	For use on gliders with wing loading of 8 p.s.f. or less, provided the placarded never-exceed speed is 135 m.p.h. or less.
Glider fabric cotton.	A. A. F. No. 16128. AMS 3802.	55 pounds per inch warp and fill.	4 pounds warp and fill.	39 pounds per inch.	80 minimum warp and fill.	Alternate to AMS 3802–A.
Aircraft linen.	British 7F1 ____	----------	----------	----------	----------	This material meets the minimum strength requirements of TSO–C15.

Fig. 41. Textile fabric used in aircraft covering.

For holes up to eight inches in size, make the fabric patch of sufficient size to provide a lap of at least two inches around the hole. On holes over eight inches in size, make the overlap of the fabric around the hole at least one-fourth the hole diameter with a maximum limit of lap of four inches.

TESTING FABRIC COVERING. Tensile testing of fabric is a practical means for determining whether a fabric covering has deteriorated to a point where re-covering is necessary. The testing may be carried out in accordance with the procedures set forth in industry or Government Specifications. In all cases the specimens should be tested in the undoped condition. The use of acetone or dope thinner is suggested as a means of removing the dope.

STRENGTH CRITERIA. The maximum permissible deterioration for used aircraft fabric based on a large number of tests is 30%. Fabric which has less than 70% of the original required tensile strength would not be considered airworthy.

REPLACEMENT OF FABRIC. Fabric may be replaced with metal sheet provided that the increase in weight does not cause the center of gravity travel to exceed its limits, that the gross weight of the aircraft remains within the operating limits, and that the original structural integrity of the aircraft is not impaired by the alteration.

In view of the complexity of this modification it would be advisable to obtain FAA engineering approval of the alteration prior to its completion.

C. AIRCRAFT FINISHES

NATIONALITY AND REGISTRATION MARKS. The identification of each aircraft is *marked,* and the marking is displayed according to FAR's.

Fixed-wing aircraft have identification marks displayed horizontally on the vertical tail surfaces or on the sides of the fuselage.

If identification marks are displayed on the vertical tail surfaces, both surfaces of a single vertical tail or the outer surfaces of a multivertical tail must be marked.

If identification marks are displayed on the fuselage surfaces, both sides of the fuselage must be marked between the trailing edge of the wing and the leading edge of the horizontal stabilizer. If engine pods or other appurtenances are located in this area and are an integral part of the fuselage side surfaces, the marks are placed on such pods or appurtenances.

The required identification marks are of equal height of not less than 12 inches.

Identification marks shall be ⅔ as wide as they are high with the exception of the number "1" which shall be 1/6 as wide as it is high.

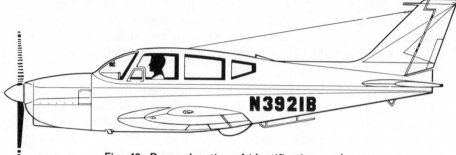

Fig. 42. Proper location of identification marks.

Markings shall be formed by solid lines of a thickness equal to 1/6 of the character height. The spacing between the numbers and letters shall not be less than ¼ of the character width.

Markings shall contrast in color with the background color of the airplane. Markings of any other nature such as emblems, insignias, etc., shall not be placed in any location or manner that may interfere with the nationality and registration marks.

AIRCRAFT DOPING. Dopes are generally supplied at a consistency ready for brush coats. For spraying operations practically all dopes require thinning. Thinning directions are usually listed on the container label. The amount of thinner to be used will depend on the dope, atmospheric conditions, the spraying equipment, the spraying technique of the operator, and the type of thinner employed. The thinning of dopes influences the drying time and tautening properties of the finish and it is necessary that it be done properly.

BLUSHING. Blushing is very common when doping under humid conditions which causes rapid evaporation of thinners and solvents. This lowers the temperature on the surface, causing condensation of moisture and producing the white appearance known as blush.

A blushed finish has very little protective or tautening value. Where the relative humidity is such that only a small amount of blushing is encountered in doping, this condition can be eliminated by thinning the dope with a blush-retarding thinner. A slight increase in room temperature is also helpful.

Under very adverse humidity conditions it is best to temporarily suspend doping operations as the use of large amounts of blush-retarding thinner is not advisable. It gives undesirable drying properties to the fabric and doping procedures.

The total number of coats of dope should not be less than that necessary to result in a taut and well-filled finish job. A guide for finishing fabric-covered aircraft follows:

(1) Two coats of clear dope, brushed on and sanded after the second coat.

(2) One coat of clear dope, either brushed or sprayed, and sanded.

(3) Two coats of aluminum pigmented dope, sanded after each coat.

(4) Three coats of pigmented dope (the color desired), sanded and rubbed to give a smooth, glossy finish when completed.

(5) Care should be taken not to sand heavily over the center portion of pinked tape and over spars in order not to damage the rib-stitching cords and fabric.

TECHNIQUE OF DOPING. Brush on the first two coats of dope and spread on the surface as uniformly as possible, thoroughly working it into the fabric. The first coat should produce a thorough and uniform wetting of the fabric. Apply succeeding coats with only sufficient brushing or spraying to spread the dope smoothly and evenly.

Apply surface tape and reinforcement patches with the second coat of dope.

With the second coat of dope, install drainage grommets on the underside of airfoils at the trailing edge and as close to the rib as practicable. After the doping scheme is completed, open the drain holes by cutting out the fabric with a small-bladed knife. Do not open drainage grommets by punching.

In cold weather dopes become viscous, and will pull and rope under the brush, and if thinned sufficiently to spray, lack body when dry. Prior to use, allow dopes to come to a temperature approximately that of the dope room, 24°C. (75°F.).

Orange peel and pebble effect result from insufficiently thinned dope or when the spray gun is held too far from the surface being sprayed.

Runs, sags, laps, streaks, high and low spots are caused by improperly adjusted spraying equipment or improper spraying technique.

Blisters may be caused by water or oil entering the spray gun. Drain air compressors, air regulators, and air lines daily.

Pin holes may be caused by not allowing sufficient time for drying between coats, or after water sanding, or they may be due to insufficiently reduced dope.

Wet areas on a doped surface indicate that oil, grease, soap, etc., had not been properly removed before doping.

D. SHEET METAL STRUCTURES

Extensive repairs to damaged stressed skin on monocoque types of aluminum alloy structures should be made at the factory of origin or by a repair station rated for this type of work. The monocoque type of fuselage utilizes circular bulkheads, horizontal stringers, and a stressed metal covering with the stressed skin contributing strength to the entire structure.

Regardless of the circumstances, aluminum structure repair work should be undertaken only by a certified mechanic thoroughly experienced in this type of work.

The use of annealed 17S (2017) or 24S (2024) alloys for any structural repair of an aircraft is not considered satisfactory on account of their poor corrosion resisting properties.

DRILLING OVERSIZED HOLES. Great care should be exercised to avoid drilling oversize holes or otherwise decreasing the effective tensile area of wing spar capstrips, wing, fuselage, or fin longitudinal stringers, or other highly stressed tensile members.

SELECTING REPLACEMENT PARTS. In selecting the alloy, it is usually satisfactory to use 24S (2024) in place of 17S (2017) since the former is stronger. Hence, it will not be permissible to replace 24S (2014) by 17S (2017) unless the deficiency in strength of the latter material has been compensated by an increase in material thickness or the structural strength has been substantiated by test or analysis.

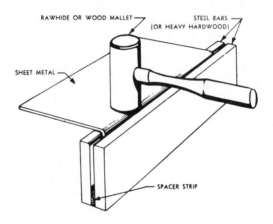

RAWHIDE OR WOOD MALLET   STEEL BARS (OR HEAVY HARDWOOD)

SHEET METAL

SPACER STRIP

Fig. 43. Forming of sheet metal.

FORMING PARTS. Bend lines should preferably be made to lie at an angle to the grain of the metal (preferably 90°). Before bending, all rough edges should be smoothed, burrs removed, and relief holes drilled at the ends of bend lines and at corners to prevent cracks from starting. For material in the heat-treated condition, the bend radius should be large.

QUENCHING. The quenching of 17S (2017) or 24S (2024) alloys in water above 100°F., or at any temperature after heat treatment, will not be satisfactory. For clad material, when the use of cold water will result in too great a distortion of the finished part, the use of oil, hot water, water spray or forced air draft is satisfactory, provided the parts will be subject to severe corrosion in service. Quenching in still air is not satisfactory.

RIVETS AND RIVETING. All protruding head rivets, (roundhead, flathead, and brazier-head) may be replaced by rivets of the same type or by AN-470 Universal head rivets. Flushhead rivets should be used to replace flushhead rivets.

Rivet edge distances should not be less than: (1) Single row—two times the diameter of the rivet and spacing not less than three times the diameter of the rivet, (2) Double row—the minimum shown in an appropriate rivet table, and (3) Triple row—the minimum shown in an appropriate rivet table.

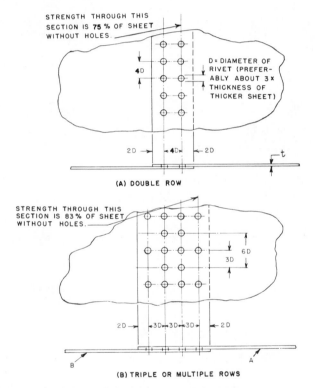

Fig. 44. Rivet hole spacing and edge distance for single-lap sheet splices.

A17S (2117) rivets may be driven in the condition received, but 17S (2017) rivets above 3/16″ in diameter, and all 24S (2024) rivets should either be kept refrigerated in the "as quenched" condition until driven or be reheat-treated just prior to driving. They would otherwise be too hard for satisfactory riveting.

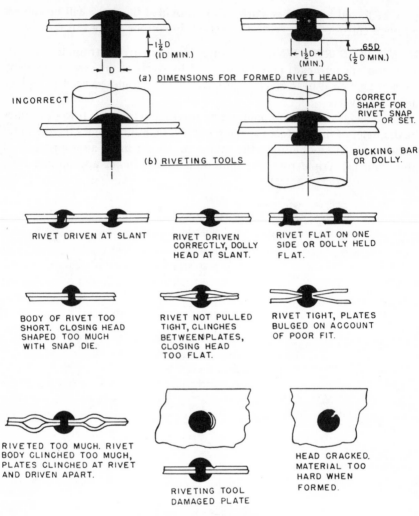

Fig. 45. Riveting practice and imperfections.

Hollow rivets should not be substituted for solid rivets in load-carrying members without specific approval of the application by a representative of the FAA.

RIVET REPAIR METHODS. Rivet holes should be drilled round, straight, and free from cracks. The rivet set snap used should be cupped slightly flatter than the rivet heads. Rivets should be driven straight and tight, but not overdriven or driven while too hard as the finished rivet must be free from cracks. Information on special methods of riveting, such as flush riveting, usually may be obtained from manufacturer's service manuals.

Round or streamline tubular members may be repaired by splicing in accordance with approved methods. When solid rivets go completely through hollow tubes, their diameter should be at least one-eighth the outside diameter of the outer tube.

To determine rivet size and pattern for a single-lap joint, use rivet diameter of approximately three times the sheet thickness. 3×0.032 equals 0.096 inch. Use of one-eighth A17S(2117)-T3 rivets (5/32 A17S(2117)-T3 would also be satisfactory).

Members which are slightly bent may be straightened cold and examined with a magnifying glass for injury to the material. The straightened parts should then be reinforced to an extent depending upon the condition of the material and the magnitude of any remaining kinks or buckles.

Local heating should never be applied to facilitate bending, swaging, flattening, or expanding on heat-treated aluminum alloy members, as it is difficult to control the temperatures closely enough to prevent possible damage to the metal and it may impair its corrosion resistance.

The diameter of rivets in stringers should preferably be between two and three times the thickness "t" of the leg, but should not be more than one-fourth the width "w" of the leg.

STEEL FITTINGS. Torn, kinked, or cracked fittings should be replaced and not repaired. Elongated holes in fittings which were designed without bushings should not be reamed oversize but such fittings should be replaced unless the method of repair is approved by a representative of the FAA.

Holes should not be filled with welding rod. Acceptable methods of repairing elongated or worn bolt holes in landing gear, stabilizer, interplane or cabane strut ends only, not originally equipped with pin plates, must be in conformity with the Administrator's standard practices.

REPAIR OF TUBULAR MEMBERS. Prior to repairing tubular members, the structure surrounding any visible damage should be carefully examined to insure that no secondary damage remains undetected. Secondary damage may be produced in some structure remote from the location of the primary damage by the transmission of the damaging load along the tube. Damage of this nature usually occurs where the most abrupt change in direction of load travel is experienced. If this damage remains undetected, loads applied in the normal course of operation may cause failure of the part.

Unless otherwise noted, welded steel tubing may be spliced or repaired at any joint along the length of the tube. Particular attention should be paid to proper fit and alignment to avoid eccentricities.

Dents at a steel tube cluster joint may be repaired by welding a specially

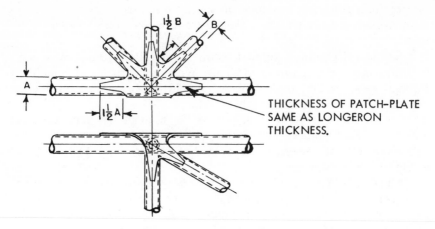

THICKNESS OF PATCH–PLATE
SAME AS LONGERON
THICKNESS.

Fig. 46. Patch plate before forming and welding.

formed steel patch plate over a dented area and surrounding tubes. Trim the reinforcing plate so that the fingers extend over the tubes a minimum of 1.5 times the respective tube diameter.

Dented, bent, or otherwise damaged tubular members can be repaired by using a split sleeve reinforcing, after first carefully straightening the damaged member. In the case of cracks, drill No. 40 (0.098) stop holes at the end of the crack.

In the repair of welded sleeves, the filling of dents or cracks with welding rod in place of reinforcing the member is not acceptable.

WELDED-PATCH REPAIR. Dents or holes in tubing may be repaired by a welded patch of the same material and one gage thicker provided:

(1) Dents are not deeper than one-tenth of, or longer than, the tube diameter, and do not involve more than one-fourth of the tube circumference; dents are free from cracks, abrasions, and sharp corners; and the dented tubing can be substantially reformed without cracking before application of the patch.

(2) Holes are not longer than tube diameter and involve not more than one-fourth of the tube circumference.

(3) No part of the patch is permitted in the middle third of the tube, and should not overlap a tube joint.

SPLICING INNER SLEEVE. If the inner sleeve fits very tightly in the replacement tube, chill the sleeve with dry ice or in cold water. If this is insufficient, polish down the diameter of the sleeve with emery cloth. Weld the inner sleeve to the tube stubs through the one-eighth-inch gap between the stubs, completely filling the gap forming a weld bead over the gap.

WELDING ENGINE MOUNTS. All welding on an engine mount should be of the highest quality, since vibration tends to accentuate any minor defect. Engine mount members should preferably be repaired by using a large diameter replacement tube telescoped over the stub of the original member and using fishmouth and rosette welds.

AXLE ASSEMBLY. Representative types of axle assemblies may be non-repairable. It will always be necessary to ascertain whether or not the members are heat-treated. Axle assemblies are generally not repairable because of the following reasons:

(1) The axle stub is usually made from a highly heat-treated nickel alloy steel and carefully machined to close tolerances. These stubs should be replaced if damaged.

(2) The oleo portion of the structure is generally heat-treated after welding and is perfectly machined to assure proper functioning of the shock absorber. These parts would be distorted after machining.

In general it will be found advantageous to replace damaged wing brace struts made either from round or streamlined tubing by new members purchased from the original manufacturer. However, there is no objection from an airworthiness point of view to repairing such members in the proper and approved manner.

Steel brace struts may be spliced at any point along the length of the strut provided the splice does not overlap any part of an end fitting.

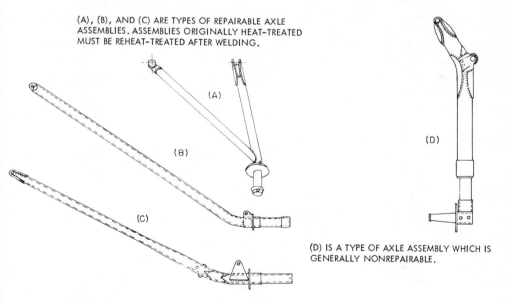

(A), (B), AND (C) ARE TYPES OF REPAIRABLE AXLE ASSEMBLIES. ASSEMBLIES ORIGINALLY HEAT-TREATED MUST BE REHEAT-TREATED AFTER WELDING.

(A)

(B)

(C)

(D)

(D) IS A TYPE OF AXLE ASSEMBLY WHICH IS GENERALLY NONREPAIRABLE.

Fig. 47. Types of repairable and nonrepairable axle assemblies.

STAINLESS STEEL STRUCTURES. Structural components made from stainless steel, particularly the "18-8" variety (18% chromium, 8% nickel), joined by spot welding, should be repaired only at the factory of origin or by a repair station designated by the manufacturer and approved by the FAA to perform this type of work. An exception to the welding procedure is that the repair can be made using bolted or riveted connections which are specifically approved by the FAA.

WINDSHIELDS AND ENCLOSURES. The following data is applicable to plastic windshields, enclosures, and windows for nonpressurized airplanes. For pressurized airplanes the plastic elements should be replaced or repaired only in accordance with the manufacturer's recommendations.

Two types of plastics are commonly used in transparent enclosures of aircraft: acrylic and cellulose acetate. Replacement panels should always be of material equivalent to that originally used by the manufacturer of the aircraft.

Installation procedures for both are generally the same. In installations involving bolts or rivets, the holes through the plastic should be slotted and oversize by ⅛" diameter and centered so that the plastic will not bind or crack at the edge of the holes.

Extensively damaged transparent plastics should be replaced rather than repaired since even a carefully patched part is not the equal of a new section, either optically or structurally.

Plastic should be cleaned by washing with plenty of water and mild soap, using a clean, soft, grit-free cloth sponge, or bare hands. Do not use gasoline, alcohol, benzine, acetone, carbon tetrachloride, fire extinguisher, deicing fluids, lacquer thinners, or window cleaning sprays. These will soften the plastic and cause crazing. Plastics lack the surface hardness of glass and care must be taken to avoid scratching or otherwise damaging the surface.

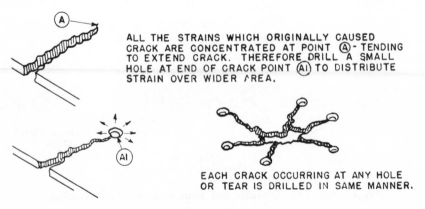

ALL THE STRAINS WHICH ORIGINALLY CAUSED CRACK ARE CONCENTRATED AT POINT (A) - TENDING TO EXTEND CRACK. THEREFORE DRILL A SMALL HOLE AT END OF CRACK POINT (AI) TO DISTRIBUTE STRAIN OVER WIDER AREA.

EACH CRACK OCCURRING AT ANY HOLE OR TEAR IS DRILLED IN SAME MANNER.

Fig. 48. Stop-drilling cracks in plastics.

E. WELDING AND WELDED STEEL STRUCTURES

Oxyacetylene or arc welding may be utilized for repair of aircraft structural elements. Some aircraft structures are fabricated from non-weldable alloys. In general, the more responsive an alloy is to heat treatment, the less suitable it is for welding. This is due to its tendency to become brittle and lose its ductility in the welded area. The following steels are readily weldable: plain carbon, nickel steels of the SAE 2300 series, chrome-nickel alloys of the SAE 3100 series, chrome-molybdenum steels of the SAE 4100 series, and low nickel-chrome-molybdenum steel of the SAE 8600 series.

The elements to be welded should be properly held in place by welding jigs or fixtures which are sufficiently rigid to prevent misalignment due to expansion and contraction of the heated material.

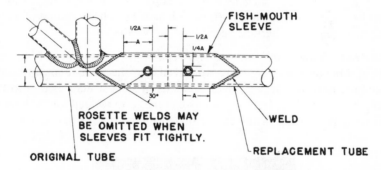

Fig. 49. Splicing by outer sleeve method—replacement by welded outside sleeve.

The parts to be welded should always be cleaned by wire brushing or other similar methods. When a wire brush is used, care should be taken never to use a brush of dissimilar metal such as brass or bronze. The small deposit left by a brass or bronze brush will materially weaken the weld and may cause cracking and subsequent failure of the weld. In case members were metallized, the surface metal may be removed by careful sandblasting followed by a light buffing with emery cloth.

No welds should be filled with solder, brazing metal, or any other filler in an effort to make a smooth appearing job as such treatment causes a loss in strength. Never weld over a weld if it can be avoided because continual reheating may cause the material to lose its strength and to become brittle. Also, never weld a joint which has been previously brazed.

Torch tip size is dependent upon the thickness of the material to be welded.

Rosette welds are generally employed to fuse an inner reinforcing tube (liner) with the outer member. Where a rosette is used, the hole should be made in the outside tube and be of a sufficient size to insure fusion of the inner tube. A hole diameter of approximately one-fourth the tube diameter of the outer tube has been found to serve adequately for this purpose. In cases of tightly fitting sleeves or inner liners, the rosettes may be omitted.

Members which depend on heat-treatment of their original physical properties should be welded using a welding rod suitable for producing heat-treated values comparable to those of the original members. Such members should be reheat-treated to the manufacturer's specifications after welding.

Airplane parts that depend for their proper functioning on strength properties developed by cold working should not be welded. For example: streamlined wires and cables.

Brazed or soldered parts should not be welded, as the brazing mixture or solder will penetrate the hot steel and weaken it.

Alloy steel parts such as aircraft bolts, turnbuckle ends, axles, and other heat-treated alloy steel parts (which have been heat-treated to improve their mechanical properties) should not be welded.

BRAZING. Brazing may be used for repairs to primary aircraft structures only if brazing was originally approved for the particular application. Brazing is not suitable for repair of welds in steel structures due to lower strength values of the brazed joint as compared to welded joints. However, brazing may be used in the repair of secondary structures.

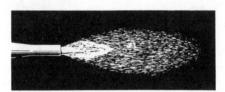

CARBURIZING FLAME
EXCESSIVE ACETYLENE. Three distinct zones. Brilliant white inner cone, whitish intermediate cone, and bluish outer envelope.

OXIDIZING FLAME
EXCESSIVE OXYGEN. Similar to neutral flame; shorter, neck-in, and acquires a purplish tinge.

NEUTRAL FLAME
BALANCED MIXTURE. Brilliant white cone surrounded by larger "envelope flame" of pale blue color.

Fig. 50. Welding flames.

F. ASSEMBLY AND RIGGING
Fixed-wing and helicopters

The Airframe mechanic must be familiar with aerodynamics and theory of flight in order to rig an airplane properly. For this reason, the FAA examinations include questions on the subject. The mechanic applicant also must be familiar with such terms as aspect ratio, longitudinal, vertical, and lateral axes, dihedral, wing heaviness, wash-out, wash-in, flaps, trim tabs, etc.

The earth's atmosphere is the medium in which aircraft move and by which they are supported.

The atmosphere is, in reality, an envelope of air which entirely surrounds the earth and rests upon the earth's surface. Air differs from the solid land and the liquid seas since it is a gas, or more properly, a mixture of gases. It has mass and weight, but does not have shape. The mass and weight of air changes constantly with seasons of the year, geographic locations, temperature and altitude variations, etc. These changes have a positive effect on aircraft performance, flight characteristics, and engine power output. Air mass and weight becomes less as altitude increases. This has a tendency to decrease aircraft performance at higher altitudes and requires the use of supplemental power units such as blowers, superchargers, etc., to offset the loss of engine power. In addition, turbo-props, turbo-jets, and the compounding of piston engines has been found to offset the loss of power during high altitude operation.

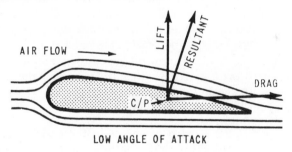

LOW ANGLE OF ATTACK

Fig. 51. Lift and drag on an airfoil.

LIFT OF THE WING. A properly designed wing produces lift because the streamline air passing above and beneath the wing reaches the trailing edge at the same time. Due to the greater camber (curvature) of the upper wing surface, the streamline air must travel farther and thus faster in the same time. This results in a decreased pressure above the wing with the higher pressure on the bottom of the wing. This difference in pressure above and below the wing results in an upward or lifting action on the wing (airfoil).

DRAG ON THE WING. The drag on a wing is a force which acts opposite to the direction of motion of the airplane and is made up of two components, profile and induced drag. Profile drag is the resistance, or skin friction, due to the viscosity (stickiness) of the air as it passes over the surface of the wing and other parts of the airplane. This works in combination with form drag which is due to the eddying and turbulent wake of air left behind. Profile drag may be thought of as a resistance, such as is always encountered when an object is thrust through a viscous medium.

Induced drag is the direct result of the force resulting from the downward velocity imparted to the air. This drag component will be parallel to the direction of motion of the wing. Induced drag value varies directly with the lift created by the wing. The greater the amount of lift created, the greater will be the induced drag.

LIFT AND DRAG COMBINED WITH THRUST AND GRAVITY (WEIGHT). From the above it can be seen that lift creates an upward action and drag creates a rearward action on the wing. Opposing lift is gravity or the weight of the aircraft. Opposing drag is the thrust created by the engine/s. These four forces in flight are common to all aircraft and vary according to the performance desired. The aircraft will climb when lift is greater than gravity and will descend when gravity is greater than the lift. Acceleration, or an increasing speed occurs when thrust is greater than drag and deceleration occurs when drag is greater than thrust.

The pilot of an aircraft has complete control of these four forces by use of the throttle and the motion of the stick or wheel. Increasing the throttle gives added thrust and acceleration, overcoming drag. Decreasing the throttle gives less thrust with deceleration and drag has the greater value. Moving the stick or wheel rearward increases lift momentarily causing the airplane to climb as lift overcomes weight. Depressing the stick or wheel forward decreases lift causing the airplane to descend as gravity or weight becomes greater than the lift of the wing/s.

CONTROL SURFACES. Control elements in the cockpit or flight deck are connected directly with the primary control surfaces through a combination of linkages and cables. Primary control surfaces are known as the rudder, elevator, and ailerons. They are controlled by rudder pedals and stick or wheel in the cockpit. Basically, control surfaces are airfoils and when moved up or down or to one side or the other they create an aerodynamic force capable of moving that portion of the aircraft. Since all aircraft are in balance around a center of gravity, the portion moved by any control surface can create an opposite movement on the other side of the center of gravity resulting in the attitude desired by the pilot. For example, when the pilot moves the wheel rearward (as in takeoff) the elevator moves to an upward attitude. This causes the tail to move downward resulting in the nose assuming an upward attitude. The wing inclined at an upward angle creates

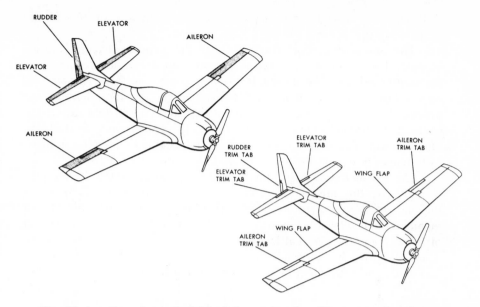

Fig. 52. Location of principal (shaded areas) and auxiliary control surfaces.

a higher lift resulting in the airplane climbing or ascending. If this principle is used as a basis, the attitude of the airplane can be easily noted when moving any of the primary control surfaces in either of their possible directions. Ailerons and elevator up or down and the rudder to the right or left, singly or in combination, permits the maneuvering of the airplane in all desired attitudes.

CONTROL AND TRIM TABS. Tabs, regardless of type, are attached rigidly or hinged to the trailing edge of the primary control surfaces. Depending on the type of aircraft, tabs can be installed on one or all three primary control surfaces. Tabs, which are small airfoils, create a movement of the surfaces that affect either a trimming of the airplane (flight at proper attitudes) or they aid the pilot in the actual movement of the control surfaces. The latter is an important consideration on large aircraft where movement of the large surfaces must be accomplished by forces other than the ability of the pilot.

Fixed trim tabs are installed to correct abnormal airplane attitudes such as a wing flying high or low, nose heaviness, etc. Made of metal plate material, they are affixed to the trailing edge of a control surface and bent in the direction that will force the control surface to an attitude slightly off center. This procedure corrects the particular rigging problem. When used on ailerons to correct a wing high or low condition, tabs eliminate the former necessity of warping (wash-in and wash-out) the wings. Washing-in or washing-out a wing is a procedure of changing the angle of incidence of a wing to increase or decrease the amount of effective lift at the wing tip.

The use of trimming tabs to balance the airplane (balance tabs) is in general use. The balance tab is a small adjustable surface hinged at the trailing edge of the moveable primary control surface. The air load on the tab produces a moment (or force) about the control hinge and deflects the control moment but without the exertion of effort on the part of the pilot. The tabs may be adjustable on the ground only, as is usually the case with aileron tabs, or they may be adjustable by the pilot in flight, as is generally the case with rudder and elevator control tabs.

The Flettner control is a larger trimming tab adjustable by the pilot in flight. By this means the rudder control may be adjusted to relieve the pilot of the constant strain of correcting for an unbalanced yawing moment, or the elevator tab may be used to trim the longitudinal balance of the airplane for different loading conditions or speed of flight. The use of the elevator tab is especially important in large aircraft due to changing fuel loads and movement of passengers during flight.

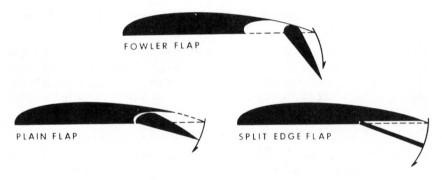

Fig. 53. Types of wing flaps.

WING FLAPS. Flaps are airfoils, or a portion of an airfoil, attached to the trailing edge of the wing and inboard from the ailerons. In the *up or retracted attitude* the flap streamlines with the wing and has no effect on the flying characteristics of the airplane. In the *down or effective position,* the flap increases the wing camber with a resultant increase in lift. They are often termed high lift devices because of this added lift advantage. In landing with the flaps down, the angle of glide is steepened and the length of the flight path is materially decreased. On takeoff with flaps down, the takeoff run is decreased and the climb out is steepened to a large extent giving added altitude and safety over obstructions.

Other advantages in the use of flaps are: lower takeoff and landing speeds, use of shorter landing fields, increased stalling angle, effective air

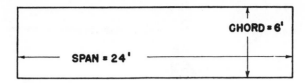

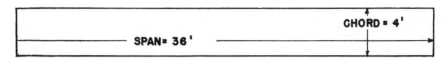

Fig. 54. Aspect ratio.

brakes within limits, and greater safety operating in and around what otherwise might be considered small airports for the particular aircraft involved.

ASPECT RATIO. The span of an airfoil (wing) is the maximum lateral dimension normal to the relative wind, or the distance between the wing tips. The span therefore determines the amount of air acted upon and hence is an important factor in the production of lift by a given wing area. For an airfoil of rectangular plan form the ratio of the span to the chord is termed aspect ratio. Thus an airfoil having a span of 30 feet and a chord of six feet (light airplanes) would have an aspect ratio of five. Very seldom is an airfoil rectangular in plan form due to the employment of tapered wings, center section cut-outs, and various tip shapes. In these cases the span squared divided by the area will give the aspect ratio. In general, it can be said that a relatively short and wide wing will have a low aspect ratio and a long narrow wing will have a high aspect ratio. A wing of low aspect ratio will be somewhat inefficient due to an unfavorable lift-drag ratio. A wing of high aspect ratio will have better efficiency due to a favorable lift-drag ratio. Structural considerations limit the higher aspect ratios as the wing length must be compatible to root attachment considerations.

STABILIZERS. The stabilizers create an attachment for the rudder and elevator and their surfaces presented to the air flow tend to steady or stabilize the aircraft's flight. However, a limit must be considered in the design of each type of aircraft since too much stabilizer area would be a disadvantage due to drag considerations and excessive weight. The vertical stabilizer is generally offset from the longitudinal axis to counteract the affect of propeller torque.

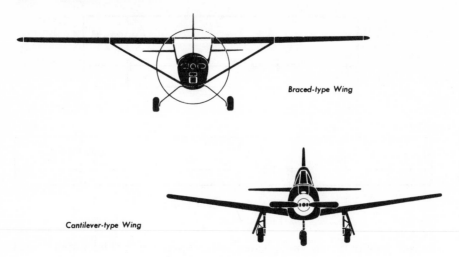

Braced-type Wing

Cantilever-type Wing

Fig. 55. Wing structures.

CANTILEVER WING. The term cantilever implies a structure that is supported at its root end and has no external bracing to aid in its support. The modern airplane of the monoplane type is typically a cantilever example as the one wing is supported entirely at the wing root with no bracing or struts to supplement its support. In some lighter aircraft, struts for bracing of the wing is still a design practice. The greatest advantage of the cantilever wing is the "cleanness" it gives by eliminating external bracing. External bracing or external units of any type creates parasite drag. This decreases the efficiency and performance of the airplane.

STALLING OF THE AIRPLANE IN FLIGHT. To properly understand the stalling characteristics of an airplane and how to recover from a stalled attitude is the pilot's greatest asset to safety in flight. Earlier the creation of lift and its effect on flight was discussed. In effect, it was brought out that so long as the wing created lift equivalent to the weight of the airplane level flight could be accomplished. Should the wing create more lift than the weight of the airplane a climb would occur. Lift is dependent on two factors, that of a streamline air flow over the wing and a minimum speed of the airplane. Understanding these flight principles can prevent accidental stalls in flight or near the ground. In either case, accidental stalls may be hazardous as gravity has an immediate affect on the airplane and the pilot's ability to maintain altitude and control. However, the landing maneuver is a normal stall maneuver. The airplane settles to the ground as the weight of the airplane becomes greater than lift at the point of touchdown.

MONOPLANE VERSUS BIPLANE. The monoplane (one wing) has many advantages over the biplane (two wings) especially from an aerodynamic efficiency standpoint. The monoplane is "cleaner" and has fewer protuberances from the fuselage, which means less parasite drag.

Biplanes are now used mostly for crop dusting. The mechanic is expected to understand the characteristics of the biplane and its rigging principles.

The upper and lower wing are attached to the main center sections. During the wing attachment process, landing and flying wires, in addition to wing struts, are utilized to form a truss-like structure. The landing wires are attached at the inboard section of the top wing and run to the outboard end of the lower wing. The flying wires are opposite, running from the inboard section of the bottom wing to the outboard section of the upper wing. The landing and flying wires cross each other at the mid-section between the upper and lower wings.

In rigging the biplane a mechanic adjusts the landing and flying wires in conjunction with the wing struts. The arrangement of the struts form the letter "N" and are referred to as the "N" struts.

The angle of incidence of the upper and lower wings of the biplane varies. The difference is referred to as decalage and may have a value of one to three degrees. *Example:* The lower wing has an angle of incidence of two degrees and the upper wing is set at four degrees. The decalage, or difference, is two degrees.

Decalage is one of the advantages of the biplane since the upper wing (having the greater angle of incidence) will stall before the lower wing. Smoother stall characteristics are apparent and there is less chance for the pilot to enter an accidental stall maneuver.

Stagger in the biplane is accomplished by having one or the other of the two wings in a forward or extended position. Positive stagger is a condition whereby the upper wing is set ahead of the lower wing. Negative stagger occurs when the lower wing is set ahead of the upper wing. Stagger of the wings has several advantages such as better vision for the pilot, easier access to the cockpit, minimizes blind spots and improves weight and balance considerations. Stagger also minimizes interplane interference, a condition where the air moving under the upper wing comes into contact with air moving over the lower wing creating turbulence, eddies, etc.

The monoplane, with its one wing and no external bracing, eliminates the inherent problems of the biplane and has a somewhat simplified rigging procedure.

PROPELLER TORQUE. The clockwise rotation of the propeller/s (as viewed from the cockpit) sets up a counter-clockwise moment that tends to rotate the fuselage. In order to reduce this condition to a minimum,

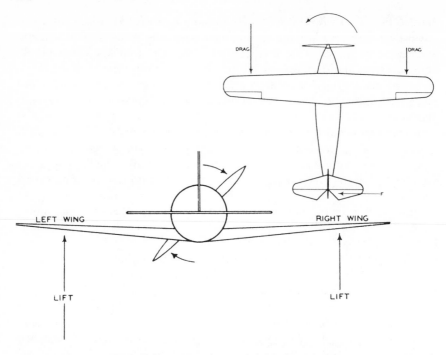

Fig. 56. Unbalanced forces due to engine torque.

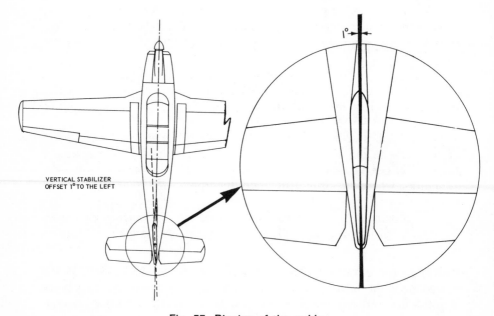

Fig. 57. Rigging of the rudder.

especially on takeoff where torque reaction is maximum, the left wing is washed-in to create more lift. This offsets the turning moment of the fuselage but increases the drag on the left wing causing a left turn tendency in the airplane. To offset this left turn tendency the vertical stabilizer is installed with its leading edge to the left of the longitudinal axis. A fixed tab may be installed on the trailing edge of the rudder, and the pilot uses right rudder action on takeoff. After rigging has been accomplished, offsetting propeller torque is of no or little consequence to the pilot in flight.

STABILITY OF THE AIRPLANE. Airplane stability is of utmost consideration and must be incorporated in the airplane from its design stages to the actual production. Though stability may be broken down into many classifications, there are two of major consequence. These are static stability and dynamic stability.

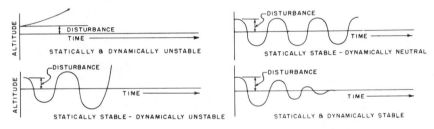

Fig. 58. Static and dynamic stability of the airplane.

Static stability is that property of the aircraft which causes it, when its state of steady flight is disturbed, to develop forces and moments tending to restore its original condition. These restoring forces and moments set up oscillations that are constant in their original creation.

Dynamic stability in the aircraft dampens the oscillations set up by the restoring forces and moments and gradually returns the aircraft to its original attitude. *Example:* A downdraft causes the right wing to assume a wing heavy attitude. Static stability immediately sets up oscillations to return the wing to its original state. These oscillations are dampened out by dynamic stability. In effect the oscillations become smaller until the wing assumes its original level attitude. This example is typical of the reactions following any disturbance causing wing low or wing high conditions in flight and also apply to inadvertent yawing and pitching of the airplane.

Sweepback built into the leading edge of wings is one design method to dampen out unnecessary yawing. Dihedral of the wings (upward angle from the horizontal) is a means of dampening out lateral disturbances in flight. Large stabilizer areas add much to the overall stability of aircraft, giving a stabilizing or keel reaction to disturbances.

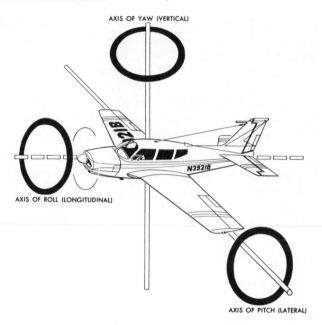

AXIS OF YAW (VERTICAL)

AXIS OF ROLL (LONGITUDINAL)

AXIS OF PITCH (LATERAL)

Fig. 59. Axes of the aircraft.

REFERENCE AXES OF THE AIRPLANE. The motion of the airplane may be studied with reference to the three fixed axes of the airplane, with their origin at the center of gravity.

Although invisible to the eye, an understanding of the airplane axes is important to understanding flight characteristics.

The longitudinal axis runs parallel to the relative wind (horizontally from nose to tail) and through the center of gravity. When rolling or banking, the airplane moves around the longitudinal axis. When the airplane is pitching (nose up or down) it is being disturbed along the longitudinal axis.

The lateral axis runs through the center of gravity and is perpendicular to the longitudinal axis. Positioned from one wing tip to the opposite wing tip the lateral axis is horizontal when the airplane is in level flight attitude. When pitching, the airplane moves about the lateral axis. When rolling or banking, the airplane is being disturbed along the lateral axis.

The vertical axis is vertical in position when related to the other two axes and is perpendicular to the longitudinal and lateral axis. Also running through the center of gravity, the vertical axis is in a vertical or upright position when the airplane is in level flight condition. When yawing (nose right or left), the airplane moves about the vertical axis.

In normal flight attitudes it can be easily understood that the airplane may be moving about one or more of these axes at any given time. In a climbing turn to the left, the climb creates movement about the lateral axis, the bank attitude causes movement about the longitudinal axis, and the turn creates movement about the vertical axis. Since all axes converge at the center of gravity, it is obvious that all attitude movements of the airplane are about the center of gravity. This is an important consideration, especially during weight and balance computations.

Helicopters

HELICOPTER THEORY OF FLIGHT. The following is a brief review of terms, etc., that must be understood in order to learn how a helicopter sustains flight and the means by which it is controlled. This is important for the mechanic in helicopter rigging and assembly.

An *airfoil* is any surface that obtains a useful reaction from the air through which the airfoil moves. The useful reaction is the creation of lift and/or thrust depending on the use of the particular airfoil. Examples of airfoils are wings, rotor blades, tail rotors, control surfaces, propellers, etc. Airfoils on helicopters consist basically of the main rotor blades and tail rotor blades.

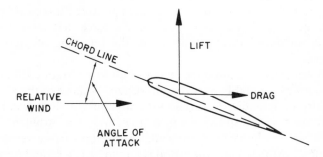

Fig. 60. Angle of attack is angle between relative wind and chord line.

The *chord* of an airfoil is an imaginary straight line from the leading (front) edge to the trailing (rear) edge of an airfoil.

Relative wind is the direction of the airflow with respect to an airfoil.

The helicopter rotor blade *pitch angle* is the acute angle between the blade chord line and a reference plane determined by the main rotor hub.

The *angle of attack* is the angle between the chord line of the airfoil and the direction of the relative wind. The angle of attack should not be confused with the pitch angle of the rotor blades.

Lift is derived from an airfoil through what is known as Bernoulli's Principle of venturi effect. The upper surface of an airfoil usually has a greater curvature than the lower surface. As air flows over the upper surface the air velocity is increased with a corresponding decrease in pressure along the upper surface. The air flows over the lower surface at a decreased

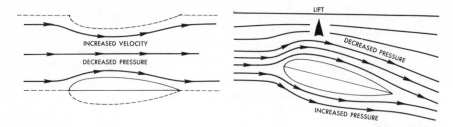

Fig. 61. (Left) Bernoulli's principle: increased velocity produces decreased pressure. (Right) Lift is produced by a combination of decreased pressure above the wing and increased pressure beneath.

velocity with a corresponding increase in pressure. This difference in pressure (lower on the upper surface and higher on the lower surface) results in an upward action known as lift. The wing of an airplane is normally an unsymmetrical airfoil (top surface has more curvature than lower surface). The rotor blade of a helicopter is normally a symmetrical airfoil (top and lower surface have the same curvature).

Whenever airfoils (rotor blades, wings, etc.) produce lift they are subject to a drag force. *Drag* is the term used for the force that tends to resist movement of an airfoil through the air or the retarding of inertia and wind resistance.

Stall. When the angle of attack increases up to a certain critical point, the air can not flow smoothly over the top surface because of the excessive change of direction required. This loss of streamlined flow results in a swirling, turbulent airflow and a large increase in drag. The turbulent airflow also causes a sudden increase in pressure on the top surface resulting in a large loss of lift. At this point, the airfoil is said to be in a stalled condition.

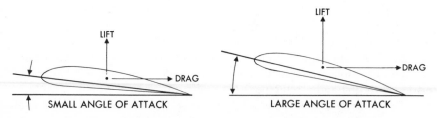

Fig. 62. As angle of attack increases, lift and drag increase.

As the angle of attack of an airfoil increases, the lift increases (up to the stall angle), providing that the velocity of the airflow (relative wind) remains the same.

The pilot of a helicopter can increase or decrease the angle of attack by increasing or decreasing the pitch angle of the rotor blades through use of the collective pitch control in the cockpit. Thus, he can increase or decrease the lift produced by the rotor blades. He must remember, however, that any increase in angle of attack will also increase drag on the rotor blades tending to slow down the rotor rotation. Additional power will be required to prevent this slowing down of the rotor.

Lift of the rotor blades varies directly with the *density of the air*. As the air density increases, lift and drag will increase. As the air density decreases, lift and drag decrease.

The higher the altitude the less dense is the air. Rotor blade, helicopter performance, engine output, etc., are all affected as altitude is increased. Hot air is less dense than cold air and moist air is less dense than dry air. It is thus obvious that the helicopter pilot must beware of hot air, high altitudes, and humid (moisture laden) air, especially when operating under critical performance conditions.

The total weight (gross weight) of a helicopter is the first force that must be overcome before flight is possible. Lift, the force which overcomes or balances the force of weight, is obtained from the rotation of the main rotor blades.

Thrust moves the aircraft in the desired direction. Drag, the retarding force of inertia and wind resistance, tends to hold it back. In vertical flight of the helicopter, drag acts downward. In horizontal flight, drag acts horizontally and opposite in direction to the thrust component. Thrust, like lift, is obtained from the main rotor.

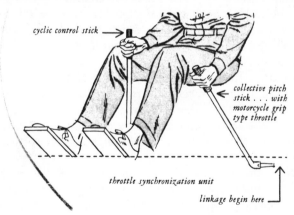

Fig. 63. Collective pitch control (stick).

HELICOPTER CONTROLS. There are four controls in the helicopter that the pilot must use during flight: (1) collective pitch control; (2) throttle control; (3) antitorque pedals (auxiliary or tail rotor control); and (4) cyclic pitch control.

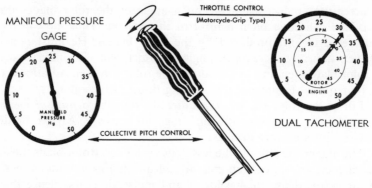

Fig. 64. Collective pitch stick is the primary control for manifold pressure; throttle is primary control for rotor RPM. A change in either control results in a change in both manifold pressure and rotor RPM.

The *collective pitch* lever or stick is located by the left side of the pilot's seat and is operated with his left hand. This lever moves up and down pivoting about the aft end and, through a series of mechanical linkages, changes the pitch of the main rotor blades. As the pitch of the rotor blades is changed, the angle of attack of each blade will also be changed. An increase in angle of attack results in more lift (with a resultant increase in rotor drag). A decrease in the angle of attack results in less lift (with a resultant decrease in rotor drag). The collective pitch control is the primary altitude control. Raising the collective pitch lever increases the rotor's lift and, through the cam linkage with the throttle, increases engine power to compensate for the additional drag. If there were no throttle compensation, the increase in rotor lift that also increases the rotor drag, would tend to slow the rotor RPM.

The *throttle* is mounted on the forward end of the collective pitch lever in the form of a motorcycle-type twist grip. The function of the throttle is to regulate the RPM of the main rotor. Twisting the throttle outboard increases the RPM and twisting it inboard decreases the RPM. The throttle must be coordinated with the collective pitch so that correct rotor RPM is maintained.

The thrust produced by the auxiliary (tail) rotor is governed by the position of the *antitorque pedals* (often referred to as rudder pedals). The pedals are linked to a pitch change mechanism in the tail rotor gear box to permit the pilot to increase or decrease the pitch of the tail rotor blades. The primary purpose of the tail rotor and its controls is to counteract the torque effect of the main rotor. In addition, the tail rotor enables the pilot to control the heading of the helicopter during hovering flight, hovering turns, and hovering patterns. In forward flight, the pedals are not used to control the heading of the helicopter. During forward flight they are used to compensate for torque to put the helicopter in longitudinal trim so that coordinated flight is maintained. The cyclic control is used to change heading by making a coordinated turn to the desired direction.

The purpose of the *cyclic pitch control* is to tilt the tip-path plane of the main rotors in the direction that horizontal movement is desired. The thrust component then pulls the helicopter in the direction of rotor tilt. The cyclic control has no affect on the magnitude of the total lift-thrust force, but merely changes the direction of this force, thus controlling the attitude and airspeed of the helicopter.

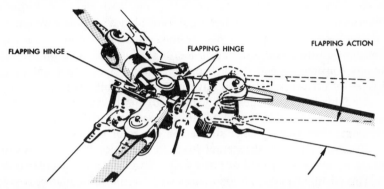

Fig. 65. Flapping action about the flapping hinges. Drag hinges can also be seen.

HELICOPTER OPERATIONAL FUNCTIONS. In a three-bladed rotor system, the *rotor blades* are attached to the rotor hub by a horizontal hinge which permits the blades to move in a vertical plane, i.e. flap up or down as they rotate.

Coning is the upward bending of the blades caused by the combined forces of lift and centrifugal force. Before takeoff, the blades rotate in a plane nearly perpendicular to the rotor mast, since centrifugal force is the major force acting on them. As a vertical takeoff is made, two major forces are acting at the same time—centrifugal force acting outward perpendicular to the rotor mast, and lift acting upward and parallel to the mast. The result of these two forces is that the blades assume a conical path instead of remaining in the plane perpendicular to the mast. Coning results in blade bending in a semi-rigid rotor. In an articulated rotor the blades assume an upward angle through movement about the flapping hinges.

When a helicopter is in a hovering position close to the ground, the rotor blades will be displacing air downward through the disk faster than it can escape from beneath the helicopter. This builds up a cushion of denser air between the ground and the helicopter. This cushion of air, referred to as *ground effect,* aids in supporting the helicopter while hovering. It is usually effective to a height of approximately one-half the rotor disk diameter. At approximately 3 to 5 MPH groundspeed, the helicopter will leave its ground effect.

Autorotation is a term used for the flight condition during which no engine power is supplied and the main rotor is driven only by the action of the relative wind. It is the means of safely landing a helicopter after engine failure or certain other emergencies.

Helicopter strength is measured basically by the *total load* the rotor blades are capable of carrying without permanent damage. The load imposed upon the rotor blades depends largely on the type of flight. The rotor blades must support not only the weight of the helicopter and its contents (gross weight), but also the additional loads imposed during maneuvers. In straight and level flight, the rotor blades support a weight equal to the helicopter and its contents. So long as the helicopter is moving at a constant altitude and airspeed in a straight line, the load on the blades remains constant. When the helicopter assumes a curved flight or path, all types of turns (except when hovering), flares, pullouts from dives, etc., the actual load on the blades will be much greater because of the centrifugal force produced by the curved flight. This additional load results in the development of much greater stresses on the rotor blades.

The load factor is the actual load on the rotor blades at any time, divided by the normal load or gross weight. Any time a helicopter is flown in a curved flight path, the load supported by the rotor blades is greater than the total weight of the helicopter. The tighter the curved path—that is, the steeper the bank—or the more rapid the flare or pullout from a dive, then the greater the load supported by the rotor; therefore, the greater the load factor.

The *transmission system* transmits engine power to the main rotor, tail rotor, generator, and other accessories.

The engine of a helicopter must operate at a relatively high speed while the main rotor turns at a much lower speed. This speed reduction is accomplished through reduction gears in the transmission system and is generally somewhere between 6 to 1, and 9 to 1. This means between 6 and 9 engine RPM's to 1 main rotor RPM. In a helicopter with a 6 to 1 ratio, if the engine turns at 2700 RPM, the main rotor turns at 450 RPM. When the rotor tachometer needle and the engine tachometer needle are superimposed over each other, the ratio of the engine RPM to the rotor RPM is the same as the gear reduction ratio.

Because of the much greater weight of a helicopter rotor in relation to the power of the engine than the weight of a propeller in relation to the power of the engine in an airplane, it is necessary to have the rotor disconnected from the engine by a clutch to relieve or reduce the starter load. *The clutch*, between the engine and rotor, allows the engine to be started and gradually assume the load of driving the heavy rotor system. Normally, the clutch does not provide disengagement of the engine from the rotor system for autorotation.

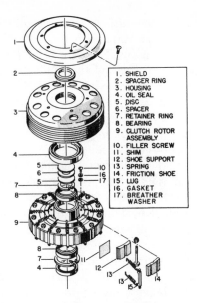

1. SHIELD	
2. SPACER RING	
3. HOUSING	
4. OIL SEAL	
5. DISC	
6. SPACER	
7. RETAINER RING	
8. BEARING	
9. CLUTCH ROTOR ASSEMBLY	
10. FILLER SCREW	
11. SHIM	
12. SHOE SUPPORT	
13. SPRING	
14. FRICTION SHOE	
15. LUG	
16. GASKET	
17. BREATHER WASHER	

Fig. 66. Helicopter clutch mechanism.

The *freewheeling coupling* provides for autorotative capabilities by automatically disconnecting the rotor system from the engine when the engine stops or slows below the equivalent of rotor RPM. When the engine is disconnected from the rotor system through automatic action of the freewheeling coupling, the transmission continues to rotate with the main rotor thereby enabling the tail rotor to continue at its normal rate. This permits the pilot to maintain directional control during autorotation.

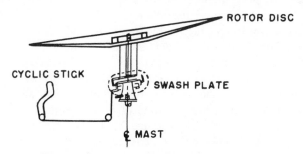

Fig. 67. Relation of swash plate to rotor.

The *swash plate* consists of two primary elements through which the rotor mast passes. One element is a disk, linked to the cyclic pitch control. This disk is capable of tilting in any direction but does not rotate as the rotor rotates. This nonrotating disk, often referred to as the "stationary star," is attached by a bearing surface to a second disk, often referred to as the "rotating star," which turns with the rotor and is mechanically linked to the rotor blade pitch horns. The rotor blade pitch horns are placed approximately 90° ahead of or behind the blade on which they control the pitch change. Blade pitch decrease takes place 90° ahead of cyclic stick position, and blade pitch increase takes place 90° after passing cyclic stick position. Thus, maximum downward deflection of the rotor blades occurs in the same direction as cyclic stick displacement and maximum upward deflection occurs in the opposite direction.

There are three fundamental types of *main rotor systems:* (1) fully articulated rotors; (2) semirigid rotors; and (3) rigid rotors.

A fully articulated rotor system generally consists of three or more rotor blades—each rotor blade is attached to the rotor hub by a horizontal hinge, called the flapping hinge, which permits the blades to flap up and down and move up and down independently of the others. The flapping hinge may be located at varying distances from the rotor hub and there may be more than one. The position is chosen by each manufacturer primarily with regard to stability and control. In this system each blade can flap, drag, and feather independently of the other blades.

In a semirigid rotor system, the rotor blades are rigidly interconnected to the hub, but the hub is free to tilt and rock with respect to the rotor shaft. In this system, only two-bladed rotors are used. The rotor (both blades) flaps as a unit; that is, as one blade flaps up, the other blade flaps down an equal amount. The hinge that permits the flapping or seesaw effect is called a teetering hinge. The rocking hinge is perpendicular to the teetering hinge and parallel to the rotor blades. This hinge allows the head to rock in response to tilting of the swash plate by cyclic pitch control, thus changing the pitch angle an equal amount on each blade—decreasing it on one and increasing it on the other.

In the rigid rotor system the blades, hub, and mast are rigid with respect to each other. In this system, the blades cannot flap or drag but can be feathered.

Rotor blade tips rotate in a common plane. *Tracking the blades* is a mechanical procedure used to bring the blades to such an attitude that they will each follow precisely the same plane of rotation.

There are several methods by which blades can be "tracked" by the maintenance personnel. One method generally used utilizes a canvas "flag" attached to the end of a pole. A different colored chalk is placed at the tip of each rotor blade. The engine is started and the rotor operated at a certain specified RPM. At each given RPM, the flag is held upright in such manner that the canvas can be moved toward the rotor blades. When the canvas strikes the rotor blades the different colored chalk will transfer from the blade tips to the canvas. On inspection of the canvas it can then be determined if all chalk colors are within a certain prescribed

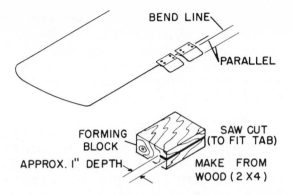

BEND LINE

PARALLEL

FORMING
BLOCK

SAW CUT
(TO FIT TAB)

APPROX. I" DEPTH

MAKE FROM
WOOD (2 X 4)

Fig. 68. Trim tab used in helicopter rotor blade tracking.

limit. If the chalk marks are outside of a prescribed limit it can be determined which blade or blades are out of track and how much. If out of limits, it will be necessary to rig or adjust the rotor blade mechanism to bring all blades within tracking limits. This is somewhat a "trial and error" method and several adjustments may have to be made before the blades are within tracking limits for the particular helicopter.

Another tracking method utilizes a strobe light. Reflectors are placed on the rotor tips. The engine is started and the rotor is operated at varying RPM. The strobe light played on the reflectors gives a visual pattern that can easily determine if the blades are within tracking limits or must be rigged or adjusted to a specified limit. Small trim tabs, installed on the trailing edge of main rotor blades, also can be utilized to correct improper blade tracking.

Trim controls are necessary for compensating control pressures caused by variable in-flight conditions.

Friction locks adjust hand pressures of controls to individual requirements, plus positive locking of controls in any position.

Clutch engagement lever provides smooth, gradual engagement of the rotor system to the engine.

HELICOPTERS AND AIRPORT OPERATIONS. The following are basically safety rules that should be remembered by both flight and ground personnel during normal operations: (a) do not air taxi over any parked aircraft; (b) do not air taxi near an aircraft taxiing or warming up; (c) do not hover at length because of the dust and foreign objects that will be blown about; (d) use caution when hovering on the downwind side of buildings or obstructions because of wind gusts; (e) the greatest hazard on the ground is the tail rotor which revolves at such high speeds that it is practically invisible; (f) persons unfamiliar with helicopters will tend to rush up to them whenever one lands; unless pre-arranged, all persons (unauthorized) should be warned away from them until the rotors have stopped; (g) when possible, always land and taxi into the wind; (h) surfaces of helicopter landing areas should be of a type that will not raise dust; (i) should a dusty landing area be unavoidable, land well away from people or buildings so any dust, etc., that is created will not be objectionable; and (j) helicopters should not be operated near a crowd of people, livestock, etc.

Briefing of ground maintenance and ground handling personnel is a must. Some important ground safety precautions are: (a) remain well clear of a helicopter until the rotor/s have stopped; (b) authorized maintenance personnel, when working in the vicinity of an operating helicopter shall use extreme caution and be sure the person in the cockpit is aware of their presence; (c) always approach a helicopter from the side or front, *never from the rear*; (d) persons departing should always depart toward the front or side, *never toward the rear*; and (e) fueling personnel (truck operators, etc.) should remain well clear of the main and tail rotor blades and should not approach the helicopter until the rotor/s are stopped.

The helicopter, like all types of aircraft, depends on good engineering, construction, materials, and proper maintenance for its efficient operation and service. Instances of failure may be traced to nicks, scratches, corrosion, or other damage to the surface of the metal. It is highly important that maintenance personnel exercise care to avoid tool marks or other damage to metals to prevent fatigue failure.

Each structural member has been designed to perform a specific function or to serve a definite purpose. In the repair of helicopters, like all other types of aircraft, the prime objective is to restore the injured or damaged part to its original condition. Very often, replacement is the only way in which this can be done. When repair of a damaged part is possible, however, the first step in this procedure is for the mechanic or technician to carefully study the part to fully understand its purpose, function, and strength requirements. Competent and satisfactory repair will follow as a natural consequence.

Fixed-wing and helicopters

WING LOADING. Wing loading is important in the design and structural fabrication of the airplane. It is a factor that determines how many pounds must be supported by each square foot of the wing at any time and under varying load conditions. Wing loading is found by dividing the total area of the wing (in square feet) into the total weight of the airplane at the time of computation. *Example:* Gross weight of an airplane is 95,000 pounds; wing area is 1,681 square feet. Wing loading equals 95,000 divided by 1,681 or 56.52 pounds per square foot.

POWER LOADING. Power loading is similar in application to wing loading. However, it is a factor that determines how much weight in pounds must be thrust forward by each horsepower of the engine/s. It is found by dividing the total horsepower (of all engines combined) into the total weight of the airplane at the time of computation. *Example:* Gross weight of the airplane is 95,000 pounds, total combined horsepower is 8,482 (four engines). Power loading equals 95,000 divided by 8,482 or 11.2 pounds per horsepower.

TEST FLYING AFTER MAINTENANCE AND/OR REPAIR. When an aircraft has undergone any repair or alteration which may have appreciably changed its flight characteristics or substantially affected its operation in flight, such aircraft, prior to carrying passengers, shall be test flown by at least a private pilot appropriately rated for the aircraft. A notation to that effect should be entered by such pilot in the aircraft log.

AIRWORTHINESS DIRECTIVES. These directives are issued by the FAA when it becomes evident that changes in aircraft design or installations are required to enhance an aircraft's airworthiness. These directives are mandatory.

RIGGING THE AIRPLANE. In the final analysis, flight characteristics of the aircraft depend on how well it has been rigged by the mechanic. The mechanic must be familiar with the airplane prior to rigging procedures and should strive to develop the best flight characteristics possible.

There are many variables the mechanic must work with, but still he strives for perfection, especially to have the aircraft fly straight and level (at cruising speed) after the pilot has properly trimmed it out in flight. This should be the rule rather than the exception, and is referred to as "flying at cruising speed with hands and feet off the controls."

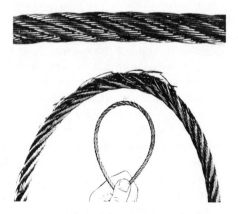

Fig. 69. Cable inspection technique.

CONTROL CABLES AND TERMINALS. Control cables and wires should be replaced if injured, distorted, worn or corroded, even though the strands are not broken. Cable sections can be spliced using standard procedures. Cable tension should be checked after installation to insure proper rigging.

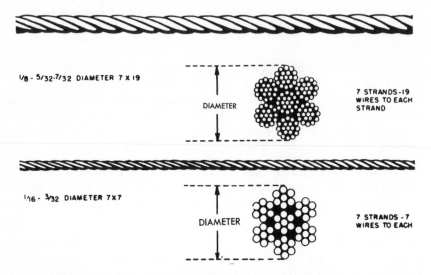

1/8 - 5/32-7/32 DIAMETER 7 X 19

DIAMETER

7 STRANDS-19 WIRES TO EACH STRAND

1/16 - 3/32 DIAMETER 7X7

DIAMETER

7 STRANDS - 7 WIRES TO EACH

Fig. 70. Woven splicing by the 5-tuck method.

When splicing cables, the cable, thimbles, shackles, turnbuckles, bolts, and other parts should be of the same size, material, and quality as the original parts or of such quality as the original parts or of such size that the repaired cable will be of strength equivalent to the original. If facilities and supplies are limited, repair may sometimes be accomplished using thimbles, bushings, and turnbuckles in place of original terminals. When this is done, flexible cable 7×7 and 7×19, having a diameter of three thirty-seconds inch or over, can be woven-spliced by means of the 5-tuck method. Flexible cable less than three thirty-seconds inch in diameter and nonflexible carbon steel 19 wire cable can be wrap-soldered.

Cables should be cut to length only by mechanical means. The use of a torch in any manner is not permitted. Wire and cable should never be subjected to excessive temperature. Soldering bonding braid to control cable is not considered satisfactory.

If cables are made from tinned steel, the cable should be coated with rust preventive oil. It is to be noted that corrosion-resistant steel cable does not require this treatment for rust prevention.

The five-tuck woven splice terminals can be utilized on 7×7 flexible and 7×19 extra flexible cables of three thirty-seconds inch diameter or greater. This type of terminal will develop only 75% of the cable strength and should not be used to replace swaged or other high efficiency terminals unless it is definitely known that the design load for the cable is not greater than 75% of the cable minimum breaking strength.

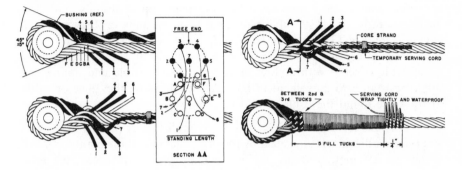

Fig. 71. Preparation of a woven cable splice.

In the wrap-soldered splice, the solder flux should be a compound of stearic acid (there should be no mineral acid present) and resin, with a composition of 25% to 50% resin. A warming gluepot to keep the flux in fluid state is desirable.

SAFETYING OF TURNBUCKLES. All turnbuckles should be safetied with safety wire using either the double or single wrap method, or with any appropriate approved special safetying device. Safety wire should never be reused. The turnbuckle should be adjusted to the correct cable tension so that no more than three threads are exposed on either side of the turnbuckle barrel. Turnbuckles should never be lubricated.

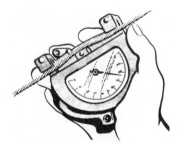

Fig. 72. Tensiometer.

RIGGING TENSION OF CABLES AND WIRES. Manufacturers' recommendations must be carefully followed in the rigging tension on control cables and brace wires. Should no specific tension information be available to the mechanic, he should use the rule of rigging cables and wires neither too tight or too loose. The tensiometer is a standard aid or tool in the rigging of aircraft cables and wires. Control cables rigged too loose will tend to vibrate and when too tight cause difficult control on the part of the pilot.

For identification purposes cables are made up of strands and wires in the strands. A 7×19 cable will have seven strands of 19 wires each giving a total of 133 wires. A 7×7 cable will have 49 wires total.

Frays in control cables can be detected easily by running your hand (preferably covered with a cloth or glove) over the cable and noting any slight resistance to movement of the hand or snags on the glove or cloth. If frays are found the cable should be replaced.

LEVELING THE AIRPLANE. This is accomplished by placing the level/s at points on the aircraft specified by the manufacturer. If the manufacturers' information is not available and no leveling points are noted, the level can be placed on the longitudinal and lateral braces at the front of the cabin section.

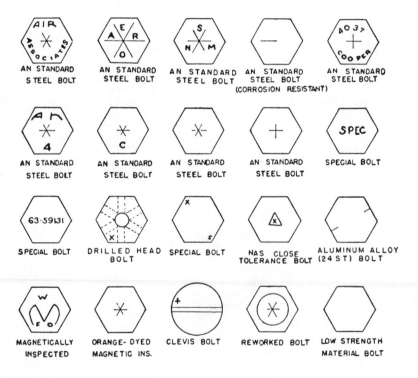

Fig. 73. Bolt identification.

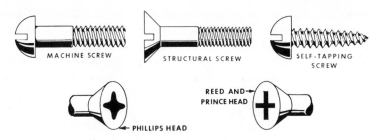

Fig. 74. Screws.

TRAMMING A WING. Tramming or lining up a wing is accomplished by adjusting all component parts to the manufacturer's recommendations. After tramming, should the spars not be properly aligned, it will be necessary to loosen the wires (drag, anti-drag, and brace) and retram the wing.

BOLTS, SCREWS, AND MISCELLANEOUS FASTENERS. AN-type aircraft bolts can be identified by the code markings on the bolt heads. The markings generally denote the bolt manufacturer, material of which the bolt is made, and whether the bolt is a standard AN-type or a special purpose bolt.

Bolts inspected magnetically (Magnaflux) or by fluorescent means (Zyglo) are identified by means of colored lacquer or a head marking of a distinctive type.

All bolts and/or nuts, except self-locking nuts, should be suitably locked or safetied. Cotter pins and safety wire should not be reused.

Bolts and nuts should be clean, dry, and thoroughly degreased before installation. Threads should not be oiled. Nuts should be torqued in all important places, such as wing-joint, engine-support, and landing-gear bolts.

The hex-head aircraft bolt is an all-purpose structural bolt used for general applications involving tension or shear loads. Alloy steel bolts smaller than No. 10-32 and aluminum alloy bolts smaller than $\frac{1}{4}''$ diameter should never be used in the primary structure.

Self-tapping screws of the AN-504 and AN-506 type are used for attaching minor removable parts such as nameplates. AN-530 and AN-531 are used in blind applications for the temporary attachment of sheet metal for riveting and the permanent assembly of nonstructural assemblies. AN-535 is a plain head self-tapping screw used in the attachment of nameplates or in sealing drain holes in corrosion-proofing tubular structures and is not intended to be removed after installation. Self-tapping screws should never be used to replace standard screws, nuts, bolts, or rivets in the original structure.

Self-locking nuts are acceptable for use on certificated aircraft subject to the restrictions on the pertinent "Manufacturer's Recommended Practice Sheets." Self-locking nuts are used on aircraft to provide tight connections which will not shake loose under severe vibration. Two types of self-locking nuts are currently in use—the all-metal type and fiber or nylon lock type. Self-locking nuts should not be used at joints which subject either the nut or bolt to rotation. They may be used with anti-friction bearings and control pulleys, provided the inner race of the bearing is clamped to the supporting structure by the nut or bolt. Nuts which are attached to the structure should be attached in a positive manner to eliminate rotation or misalignment when tightening the bolts or screws.

Fiber or nylon lock nuts are constructed with unthreaded fiber or nylon and have a smaller diameter than the nut, and when a bolt or screw is entered, it taps into the insert, producing a locking action. After the nut has been tightened, round or chamfered end bolts, studs or screws should extend at least the full round or chamfer through the nut. Flat end bolts, studs or screws should extend at least 1/32 inch through the nut. When fiber-type self-locking nuts are reversed, care should be exercised that the fiber has not lost its locking friction or become brittle.

RIGGING AND ASSEMBLY. General in nature, the following text centers around typical rigging information and procedures.

A twisting movement in a recently recovered control surface would indicate the fabric is too tight.

The angle of incidence can be measured with the aid of a straight edge and a bubble protractor.

Vibration of rigging wires is an indication of excessive looseness.

Maximum weight limits are assigned to baggage and other compartments of the airplane. This is necessary since weight distribution affects aircraft balance and stability.

Dihedral of the wings can be determined by the use of a dihedral stick and a protractor level.

The center of lift and center of gravity relationship is important to the flight characteristics of the airplane. Relative to each other the center of lift must always be behind the center of gravity. In the event of a stall or engine failure, the airplane will be noseheavy, thereby increasing the airspeed. This permits control of the airplane a maximum length of time during mechanical or flying difficulties.

If not otherwise corrected, the removal of landing gear streamlining will cause the airplane to be noseheavy. This is due to the additional drag created at time of removing the streamline units.

When correcting rigging problems with fixed control surface tabs, the mechanic most generally will have to use the trial and error method until the tab correction solves the problem. Blocks of wood should be used when bending a fixed tab for protection against metal fatigue and cracks. Never bend a fixed tab with the metal jaws of a tool in direct contact with the tab metal.

Excessive positive stagger in a biplane will cause tail heaviness due to upward action on the nose of the biplane.

Drag and anti-drag wires in the wing structure must be rigged to the proper tension. If too tight they may distort the shape of the wing (destroy symmetry) and if too loose will decrease the strength of the wing structure and possibly cause a metallic ring or noise within the wing.

Wing heaviness can be corrected by bending the aileron tab on the heavy side upward. Tail heaviness can be corrected by bending the elevator tab upward.

In tramming a wing, the mechanic should work from the root outboard.

When washing-in a wing (increasing lift) or increasing the angle of incidence, the mechanic must remember that drag at that location will also increase. This may call for further rigging to eliminate any turning effect that is created by the increase in drag.

Generally the left wing is rigged to have more lift than the right wing in order to offset propeller torque.

Micarta fairleads have excellent wearing qualities and decrease cable wear to a great extent.

Ailerons are generally rigged with a small amount of droop in them. This permits their streamlining with the trailing edge of the wing while in flight.

Control surfaces can be statically balanced by attaching weights ahead of the hinge line. Weights should never be placed behind the hinge because of the resultant control surface flutter. Flutter of a control surface may also be caused by loose hinge fittings. The rudder should be rigged in the neutral position. The rudder pedals should be even during the rigging process.

Bonding of aircraft (connectors between metal parts) eliminates sparks and arcing that build up due to static electricity in the airplane structure. Bonding is accomplished by using clamps or flexible wires (pigtails). Surfaces to which bonding is attached should be free from dirt, paint, or other nonconducting material.

Aircraft tires should be inflated, deflated, and again inflated prior to installation. This will insure proper relation between the tube and carcass with the carcass tightly against the wheel flanges. The axle nut of an airplane wheel should be tightened just enough to eliminate any wheel drag or wheel side play.

Fairings should be carefully inspected for looseness, as they will cause turbulence and affect the airplane flight characteristics.

G. INSPECTIONS

One-hundred-hour inspection. An inspection of an aircraft within each 100 hours of time in service and is a complete airworthiness inspection of such aircraft and its various components and systems in accordance with procedures prescribed by the Administrator.

Annual inspection. An inspection of an aircraft required once each 12 calendar months and is a complete airworthiness inspection of such aircraft and its various components and systems in accordance with procedures prescribed by the FAA.

Progressive inspection. A continuing airworthiness inspection of an aircraft and its various components and systems at scheduled intervals in accordance with procedures prescribed by the Administrator.

MAJOR REPAIR AND ALTERATION FORM FAA-337. The form FAA-337 should be executed in duplicate. After the major repair or alteration has been examined, inspected, and approved or rejected, the original should be given to the aircraft owner for retention as part of the permanent aircraft maintenance records—the FAA retains one copy. Though not required, it is also a good practice for the maintenance person responsible for the repair or alteration to retain a copy for his records.

MAINTENANCE RELEASE. The release must contain the identity of the aircraft or component. In the case of an aircraft, it should consist of at least the following: (a) make of aircraft; (b) model; (c) serial number; (d) nationality and registration mark; and (e) location of repair. When the repair is to a spare component such as a wing, landing gear, propeller, powerplant, or appliance, the identification must include manufacturer's name, name of component, model, and serial number, if any. The following statement should also be included: "The aircraft and/or component identified above was repaired and inspected in accordance with current Federal Aviation Regulations and was found airworthy for return to service."

The maintenance release may be combined with the "customer's work order" to provide one document so long as it contains the required information.

RADIO ANTENNAS. It is satisfactory to use one antenna for transmission and reception of communications provided such antenna is a satisfactory compromise for the frequencies to be used.

Fixed and trailing wire antenna installations must be tailored to fit the particular type of aircraft. Other types of antennas are more compact and can be installed as complete units on various types of aircraft.

If an antenna is attached at the trailing edge of wings and leading edge of horizontal stabilizers, the attachment must be secured to lugs *firmly*

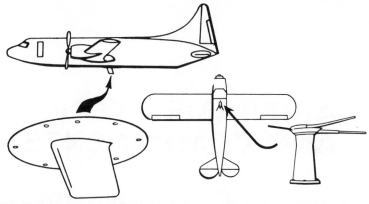

Fig. 75. Preferred position (left) for distance measuring equipment and/or ATC radar beacon antennas. Preferred VOR antenna location (right) for maximum signal pickup with minimum ignition interference.

fixed to the structure. The lugs must be welded, riveted, clamped, or bolted, whichever method is most suitable to the structure.

Antennas for omnirange (VOR) and instrument landing system (ILS) localizer receivers must be located at a position on the airplane to obtain maximum sensitivity for the desired signals and minimum response to undesired signals such as VHF energy radiated by the engine ignition system. The best location for the VOR-localizer receiving antenna on most small airplanes is over the *forward part of the cabin.* Do not install antenna on primary structures when installing on bottom.

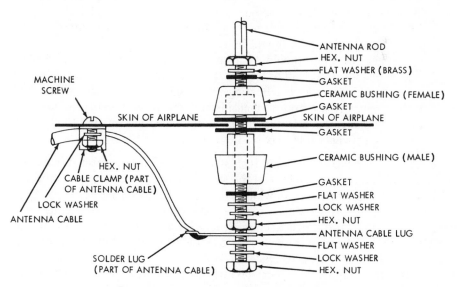

Fig. 76. Typical whip antenna installation.

3. AIRFRAME SYSTEMS AND COMPONENTS

A. AIRCRAFT LANDING GEAR SYSTEMS

LANDING GEAR EQUIPMENT. Wheels should be inspected at periodic intervals for cracks, corrosion, dents, distortion, and faulty bearings. In split-type wheels, bolt holes which may have become elongated due to some play in the through-bolt should be reconditioned by the use of Rosan inserts or other suitable means. The bolts should also be inspected for wear; if excessive wear is evident the bolts should be replaced. In bolting the wheel halves together, care should be taken to have the nuts properly tightened. These should be inspected periodically to be sure that they are tight so that there will be no movement between the two halves of the wheel. Grease-retaining felts in the wheel assembly should be kept in a soft, absorbent condition. If they become hardened they should be cleaned in gasoline; if this fails to soften them, they should be replaced.

Wheels that wobble excessively due to deformation resulting from a severe side load impact should be replaced.

Wheel bearings should be inspected periodically to detect damage caused by maladjustment or foreign material. Damaged or excessively worn parts should be replaced. Bearing cones should be packed with a high melting-point grease prior to their installation. When assembling the wheel to the airplane, the axle nut should be tightened just enough to eliminate any drag or wheel side play on the axle.

TIRES. For maximum safety, it is essential that tires be inspected frequently for cuts, worn spots, bulges on the sidewalls, and foreign bodies imbedded in the treads. Any repairs which are necessary should be made.

Tire repairs should be made by the original tire manufacturer or a reliable local tire repair agency. Specifications state that tires should not be repaired if any of the following conditions are found during inspection:

(a) Flex breaks or evidence thereof.

(b) Bead injuries which extend into more than three plies of a tire having

154

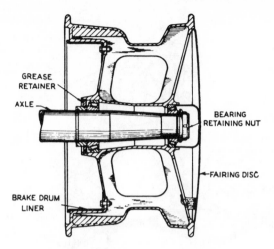

GREASE
RETAINER

AXLE

BEARING
RETAINING NUT

FAIRING DISC

BRAKE DRUM
LINER

Fig. 77. Wheel and axle assembly.

16 or more plies or into 25% of the plies on all other tires.
(c)   Evidence of separation between plies or around bead wire.
(d)   Injuries requiring inside reinforcement, including all casings needing section repairs.
(e)   Kinked or broken beads.

In treading tires, three basic methods of applying new tread stock to a tire are in use. They are known as top capping, full capping, and retreading. In top capping, the tread is buffed across the top of the tire and new tread material known as capping stock is applied. In the full capping process, the buffing is usually carried down on the shoulders to the sidewall ring to receive the new tread material known as camelback. This material is similar to capping stock except that it is wider and the abrupt bevel is replaced by longer, tapering sides. The third method is known as retreading (sometimes called remolding). The old rubber is rasped from bead to bead and replaced with new camelback and sidewall material. This results in a new tire appearance. A later development is the flex-to-flex retread which can be classified as being between the full cap and retread process. In this method, a very wide camelback is used and new sidewall material is then applied which reconditions or covers the flex area. This new sidewall material does not extend down to the bead as it does in the full retread method.

The number of times a tire may be treaded can be determined only by a thorough inspection which would disclose any sidewall bruises, ply separation, broken bead wire, or other defects that indicate the carcass is not sound enough to justify an additional tread. When a carcass has been recapped three times (or retreaded) its airworthiness may be questioned

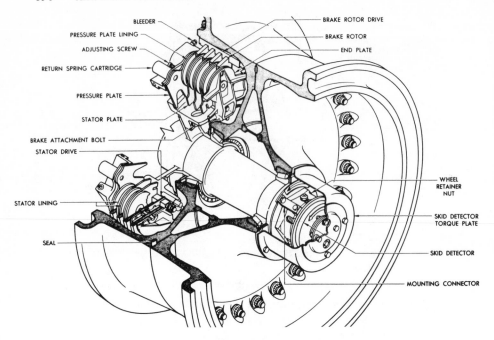

Fig. 78. Main landing gear and brakes.

although there have been tires which gave satisfactory performance with more than three such treads.

HYDRAULIC BRAKES. The safety of the aircraft and its cargo, passengers, etc., depends more often on the proper application of the brakes than on any other unit except the flight surfaces. Not only must the brakes stop the aircraft, but they must bring it to a stop within certain very definite limits of space and must be able to decelerate it in about one-third or one-half the time than the power plant required to accelerate it. Besides being efficient, they must operate easily and with a wide margin of safety. They must be designed so that braking force can be applied equally at both wheels with practically no chance for human error in the application of pressures. Brakes must also be very durable and tough.

Braking action consists of the application of a controlled force to stop a moving airplane on the ground or to keep it stationary. When the braking force is applied it develops *friction,* which is the resistance of relative motion between two surfaces in contact. Thus, by forcing a stationary surface, the rubbing action between the two surfaces will slow down the moving surface.

Brake mechanisms may be of various designs. They all require a rotating unit and a non-rotating unit, each containing one of the braking surfaces that are rubbed together to achieve the braking action. The rotating unit on airplanes may consist of brake drums or one or more disks either

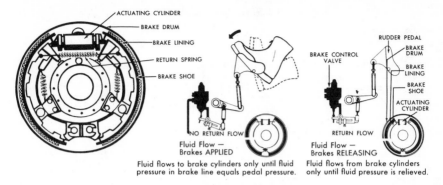

Fig. 79. Expander shoe brakes.

splinted or secured rigidly to the wheel. The non-rotating unit may consist of brake shoes, disks, or pucks, and linkage necessary to apply the shoes to the drums or to squeeze the disks together or the pucks against a single rotating disk, and the accessories to which these units are attached.

Brakes are classified as the *expanding-shoe,* the *spot,* the *multiple disk,* or the *expander-tube* type.

The hydraulic brake pressure necessary for operating the brake mechanisms may be provided by pressure bled away from the main hydraulic system, or they may be actuated by an independent system.

In the independent system, a master cylinder is employed in which a charge of fluid is trapped by piston action, then compressed by foot pressure. The pressure thus created in the master cylinder is transmitted along the lines to the brake cylinder.

In the integral system a unit, called the power brake control valve, bleeds away part of the main hydraulic system at reduced pressure, this pressure is then released to the brake cylinder as needed.

The independent system is simple and quite light in weight; therefore, it is commonly used on light aircraft and slow-moving types of larger aircraft. With its simplicity also goes reliability, and since this system is very reliable, no emergency system is involved.

The integral system makes available higher operating pressures than does the independent system. The pressure is a fraction of the main hydraulic pressure as bled from the pressure manifold of the main hydraulic system by the power brake control valve. In some cases, the pressure delivered by the power brake control valve to the brake actuating mechanism is too high. In this case, a second unit known as a brake debooster is used—the brake debooster pressure is delivered to it by the power control valve. It is important to remember that pressure for the integral system comes from the main hydraulic pump, and that in case of failure of the engine-driven pump, you usually have an accumulator and hand-pump with which

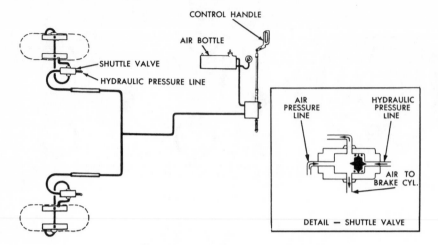

Fig. 80. Brake emergency air system.

you can charge it. Therefore, an emergency brake system, or a brake-charging system, is possible to incorporate. In addition, the brake control valve is often designed so that the right and left brakes can be operated either independently or together. When operated together, pressures are applied equally to both brakes automatically, thus resulting in equal braking pressures at both wheels.

THERMAL EXPANSION in the power system is absorbed by the general system and brake system accumulators, and will be evidenced on the gages by somewhat higher pressures. Pressures in excess of approximately 1,500 PSI are relieved to the general system return line through the thermal relief and exhaust valve.

The clearance between moving and stationary parts of a brake should be maintained in accordance with the manufacturer's recommendations. The brake should be disassembled and inspected periodically and the parts examined for wear, cracks, warpage, corrosion, elongated holes, etc. If any of these or other faults are indicated, the affected parts should be repaired, reconditioned, or replaced, depending on the seriousness of the difficulty. This should be done in accordance with the manufacturer's recommendations.

Surface cracks on the friction surfaces of brake drums occur frequently due to high surface temperature. These may be disregarded as seriously affecting the airworthiness until they become cracks of approximately one inch in length. The brake drums or surfaces then should be replaced.

Hydraulic brake maintenance includes inspection of the entire hydraulic system from the reservoir to the brakes themselves. The fluid in the reservoir should be maintained at the recommended level with the proper

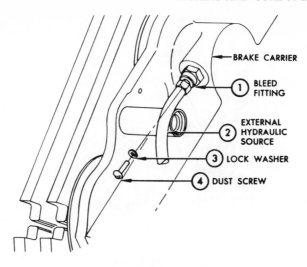

Fig. 81. Bleeding brake.

brake fluid. Flexible hydraulic hose which has deteriorated due to long periods of service should be replaced. When air is present in the hydraulic system, the system should be bled in accordance with the manufacturer's instructions. Hydraulic piston seal gaskets should be replaced when there is evidence of leakage.

Mechanically operated brake parts should be kept free of dirt and foreign matter and should work freely at all times. Excessive play in the linkage system should be kept to a minimum.

B. HYDRAULIC SYSTEMS

Modern aircraft are of such size and complexity in their operation that it is not practical to operate all the systems with power taken directly or indirectly from the engine/s. Therefore, a medium commonly known as *fluid under pressure* is used for this purpose. The study of fluids under pressure is called hydraulics, and the system which controls such fluid flow is known as the hydraulic system.

Among the aircraft components that are controlled and/or operated hydraulically are wing flaps, landing gear, cowl flaps, brakes, nose-wheel steering, and windshield wipers.

Hydraulic units that comprise the aircraft hydraulic system are as follows:

PUMPS. The hand pump serves as a substitute for the power pump during emergencies in flight or as a source of pressure for checking the hydraulic system when the airplane is on the ground. All hand pumps are of the reciprocating or piston type and may be classified as single-action or double-action.

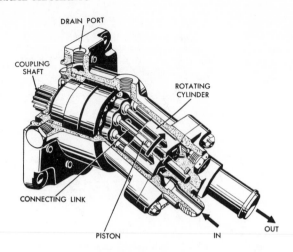

DRAIN PORT

COUPLING
SHAFT

ROTATING
CYLINDER

CONNECTING LINK

PISTON   IN   OUT

Fig. 82. Piston-type hydraulic power pump.

The primary energizing unit of the hydraulic system is the power pump. It normally delivers fluid under pressure to the actuating units. Power pumps may be of the gear or piston type.

The advantage of the piston-type power pump is its ability to develop higher pressures than other types of pumps. This pump is of the positive displacement type. One version consists of a housing, a cylinder barrel containing seven cylinders, seven pistons and their connecting rods, a drive shaft to which the piston rods are attached by ball-and-socket joints, and the universal drive linkage. The cylinder barrel housing is a head which contains two ports—the in-port and the out-port.

RESERVOIR. An adequate supply of fluid for the hydraulic system is stored in the reservoir. It is usually located at the highest point in the system so that the pumps will be primed at all times. Fluid flows from the reservoir to the hydraulic pumps, is forced by them throughout the system, and eventually returns to the reservoir. The reservoir not only supplies the operating needs of the system, but also replenishes fluid lost through leakage. In addition, the reservoir furnishes a place for the fluid to purge itself of air bubbles. Foreign matter picked up in the system may also be separated from the fluid in the reservoir. Reservoirs are always vented to the atmosphere and are generally equipped with a fluid quantity gage to be used as a guide in filling.

FILTERS. Filters may be divided into three general classes: disk filtration, screen type, and parchment filters.

Filters of the disk type (Cuno) and screen type employ the principle of edge filtration. They consist of a head, a sump or housing, and a cartridge or filtering element. Some screen type units incorporate a ball-check by-pass valve in the cartridge.

A parchment filter consists of a porous cellulose sheet. To provide an increased filtering area, the sheet is folded vertically. Fluid surrounds the filter element, and before it can reach the outlet, passes through the parchment sheet or through the relief valve. This type of filter is usually used inside the hydraulic reservoir, where the pressure is not very high.

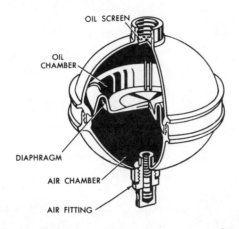

Fig. 83. Hydraulic system pressure accumulator.

ACCUMULATORS. The purpose of the accumulator is to store the fluid under pressure. This fluid can be used to supplement the power pump output during peak loads. Additionally, it may be used for limited operation of mechanisms when the power pump is not working. The accumulator also tends to damp out pressure surges in the hydraulic system. The pressure accumulator may be either of two types—the diaphragm type or the floating piston type.

During operation, an initial charge of compressed air is put into the accumulator, forcing the diaphragm upward or the piston forward. When pressure in the system builds up higher than the air pressure, fluid will be forced into the oil compartment. This fluid will push the diaphragm down or the piston forward and further compress the air. During periods of peak loads the highly compressed air will tend to force the fluid back into the system. If the power pump is not running, the air will supply a limited amount of fluid under pressure for the operation of a mechanism.

ACTUATING CYLINDERS. The purpose of the actuating cylinders is to transform pressure into mechanical force or action where linear motion is required to move some mechanism.

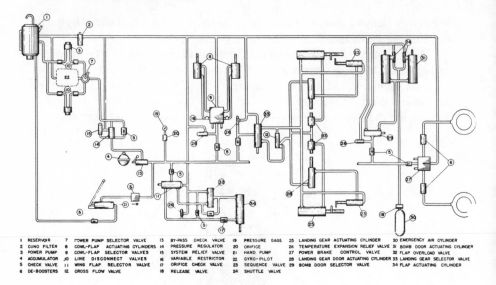

I	RESERVOIR	7	POWER PUMP SELECTOR VALVE	13	BY-PASS CHECK VALVE	19	PRESSURE GAGE	25	LANDING GEAR ACTUATING CYLINDER	30 EMERGENCY AIR CYLINDER
2	CUNO FILTER	8	COWL-FLAP ACTUATING CYLINDERS	14	PRESSURE REGULATOR	20	ORIFICE	26	TEMPERATURE EXPANSION RELIEF VALVE	31 BOMB DOOR ACTUATING CYLINDER
3	POWER PUMP	9	COWL-FLAP SELECTOR VALVES	15	SYSTEM RELIEF VALVE	21	HAND PUMP	27	POWER BRAKE CONTROL VALVE	32 FLAP OVERLOAD VALVE
4	ACCUMULATOR	10	LINE DISCONNECT VALVES	16	VARIABLE RESTRICTOR	22	GYRO-PILOT	28	LANDING GEAR DOOR ACTUATING CYLINDER	33 LANDING GEAR SELECTOR VALVE
5	CHECK VALVE	11	WING FLAP SELECTOR VALVE	17	ORIFICE CHECK VALVE	23	SEQUENCE VALVE	29	BOMB DOOR SELECTOR VALVE	34 FLAP ACTUATING CYLINDER
6	DE-BOOSTERS	12	CROSS FLOW VALVE	18	RELEASE VALVE	24	SHUTTLE VALVE			

Fig. 84. Typical pressure regulator system.

These units consist essentially of a cylinder, one or more pistons and piston rods, and the necessary seals. Actuating cylinders are usually double-acting. This means that fluid under pressure can be applied to either side of the piston to provide movement in either direction. Single-acting cylinders with a spring return are used for actuating brakes.

SWIVEL JOINTS. The purpose of swivel joints is to allow the use of rigid tubing in the plumbing of a hydraulic unit which rotates about a fixed pivot. For instance, if the landing gear actuating cylinder has to rotate during the raising and lowering of the gear, swivel joints eliminate the use of flexible tubing.

PUMP CONTROL VALVES allow the pilot to direct (manually) the output of the pump to the main hydraulic system when some mechanism is to be operated, or to the reservoir when there is no need for pressure.

POWER CONTROL VALVE. A power control valve is essentially a hand shut-off valve with an automatic turn-on feature. It is, therefore, semi-automatic in operation; that is, it must be closed manually, but it opens automatically. When the unit is open, it permits free circulation of fluid from the power pump to the reservoir and thus relieves the pump of load.

TIME-LAG CONTROL VALVE. The purpose of the time-lag control valve is to relieve the power pump of load when no mechanism is being operated. The unit consists essentially of a three-port housing containing a plunger and metering valve. One end of the plunger extends outside the housing.

This plunger is mounted in line with the return port. A piston head is mounted on the plunger. This head is equipped with a one-way cup which will allow fluid to flow by the head toward the left only. The metering valve is installed in such a way that fluid passing through it is by-passed around the one-way cup on the piston head.

PRESSURE REGULATORS. The pressure regulator, completely automatic in operation, relieves the power pump of load. When pressure in the system reaches a certain maximum, the unit opens and allows the output of the pump to return to the reservoir. If system pressure drops to a certain minimum, the unit closes and directs the output of the pump to the system.

FLOW CONTROL VALVES control the direction and the rate of flow of fluid through the hydraulic system. These valves consist of orifices, variable restrictors, check valves, line disconnect valves, and sequence, shuttle, and crossflow valves.

An orifice or variable restrictor limits the rate of flow of oil in both directions in a line. In so doing, these units cause the mechanism being operated to move more slowly.

Check valves allow free flow in one direction, no flow in the opposite direction. They may be used to trap pressure in some part of the system. Check valves may be used in either the pressure or the return line.

The orifice check valve allows free flow in one direction and restrained flow in the other direction. It is used to retard the operation of such units as flaps and landing gear in one direction.

The function of the *cross-flow valve* is to by-pass fluid from the landing gear up-line to the down-line when the gear is being extended. When the landing gear is released, its weight causes it to fall so rapidly that fluid cannot fill in behind the piston in the landing gear actuating cylinder. The weight of the landing gear also causes pressure to build up on the opposite side of the piston. The cross-flow valve permits fluid to flow from the *up* side of the piston to the *down* side, thus allowing the gear to fall more easily and with an even motion.

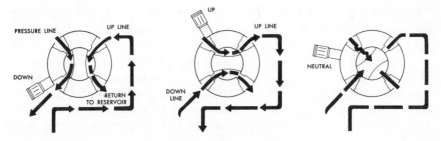

Fig. 85. Rotor-type selector valve.

SELECTOR VALVES control the direction of operation of a mechanism. They do this by directing fluid under pressure to the desired end of the actuating cylinder, and at the same time directing fluid from the opposite end of the cylinder to the reservoir. Selector valves may be of the rotor, the poppet, or the piston type. The rotor type is common on smaller aircraft, whereas the piston type is widely used on heavier aircraft.

RELIEF VALVES serve to limit or control the pressure in a hydraulic system. They are usually safety devices which prevent damage to the various parts of the system. Some units are designed to trap oil under pressure in a section of the system for holding a particular unit such as a landing flap in the retracted position when not in use or in extended position when in use.

Flap overload valves are used occasionally to prevent lowering of the flaps to the full down position when the airspeed is high enough to damage them or the flap linkage. Some systems incorporate hydraulic pressure switches instead of relief valves. These switches either turn the power pump on or off in response to changes of pressure in the system or warn the pilot when the system pressure drops below a specified value.

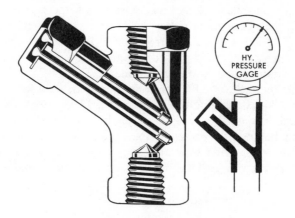

Fig. 86. Pressure gage and snubber.

PRESSURE GAGE AND SNUBBER. The hydraulic gage is mounted on the pilot's instrument panel. In large aircraft, a second hydraulic gage is mounted on the flight engineer's instrument panel.

Except for dial calibration, the hydraulic pressure gage is very similar to the oil pressure gage in that it operates on the Bourdon-tube principle. One end of the tube is attached solidly to the housing of the gage, the other end is free to move (to straighten out and recoil under fluctuating pressures) and actuate the indicator linkage and indicator. A pressure line, equipped with a snubber, tees off the hydraulic system some place after the power pumps and the relief valve, and attaches to a fitting in the base of the instrument. Hydraulic pressure gages are graduated in pounds per square-inch (PSI).

POSITION INDICATORS. Some of the units operated by the hydraulic system have position indicating devices to show the position of the unit being moved. The wing flaps, for example, have an indicator that shows their position at all times. Most airplanes have warning horns or buzzers that sound when the throttle is retarded for landing, and the landing gear is not down and locked.

HYDRAULIC SYSTEM SERVICING. Airplane hydraulic systems should be maintained, serviced, and adjusted in accordance with the manufacturer's maintenance manuals.

Hydraulic lines and fittings should be carefully inspected at regular intervals to insure airworthiness. Metal lines should be checked for leaks, loose anchorages, scratches, kinks, or other damage. Flexible hose lines should be checked for leaks, cuts, abrasions, soft spots, or other deterioration or damage. Fittings and connections should be inspected for leakage, looseness, cracks, burrs, or other damage.

When inspection shows a line to be damaged or defective, the entire line should be replaced. However, if the damaged section is localized a repair section can be inserted. In replacing lines, always use tubing of the same size and material as the original line; use the old tubing as a template in bending the new line, unless it is too greatly damaged. In this case a template can be made from soft iron wire. Soft aluminum tubing under one-fourth inch outside diameter may be bent by hand. For all other tubing, an acceptable tube-bending tool, hand or power, should be used.

Bending should be done carefully to avoid excessive flattening, kinking, or wrinkling. A small amount of flattening in bends is acceptable but should not exceed an amount such that the small diameter of the flattened portion is less than 75% of the original outside diameter.

When installing the replacement tubing, it should line up correctly with the mating part and should not be forced into line by means of the coupling nuts.

In making tube connections, only hydraulic fluid should be used as a lubricant. The connection should then be tightened, but overtightening will damage the tube or fitting and may cause failure and undertightening may cause leakage.

Minor dents and scratches in tubing can be repaired. Scratches or nicks no deeper than 10% of the wall thickness in aluminum alloy tubing only, not in the heel of a bend, can be repaired by burnishing with hand tools. Severe die marks, seams or splits in the tube should not be repaired;

such lines should be replaced. Any crack or deformity in a flare is also unacceptable and should be rejected.

A dent less than 20% of the tube diameter is not objectionable unless it is in the heel of a bend. Dents may be removed by drawing a bullet of proper size through the tube by means of a length of cable.

In the replacement of flexible lines, the hose should never be stretched tight between two fittings as this will result in over-stressing and failure at the ends under pressure. The length of hose should be sufficient to provide about 5 to 8% slack.

TEMPORARY HYDRAULIC SYSTEM. In the temporary pressure hydraulic system, the pressure is maintained only temporarily. The pressure is on only while one of the hydraulic systems is operated, and is released as soon as operation is completed. This arrangement makes it unnecessary for the engine to run constantly against pressure. Such a system saves engine power or power pump wear, and cuts down the leaks that would likely occur if the fluid lines were constantly under pressure.

Pressure for the operation of the system is built up by the engine-driven pump. Pressure is relieved in the system at a predetermined setting, usually 1,000 PSI, by a power control valve which automatically disengages—depressurizing a particular system and directing pump hydraulic pressure back to the reservoir. The unit being operated automatically stops when the piston reaches the end of its stroke for the system being operated—pressure suddenly trips the control valve. This control valve action can also be used for placing a unit in any intermediate position by turning the selector valve quickly to neutral. Pressure trapped in the cylinder and lines will hold the unit (such as the landing flaps) in position and simultaneously trip the control valve.

PRESSURE ACCUMULATOR SYSTEM. In the pressure-accumulator system, the units operated are the landing gear, wing flaps, brakes, cowl flaps, etc. The carburetor heat rise system also operates from the hydraulic system.

The main power system consists of pressure supply pumps, pressure regulating valves, pressure relief valves, pressure accumulators, fluid reservoir, and all piping with necessary restrictions and check valves to carry the pressure to the different system selector valves.

The engine-driven pumps receive a supply of fluid from the reservoir. One-way check valves are provided in both right and left engine pressure lines so that failure of one pump will not render the pressure from the other pump ineffective. Pressure passes to the relief side of the unloading and relief valves (which may be set to relieve pump pressure in excess of 1,150 PSI) and through the one-way check valve, which holds pressure built up beyond this point. Pressure is then directed through all the lines leading to the various system selector valves, where it is checked if no systems are being operated. Systems may operate to 3,000 PSI or more.

Pressure builds up in the control line routed back to the control port of the unloading valve and begins to charge the general system and brake system accumulators. Pressure going to the brake system accumulator passes through a one-way check valve incorporated in the thermal relief and exhaust valve; this prevents the pressure from returning to the general system. Although the general system accumulator starts charging at the same time, it will not charge as fast, because of fluid having to pass through a restrictor valve.

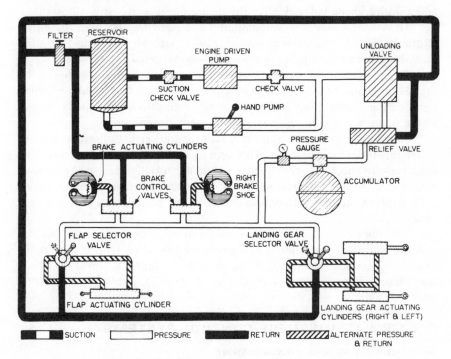

Fig. 87. Constant pressure hydraulic system.

The general system pressure will feed into the brake system whenever the brake pressure drops below that of the system pressure. As soon as the predetermined pressure (1,150 PSI) is built up in the line connected to the pressure port of the unloading and relief valve, the relief valve will open slightly to prevent the pressure from increasing further. General system pressure increases until it reaches 1,000 PSI, at which point, through the line leading to the control port of the unloading valve, it will force the unloading and relief valve completely open. The pressure trapped in the system by the one-way check valve holds the valve open to create an idling circuit which prevails until some unit of the hydraulic system is operated. When a unit is operated, the accumulator pressure will immediately start dropping. When the pressure is reduced to 800 PSI, the pump unloading and relief valve closes and cuts the pumps back into the system to direct pressure to the operating unit. After the operation is completed, pumps will continue to recharge the accumulators until the pressure is restored and the idling circuit is established. Return fluid from the operating systems flows through separate return lines.

The metering type of check valve serves the same purpose in a system as an orifice check valve. Unlike the orifice check valve, it is adjustable.

The *line-disconnect valve* eliminates the need for draining the entire hydraulic system when the engine or power pump is changed. It is used in hydraulic lines which are periodically disconnected.

The *by-pass check valve* is essentially a check valve which can be manually opened to allow free flow in both directions. It is usually installed between the accumulator and hand pump connections, so that the output of the hand pump can be diverted into the accumulator.

Sequence valves cause one hydraulic operation to follow another in a definite order or sequence. This unit is also called a timing valve or load-and-fire check valve. Sequence valves are usually used in the landing gear system, so that the nose wheel will be retracted before the main wheels during retraction and will be extended after the main wheels during extension.

The *shuttle valve* directs fluid automatically from either the normal source or the emergency source to the actuating cylinder. During emergency operation, the port connected to the normal operating line will be closed and there will be no loss of fluid even if there is a leak in the normal operating line.

C. CABIN ATMOSPHERE CONTROL SYSTEMS

Air conditioning and pressurization systems for large jet aircraft

Cabin atmosphere systems basically include air conditioning, pneumatic system, pressurization, and oxygen systems. These systems provide a cabin atmosphere that is comfortable and assures sufficient "cabin altitude" for the crew and passengers within the operable flight altitudes of the aircraft.

The design, material, and fabrication of the structure and associated assemblies must be such that pressurized areas can withstand the forces created when the cabin is pressurized to a relatively low altitude (8,000 feet plus or minus) and the aircraft is at a flight level of, say, 43,000 feet MSL.

AIR CONDITIONING AND PRESSURIZATION SYSTEMS provide ventilation, controlled temperature and regulated air pressure for the control cabin and main cabin. The cargo compartments are pressurized to the same differential pressure as the cabins, but have a low ventilation rate and will not sustain combustion. These interrelated systems are supplied with heated and compressed air from turbocompressors (air compressors) on engines No. 2 and 3 and engine bleed air from all four engines. Provision has been made for pre-conditioning the cabins with an external source prior to engine start.

PNEUMATIC SYSTEM. The pneumatic system supplies high temperature compressed air for air conditioning and pressurizing the airplane and for low pressure engine starts. The system consists of turbocompressors mounted on engines No. 2 and 3 and low pressure bleed air from all four engines. The main air supply is obtained by compressing fresh air in the turbocompressors, the air enters through a ram inlet on top of the engine nacelle. The turbocompressor is driven by bleed air from the last stage of the high pressure compressor (N_2) of its respective engine. After a turbocompressor is put into operation, it is independent and self governing. Each turbocompressor system incorporates a pressure regulator valve, a turbine inlet shutoff valve, a surge bleed valve, a speed topping controller, an overspeed trip mechanism, a fuse plug and its own oil reservoir. The oil quantity can be checked by a two-scale dip stick attached to the filler cap which is located on the right side of each turbocompressor.

An alternate air supply is obtained by bleeding low pressure air (N_1) from the intermediate compressor case of each engine. The bleed air system consists of appropriate ducting, a shutoff valve and a flow limiting venturi.

The compressed and heated air from each turbocompressor as well as the engine bleed air is routed through the left and right wing manifold and crossover pneumatic duct to the air conditioning system. The operating and utilization of turbocompressors and engine bleed air systems can be monitored by reference to tachometers, indicating lights and switches mounted on the flight engineer's upper panel.

AIR CONDITIONING SYSTEM. Compressed, high temperature air is supplied from the turbocompressors or engine air bleeds to the air conditioning system. The system consists of two units, each including a shutoff valve, primary and secondary heat exchangers, an air cycle machine, a water separator, an anti-icing control valve and thermostat, a turbofan with shutoff valve, three temperature limiting switches, and associated ducting. The anti-ice control valve and thermostat prevents freezing of the water separator by limiting the temperature across the water separator to a minimum of 35°F. The turbofan operates on the ground to provide coolant airflow across the primary and secondary heat exchangers.

The compressed air is used directly for heating or is routed through the air conditioning units depending upon the position of the main cabin and control cabin temperature valves. The air for the air conditioning unit flows through the primary heat exchanger where it is cooled or semi-conditioned. Depending upon the temperature required, part or all of this semi-conditioned air is directed through the compressor section of the air cycle machine (ACM), through the secondary heat exchanger, and then through the expansion turbine of the ACM which further cools the air.

Personal ventilation for each passenger is provided through the gasper system. This system supplies cool air through a variable area opening, adjustable by each person.

The main and control cabins have separate and independent temperature control systems. Each system consists of a temperature selector that has two modes of operation for temperature control: automatic and manual. The temperature regulator for each cabin has temperature inputs from ambient, cabin, and duct sensors and a temperature topping circuit to limit the duct temperature to 160°F. The temperature valve actuator for the main cabin positions has a six-valve mixing unit through mechanical linkages, and the actuator for the control cabin positions has a two-valve unit. These temperature valves mix the hot, cold and semi-conditioned air to maintain the selected main cabin and control cabin temperatures.

If main cabin temperature cannot be controlled in the normal manner using the temperature selector (AUTOMATIC or MANUAL), a manual override receptacle is provided in the floor of the passenger cabin. After inserting the landing gear emergency extension crank into the receptacle, the electric actuator can be disengaged and the main cabin temperature valves positioned by turning the crank. If such a temperature control problem occurs in the control cabin, there is a manually operated shutoff valve to stop the flow of air into the cabin. The valve is accessible to the crew through a door located on the right rear corner of the control cabin floor.

PRESSURIZATION SYSTEM. Compressed air is supplied by the turbocompressors or the engine air bleeds through the air conditioning and distribution systems to the pressurized compartments. Cabin pressurization is provided by controlling the rate at which the cabin ventilating air is permitted to exhaust from the airplane through three cabin pressure outflow valves. Automatic control is provided by a pneumatically operated cabin pressure regulator which allows selecting any cabin altitude from −1,000 feet to +10,000 feet within the differential pressure limit of 8.6 PSI, and also provides control of the rate of change of cabin pressures. A pneumatically operated manual pressure controller permits rate control and allows selecting any cabin pressure differential to 9.42 PSI should the automatic controller malfunction. Other system components include filters, pneumatic relays, jet pumps and cabin altitude indicating and warning devices. The pneumatic relays alleviate system response time lag. The jet pumps, bleed air operated, provide a negative pressure source for the automatic and manual controllers and for the outflow valves to relieve the surge effect at the start of cabin pressurization. The jet pumps also assist in depressurizing the cabin at landing. A cabin altitude warning horn will sound intermittently if the cabin altitude exceeds 10,000 feet. Positive pressure relief devices protect the airplane against overpressurization. The automatic controller has a differential pressure relief valve set at 8.6 PSI, and each outflow valve has a 9.42 PSI differential relief valve. Decompression panels are provided in bulkheads between pressurized sections to provide local structural protection should the airplane become rapidly depressurized at altitude. Pressure relief valves are also provided in the aft bulkhead of the aft cargo compartment to equalize pressure with the adjoining compartment. There is no automatic pressure pump device in this system.

CABIN PRESSURE OUTFLOW VALVES. Three cabin pressure outflow valves provide cabin pressure regulation, negative pressure and positive pressure relief. They are located along the bottom centerline of the airplane, one just aft of the forward cargo compartment, one near the forward end of the aft cargo compartment, and one to the rear of the aft bulkhead of the aft cargo compartment. These valves allow air to flow overboard from the cabin as determined by the automatic controller or manual controller settings. The outflow valves also function as safety valves to relieve at a 9.42 PSI positive pressure differential or at a 0.36 PSI negative pressure differential. These two conditions override the automatic or manual control signals. An independent safety device, the cabin altitude limit control, is provided in each outflow valve to limit cabin altitude to approximately 13,000 feet by closing the outflow valve. However, pressurization can be maintained only if the pneumatic system supplies sufficient airflow, and leakage is not excessive.

Air conditioning and pressurization systems for large reciprocating-engine aircraft

Air for the cabin area is obtained from scoops in each wing leading edge. The air passes through engine-driven superchargers, where the air pressure is raised as necessary for use by the air conditioning and pressurization equipment. From the superchargers, the air is ducted to the heater compartment. Here, a mixing valve, positioned by the cabin temperature control circuit, routes the air to be heated, cooled, or sent through without change. On airplanes so equipped, the air, upon leaving the heater compartment, may be cooled by a Freon refrigeration system.

The air at the proper temperature is then distributed throughout the occupied areas. The exhaust from these areas then leaves the cabin through various venturis used to ventilate compartments such as toilets, radio racks, lounges, galley waste, thermistor, etc.

Cabin pressure is controlled by a pressure regulator valve, the opening of which is varied automatically by instruments sensing cabin pressure. Both positive and negative pressure relief valves are also provided.

SUPERCHARGERS. A cabin supercharger is installed in each outboard nacelle. Engine power is supplied to the supercharger through a ratchet-type disconnect clutch from which it is transmitted to a gear box by a driveshaft. The clutches are connected by cables to clutch control levers. The clutch control levers are located on the right side of the cockpit floor and are used to declutch the superchargers.

DRIVE. Power is taken from the engine accessory pad through a disconnect clutch and drive shaft to the variable speed transmission. If for any reason the clutch is disengaged, the engine must be stopped before attempting to reengage.

LUBRICATION SYSTEM. Lubricant from a pump in the transmission is routed to an oil cooler located in the fairing aft of the belly coolant airscoop. It is then returned to the nacelle where it is filtered prior to entering the lubrication, cooling, and control circuits. Coolant air across the oil cooler is automatically controlled to maintain the desired oil temperature.

CONTROL. Automatic control of the supercharger is provided by a flow control valve which is supplied air pressures from a meter in the supercharger discharge air duct. The flow control valve uses the air pressures to operate an oil valve. The oil pressures from this valve position a piston that controls the variable speed transmission driving the supercharger. A disconnect lever is provided to the right of the first officer's seat for operation of each drive clutch.

INSTRUMENTS. An airflow instrument in the cockpit indicates whether satisfactory air pressures exist at the flow control valve. Supercharger discharge duct pressure, transmission bearing oil pressure, and oil temperature are also indicated. A warning light in the cockpit is illuminated in the event of low oil pressure.

DISTRIBUTION SYSTEM. Air is obtained from scoops located between nacelles in each wing leading edge. The scoop discharges into a plenum chamber where the moisture particles settle out. From the plenum chamber, the air goes either to the supercharger for pressurized flight or through the supercharger bypass valve during unpressurized flight. The air then passes into the fuselage through check valves that prevent the backflow of air should one supercharger be inoperative. Air from the two superchargers then enters the heater compartment where it either can be routed through a heater, through an aftercooler, or through the aftercooler and expansion turbine. The selection of the routes is made by a 3-port mixing valve (port A—turbine, B—aftercooler, C—heater) which can select air from any one of the three sources or a mixture of two. The valve is positioned in response to signals from the cabin temperature control circuit. Air discharging from the mixing valve passes into the cabin distribution system. On airplanes so equipped, it passes through the Freon refrigeration system evaporator before distribution to the main cabin. The distribution system consists of a main duct which distributes air through the floor corrugations to radiant side panels, and discharges to the cabin through openings beneath the baggage rack.

Air, in addition to being supplied to the main cabin distribution system, is supplied to individual ducts for distribution to the aft toilets and to the cockpit mixing valve. This valve also has access to air from the heater through one port of the windshield anti-icing valve. Air from the cockpit mixing valve is ducted to louvered outlets located outboard of the rudder pedals at floor level for both the captain and the first officer. The direction of the louvers is adjusted by push-pull controls. The position of the cockpit mix valve is selected by the cockpit temperature control rheostat.

Air for the individual cold air outlets and voltage regulator cooling is obtained directly from the discharge of the aftercooler, or from discharge of the turbine during turbine operation. On airplanes so equipped, cold air during Freon or turbine operation will come from the discharge of the Freon evaporator rather than the turbine discharge.

Air for anti-icing and vinyl warming of the windshield is obtained from the discharge of the heater through a windshield valve. Air from the windshield valve is discharged through the gap between the windshield panels. The windshield valve is positioned in response to the windshield selector switch settings. In order to prevent fogging of the windshield while refrigeration is required, a bypass valve and flow limiting venturi will allow a small quantity of air to go to the windshield. This bypass valve opens when the anti-icing selector switch is in the 10°-to-0° DEFOGGING position.

Air is exhausted overboard through flow-limiting devices at the cabin temperature sensing element (thermistor), at each toilet and sink, at the radio rack, and at the galley area. The remainder of the air is exhausted from the main cabin to the underfloor area. Under the floor, the air passes alongside the baggage compartments to the heater compartment and overboard through a cabin pressure control valve.

When on the ground with the engines not running and ground power supplied, air will be drawn in by a blower through a valve in the left side of the fuselage forward of the main passenger entrance door. On Freon equipped airplanes, when the cabin heater is off, the air from this blower is ducted directly into the main distribution system, bypassing the heater compartment. In flight, this blower will recirculate cabin air if Freon is required, one supercharger is inoperative, and the cabin pressure is in excess of 1 PSI.

An air conditioning truck connection is provided aft of the rear baggage door on the right side of the fuselage. Air from this source goes directly into the main cabin distribution system. At the time air from the air conditioning truck opens the valve in the side of the fuselage, the ground blower is automatically stopped.

CABIN HEATER. The cabin heater consists of a central combustion chamber where gasoline is mixed with air and burned. Air for the cabin passes on the outside of the chamber. Combustion air for the heater is supplied through a scoop located between the No. 2 nacelle and the fuselage in the leading edge of the wing. Air from this scoop discharges into a plenum chamber and is then ducted to the combustion chamber of the heater.

For ground operation of the heater, combustion air is obtained from the underside of the left wing fillet area by a blower. Fuel for the heater is obtained from the No. 2 main fuel tank, an alternate supply being available by cross-feeding from the anti-icing fuel pump. Fuel valves are installed and are controlled by thermal switches in the heater discharge airstream. A fuel pressure regulator maintains fuel pressure on the spray nozzle for correct gasoline and air mixture in the combustion chamber. A second set of controls, identical to the above, is provided. A fuel selector switch in the cockpit selects the set of controls desired. The nozzle fuel pressure is indicated on the heater control panel in the cockpit. Heater air discharge temperature is also indicated on the heater control panel.

Dual spark plugs and ignition units ignite the fuel in the combustion chamber. A switch for checking individual ignition systems is provided in the cockpit. When the heater ignition selector switch is in the NORMAL (DUAL IGNITION) position, both combustion chamber spark plugs in the heater fire simultaneously. With the selector switch in the CHECK (SINGLE IGNITION) position, one of the two spark plugs will fire when #1 FUEL is selected; the other spark plug will fire when #2 FUEL is selected.

If the duct downstream of the heater should overheat, dropout switches will cause a fuse to blow, thus stopping heater operation. A duplicate set of dropout switches is installed.

Power is supplied to the heater circuits when the cabin heater master switch is on and either the cockpit temperature control rheostat is within the last 30° of travel toward HOT or the windshield switch is in the 0°-to-40° or ANTI-ICING positions. When the heater is off, the temperature of the air at the heater discharge will be dependent upon the heat of compression generated as the air passes through the superchargers and, if needed, upon the action of the aftercooler and cooling turbine, or Freon condenser, if installed.

AFTERCOOLER. The aftercooler consists of a number of tubes through which air to the cabin, heated during compression in the superchargers, passes. Outside air flows over the tubes cooling the cabin air. At full coolant airflow, the cabin air is cooled to approximately outside air temperature. The coolant air for the aftercooler is obtained from the belly scoop located in the wing area on the underside of the fuselage. The air is ducted through the aftercooler, through a fan, and exhausted overboard through an exhaust door located in the fillet area on the underside of the left wing. In flight, the exhaust door automatically controls the aftercooler coolant airflow. For ground operation, air is drawn through the coolant circuit by the fan driven by the turbine.

TURBINE. The air from the aftercooler, cooled to only slightly above ambient temperature, enters the turbine. Here, the high-pressure air is allowed to expand and drive the turbine wheel connected to the fan in the coolant air circuit. The removal of energy from the turbine by the fan cools the air below outside temperature.

Since the superchargers are limited in the amount of pressure rise they can produce, a compression ratio limit switch is provided to prevent the pressure from exceeding prescribed values with a resultant stalling of the supercharger. As the prescribed value of pressure rise is approached, the compression ratio limit switch stops the mixing valve from forcing larger percentages of air through the turbine. If the cabin pressure is then increased, the compression ratio limit switch will signal the mixing valve to reverse and allow a smaller percentage of the air to pass through the turbine and thus prevent overloading the supercharger. In other words, as the cabin pressure is increased, the refrigeration available from the expansion turbine decreases. See paragraph on the Main Cabin Mixing Valve for further discussion of the compression ratio limit switch.

FREON SYSTEM. Freon refrigeration is accomplished by the evaporation of low-pressure Freon liquid in the evaporator, where the Freon is inside the tubes and cabin inlet air around the tubes. The compressor removes the gas being generated by the evaporating liquid and discharges it as high-pressure, high-temperature gas to the condenser. The condenser is located outside the refrigerated area. Here outside air flowing across the tubes cools the high-temperature gas, causing it to condense to a high-pressure liquid. Air used to accomplish this condensation is then discharged overboard, carrying the heat with it. The liquid Freon is piped to the expansion valve where the pressure is again dropped to that required in the evaporator. The expansion valve meters the correct amount of Freon to the evaporator where complete evaporation of the liquid occurs, accomplishing refrigeration of cabin air.

The outside airflow across the condenser is controlled by the exhaust door. On the ground, with the Freon switch on and the temperature control calling for Freon system operation, the door opens wide and the condenser blower supplies the air. When the throttles are advanced, the door goes to a trail position, where it stays for flight operation. At the time the weight is removed from the landing gear, power to the condenser blower is cut off and the blower is allowed to windmill during ram air supply. An indicator light in the cockpit illuminates when the door is extended beyond the trail position. Thus, for normal ground operation, this indicator light will be illuminated. In flight, the door position light should be off. If it does illuminate, speed should be limited to 255 knots IAS to prevent damage to the blower and the exhaust door.

Suction temperature is the temperature of the gas leaving the evaporator. Superheat is the rise in temperature of the gas above the boiling point of the liquid in the evaporator. Both suction temperature and superheat are indicated in the cockpit. Before reading these instruments, it should be ascertained whether the system is in operation or not. This is evidenced by a surge of the ammeters at the time the compressor starts. (Check by turning the Freon switch off and then on again and wait 5 seconds.) Stabilization should be reached before making observations (approximately 5 minutes after starting).

Suction temperature will vary as the inlet air temperature to the evaporator changes. An indication of the refrigeration being supplied to the cabin is the difference between the cabin and suction temperatures. With normal superheat ($0°$ to $15°F$), this difference should be at least $20°F$ and with high cabin temperatures (above $80°F$) should approach $45°F$. The larger the difference, the more refrigeration supplied to the cabin. During humid conditions, the refrigeration supplied to the cabin will be less because refrigeration is being used to remove water vapor from the air entering the cabin. When the suction temperature reaches $32°F$, the compressor is automatically unloaded and refrigeration stops. This is done to prevent ice formation on the evaporator. Should the indicator in the cockpit show less than $32°F$, the Freon switch should be turned off until the suction temperature rises again.

High superheat readings indicate that the supply of refrigerant to the evaporator is not sufficient, and the evaporator is not performing at maximum efficiency. With high superheat the system need not be turned off. Should superheat reading be high with OAT above $70°F$, the system should be serviced.

As the evaporator inlet air temperature approaches $32°F$, the superheat will rise above normal. When it reaches $32°F$ or below, the superheat should return to zero. Surges in superheat and suction temperatures followed by normal values are to be expected when the compressor restarts after an off period.

Indicator lights are provided for compressor low oil pressure, condenser exhaust door extension beyond the inflight position, and compressor (inoperative) motor overheat.

MAIN CABIN MIXING VALVE. The main cabin mixing valve is a 3-port valve. The ports are selected by an actuator which receives its signal from the cabin temperature control circuit. The actuator also has switches that start and stop the Freon system. At the discharge end of the mixing valve is a restrictor damper positioned by a separate actuator. This actuator receives its signal from the windshield anti-icing switch through the compression ratio limit switch. The damper valve is used to increase the pressure in the duct system and force larger quantities of air to the windshield when the selector is in the ANTI-ICING position. If the increase in pressure causes the supercharger to approach an overload, the compression ratio limit switch will cause the actuator to reverse, decreasing the restriction and thus lowering the duct pressure.

CABIN PRESSURE CONTROL VALVE. The cabin pressure control valve regulates the outflow of cabin supercharger air from the cabin to control cabin pressure. The outflow air is controlled by the opening or closing of the cabin pressure control valve. When the cabin pressure is controlled manually, the valve is actuated by an electric motor controlled from the cockpit. When the cabin pressure is controlled automatically, the cabin pressure control valve is actuated either electrically or pneumatically, depending on which of the cabin pressure control systems is installed on the airplane.

Oxygen systems for large jet aircraft

There are two independent oxygen systems: flight crew and passenger. The crew oxygen system provides supplemental and protective breathing oxygen through diluter demand regulators for pre-decompression and post-decompression use, and for smoke protection. The crew oxygen is supplied from a cylinder located in the forward overhead panel. The passenger oxygen system provides supplemental continuous flow oxygen for use during and after an emergency descent due to decompression. The passenger oxygen is supplied from cylinders located in the aft cargo compartment.

FLIGHT CREW OXYGEN SYSTEM. The flight crew oxygen system normally provides diluted oxygen on demand. However, 100% oxygen on demand or under pressure can be obtained when desired. The system consists of a high pressure oxygen cylinder assembly, a system shutoff valve, a pressure reducing regulator, five diluter demand regulators, five oronasal masks and five smoke goggles.

CREW AND PASSENGER PORTABLE OXYGEN. A portable oxygen cylinder assembly, consisting of a cylinder, a regulator, a full face mask, and a sling, provides oxygen for protective breathing during investigation of smoke outside of the control cabin, or for supplemental breathing when required. The cylinder pressure is 1,800 PSI (at 21°C), and is charged to a capacity of 11 cubic feet with free oxygen. The cylinder is equipped with two outlets: a threaded demand outlet and a bayonet-type 3-liter-per-minute constant flow outlet. One full face mask is provided to fit the demand outlet. The regulator includes a shutoff valve, a pressure reducing mechanism, a pressure gage, a safety plug and a charging valve. The portable oxygen cylinder assembly is located on the rear bulkhead of the control cabin.

In the passenger cabin there are twelve portable oxygen cylinder assemblies intended for first aid use, located in the hat racks at placarded positions. The capacity of the cylinders is 7.15 cubic feet of free oxygen. The cylinders are equipped with two bayonet-type outlets: one provides a constant flow of 2 LPM and the other 4 LPM.

Oxygen systems for large reciprocating-engine aircraft

CREW SYSTEM. Oxygen is supplied to the crew system from a cylinder located on the forward side of the flight compartment bulkhead behind the first officer's station (aft of the bulkhead on some airplanes). A regulator, flow indicator, and outlets are provided for the captain, first officer, and third crew member, and on some airplanes, for additional crew members. Oxygen pressure is indicated by a gage mounted on the side panel adjacent to the first officer's seat.

PASSENGER SYSTEM. Passenger oxygen is supplied to the outlets in the cabin areas by one or two oxygen cylinders. Continuous-flow-type regulators are installed in the cabin beside the supply cylinder. A gage to indicate oxygen cylinder pressure is installed at the passenger supply cylinder. The passenger oxygen system is not installed on some airplanes.

A cabin low-pressure warning horn will sound when the cabin altitude exceeds 10,000 feet.

Crew system operation:
1. Open the cylinder shutoff valve.
2. Check individual regulators for desired settings (NORMAL or 100%).
3. Close the cylinder shutoff valve when the system is not in use.

Crew system portable operation:
1. Open the cylinder shutoff valve.
2. Close the cylinder shutoff valve when the cylinder is not in use.

Passenger system operation:
1. Open the cylinder shutoff valves.
2. Plug masks into desired outlets. (Remove masks from outlets when not in use to conserve supply).
3. Close the cylinder shutoff valves when the system is not in use.

D. AIRCRAFT INSTRUMENT SYSTEMS

Instrument systems for large jet aircraft

FLIGHT INSTRUMENTS. All the flight instruments except the standby magnetic compass are located on the pilot's and navigator's instrument panels. The standby magnetic compass folds out of sight under the overhead panel. The flight instruments are: turn-and-bank indicators, rate of climb indicators, clocks, the standby magnetic compass, ram air temperature indicators, air-speed indicators, altimeters, and Machmeters, compass system with radio magnetic indicators, and an integrated instrument system consisting of pictorial deviation indicators and horizon flight director indicators.

AUTOPILOT. The autopilot may be used to control the airplane (1) in pitch by moving elevator control tabs and horizontal stabilizer, (2) in bank by moving the aileron control tabs, and (3) in yaw by moving the rudder. The autopilot moves the control surfaces through electric motors and mechanical linkages.

PITOT-STATIC SYSTEM. Measurement of total and static air pressure is accomplished through the pitot-static system consisting of pitot tubes, static pressure ports and associated tubing. The pitot-static system pressures are used in determining airspeed and Mach number. Total and static pressures are also sensed by the rudder airspeed switch, the autopilot air data sensor, the Mach airspeed warning switch, the mach trim synchrotel transmitter, and the flight recorder; while static pressure alone is supplied to the rate of climb indicator, the altimeter and the differential pressure switches for equipment cooling control. Static pressure is sensed through four static ports on the right side of the fuselage and three static ports on the left side of the fuselage. One port on the left connected by tubing to one port on the right is in turn connected to the pilot's instrument manifold. Another left and right pair, manifolded together, are connected to the copilot's instruments. A third pair is connected to the autopilot air data sensor, the rudder airspeed switch, the Mach airspeed warning switches, the Mach trim synchrotel transmitter, the cabin dual altimeter and differential pressure gage, the navigator's altimeter and true airspeed indicator, differential pressure switches and the flight recorder. A fourth static port on the right-hand side supplies static pressure to the cabin pressure automatic controller. The two pitot-tube probes used to sense total air pressure are on opposite sides of the airplane and are not interconnected. The pitot tube on the left side of the airplane supplies total air pressure to the pilot's pitot-static-operated instruments. The pitot tube on the right side of the airplane supplies pressure to the copilot's instruments. The copilot's pitot tube also supplies pressure through a shutoff valve to the autopilot air data sensor, the rudder airspeed switch, the flight recorder, the Mach airspeed warning switches, the Mach trim synchrotel transmitter, and the navigator's true airspeed indicator.

Fig. 88. Machmeter.

MACHMETER. Machmeters are provided for both the pilot and copilot. The instruments provide a continuous indication of Mach number during flight. Range of indication is .3 to 1.0. A radial pointer on each instrument moves to indicate the flight Mach number of the airplane. The limiting Mach number, never-exceed Mach number (M_{NE}) and normal operating Mach number (M_{NO}) are .906. The indicators sense pressure from the pitot-static system.

Instrument systems for large
reciprocating-engine aircraft

The DC-operated electrical instruments require 28 volts DC from the generators or ground power source. The AC-operated instruments require 115 volts AC from the inverters or alternators. Some of the electrical instruments are self-energized and operate from electric current which they generate themselves.

The mechanical instruments are operated by changes in pressure, and measure system pressures as well as vacuum, absolute, and differential pressures.

ELECTRICAL INSTRUMENTS (28-VOLT DC).

Carburetor air temperatures
Engine oil temperatures
Engine oil quantities (some airplanes)
Engine cylinder head temperatures
Cabin supercharger oil temperatures
Outside air temperatures
Pitot heat—ammeters
Propeller de-icing ammeters
Anti-icing fluid quantities
Hydraulic fluid quantities
Freon temperatures
Cabin and airfoil heater temperatures (some airplanes)

ELECTRICAL INSTRUMENTS (26-VOLT AC OR 115-VOLT AC).

Flux Gate or Tyrosyn compasses
Gyro horizons
Turn-and-bank indicators
Fuel flowmeters
Torquemeters
Fuel quantity indicators
Fuel pressures
Oil pressures
Engine oil quantities (some airplanes)
Cabin supercharger oil pressures
Wing flap positions
Cabin and airfoil heater fuel pressures
Cabin pressure regulators (some airplanes)
Cabin pressure controllers (some airplanes)
Cabin pressure change limit controls (some airplanes)
Cabin pressure rate controls (some airplanes)
Mixing valve position indicators

ELECTRICAL INSTRUMENTS (SELF-ENERGIZED).

Engine tachometers

PRESSURE INSTRUMENTS (ABSOLUTE).

Engine manifold pressures
Altimeters.

PRESSURE INSTRUMENTS (DIFFERENTIAL).

Airspeed indicators
Rate-of-climb indicators

Cabin supercharger airflows
Cabin pressure differentials (some airplanes)
Cabin pressure regulators (some airplanes)
Cabin pressure change limit controls
Cabin supercharger duct pressures

Note: On some airplanes the cabin pressure regulator and the change limit control are directly operated by differential pressure, and function to deliver AC power to the cabin pressure control amplifier.

PITOT-STATIC SYSTEM. A dual pitot system supplies the ram air pressure for both airspeed indicators on the flight instrument panels. The pitot systems are entirely separate, each consisting of a pitot tube in the nose section of the airplane and the necessary lines to the respective instrument panel. The left pitot tube supplies ram air for the captain's airspeed indicator, and the right tube supplies ram air to the first officer's (and navigator's, if installed) airspeed indicator. Each pitot head is heated by an electrical element controlled by the pitot heater switch located on the forward overhead panel. No other flight compartment control is provided for the system.

A dual, electrically heated, static system supplies the static air pressure necessary for the operation of the airspeed, rate-of-climb, and altimeter indicators located on the captain's and first officer's (and the navigator's, if installed) flight instrument panels, and also for the operation of the autopilot constant altitude control unit and the cabin pressure control instruments. Each static source located on the nose section of the airplane is heated by an electrical element controlled by the pitot-static heater selector switch.

An alternate ice-free static source is located inside the fuselage, aft of the rear pressure dome. Two valves, one outboard of the captain's and one outboard of the first officer's station, permit selection of the alternate source in case of icing or malfunctioning of the main static source. A variation in instrument readings will be noticed when the alternate static source is used.

DIRECTIONAL GYRO PRECESSION, LIMITATIONS. The directional gyro may precess as much as $3°$ in 15 minutes.

GYRO HORIZON INDICATOR, LIMITATIONS. The gyro in the gyro horizon instrument on the main instrument panel can be erected quickly by using the CAGE knob on the instrument case. If the CAGE knob is not used the gyro may take several hours to erect by itself. After power is applied the gyro can be erected by actuating and holding the CAGE knob in the CAGE position until the horizon bar stops oscillating. The gyro should be erected immediately after power is applied or at least five minutes before take-off so that any tumbling or drifting can be detected before take-off. The gyro should not be erected by use of the CAGE knob unless it has been determined that the airplane is in a level attitude.

E. COMMUNICATIONS AND NAVIGATION SYSTEMS

The following discussion is non-technical in nature and is designed only to familiarize the airframe mechanic with the basic functions of aircraft radio equipment, its location, and the maintenance problems in its installation and removal.

AUTOMATIC PILOT (AUTOPILOT) PRINCIPLES. Autopilot equipment is essential, especially in large aircraft, to relieve the pilot/s of constantly maintaining proper altitude, attitude, headings, and course. The autopilot thus permits the pilot/s to pay more attention to many other related duties which in turn adds considerably to inflight safety and passenger comfort. Generally speaking, in high performance aircraft the autopilot can do a better job of flying the aircraft under normal flight conditions than can the human pilot. The autopilot can sense changes in heading, pitch, and trim faster than the pilot and can then follow with a more precise correction. Autopilots in small aircraft perform basically the same functions. An important consideration is that the pilot of a small aircraft can devote more time to his radio, navigation, and other duties.

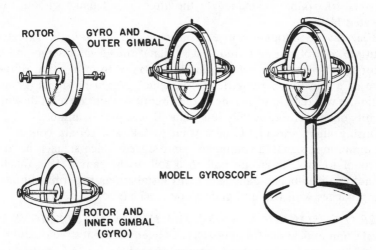

Fig. 89. Simple gyroscope.

The vertical gyro (the remote gyro for the attitude indicator) provides a good stable reference platform for the roll and pitch axes. The direction gyro (for the heading indicator) provides a stable directional reference. To the elevator, ailerons, and rudder (primary control surfaces) are attached servo motors that are capable of moving the particular control surface. Should the aircraft deviate from the attitude and heading in which the pilot has placed it when he last engaged the autopilot, synchros in the gyros signal the servo motors by way of a coordinating computer. Thus, this will correct the particular deviation by a proper movement of the particular control surface.

The autopilot may be designed around all of the three axes of the airplane or, as in small aircraft, it may be designed for only one axis such as directional or yaw corrections.

Autopilot operations require frequent checks, adjustments, etc. For instance, a trim servo is required so that the elevator trim reflects shifting or changing loads in the particular aircraft under many various flight conditions. Without this trim servo the elevator could gradually (or slowly) be held out of the exact alignment under considerable stress. This condition could cause a sudden disengagement of the servo and result in a violent climb or dive.

To provide smooth operation, rate gyros, also called sensors, are required in the pitch, roll, and the yaw axes to detect the rate at which the aircraft is deviating from its set attitude. This permits the autopilot to govern the rapidity and force of the correction and prevent over- and undercontrolling of the aircraft.

Controls must be provided to the pilot so that when the autopilot is engaged, the pilot can maneuver the aircraft and make smooth turns to selected headings.

Basically the autopilot can sense *attitude* and not *altitude*. Therefore, an altitude control can be introduced to signal the elevator servo and the elevator trim servo whenever the aircraft deviates from a given altitude. Known as a sensitive barometric sensor, which is in a no-signal (or null) condition, it will remain so as long as the aircraft remains on altitude within the limits of the sensor, usually less than 40 feet plus or minus.

During flight, should the pilot want to follow a certain course such as an omnirange signal, a computer is introduced which signals the automatic pilot to turn, intercept, and then follow the radial. This function of the automatic pilot is expanded in many installations, to permit automatic low approaches with ILS and glide slope receivers.

FEDERAL COMMUNICATIONS COMMISSION REGULATIONS. The Federal Communications Commission (FCC) requires that a radio-equipped aircraft (considered to be a mobile radio station) have an FCC radio station license posted in plain view in the cockpit or flight deck of the aircraft.

It is necessary that each pilot engaging in two-way communications from an aircraft hold a Restricted Radiotelephone Permit. In either case, the aircraft radio station license and the pilot's Permit are obtained by making application on approved forms to the FCC.

An aircraft mechanic who is authorized to move aircraft from the servicing area to loading ramps, etc., on a radio-controlled airport, must obtain the Restricted Radiotelephone Permit for communications with the control tower during taxi operations.

RADIO INSTALLATION CONSIDERATIONS. The safety of aircraft is dependent, to a large degree, on the satisfactory performance of airborne radio and electronic equipment.

Reliability of the radio systems and their performance is proportional to the quality of maintenance received and the knowledge of those who perform such maintenance. The term "radio systems" means those units of antenna, power sources, sensors, receivers, transmitters, indicators, etc., that together perform a function of communications and/or navigation.

The mechanic responsible for the maintenance and servicing of radio or allied equipment must be able to properly inspect each unit, assemblies, wiring and control systems for damage, general condition, and proper functioning to assure the continuous satisfactory operation of the systems.

Repairing, adjusting, testing, etc., of radio and electronic equipment and systems must be accomplished in accordance with the manufacturer's maintenance instructions, manuals, etc. and also in accordance with current regulations.

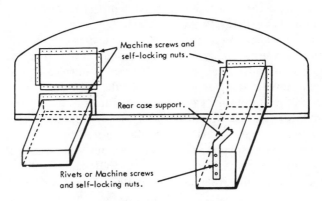

Fig. 90. Typical panel mount.

INSTALLATION AND MOUNTING OF RADIO EQUIPMENT. Radio equipment should be attached to the aircraft by means of locking devices to prevent loosening from vibration while in service. Examples of locking devices are self-locking nuts, serrated washers, cotter pins, self-locking holddown clamps, or snap-slides and holddown assembly nuts which are safety wired.

Items mounted on shock mounts should have sufficient clearance for normal vibration and swaying of the equipment without striking adjacent equipment or parts of the aircraft.

In order that occupants will not be endangered by moving equipment during minor crash landings, the equipment mounting and rack should be capable of withstanding ultimate acceleration for which the aircraft is designed.

REDUCING NOISE IN RADIO RECEIVERS. Radio equipment should be bonded to the aircraft in order to provide a low-resistance ground circuit and to minimize radio interference from static electrical charges.

Bonding jumpers should be as short as possible and installed in such a manner that the resistance of each connection does not exceed 0.003 ohms.

Aluminum alloy, tin-plated, or cadmium-plated copper jumpers should be used for bonding aluminum alloy parts. Copper, brass, or bronze jumpers should be used to bond steel parts.

The most effective method of minimizing engine ignition radio interference is to shield the ignition system. This involves enclosing in metal all parts of the circuit that might radiate electrical charges.

PREFERRED LOCATION OF ANTENNAS. Fixed and trailing wire antenna installations should be tailored to fit the particular type of aircraft. Other types of antennas are much more compact and may be installed as complete units on many types of aircraft.

Loop antennas enclosed in streamlined housings or exposed loops are satisfactory for external mounting on the aircraft. Loops may be installed internally in an aircraft when proper attention is given to avoiding interference from metallic structures and skin of the aircraft.

Antennas for omnirange (VOR) and instrument landing system (ILS) localizer receivers should be located at a position on the airplane where they will have the greatest sensitivity for the desired signals and minimum response to undesired signals such as VHF energy radiated by the engine ignition system. The best location for the VOR-ILS localizer receiving antenna on most small airplanes is over the forward part of the cabin. The rigid V-type antenna should be mounted so that the apex of the "V" points forward and the plane of the "V" is level in normal flight attitude.

RADIO INSTALLATIONS. When installing radio equipment, first consider areas or locations designated by the airframe manufacturer and use factory supplied brackets or racks. Always follow the aircraft manufacturer's installation instructions. When this is not possible and/or the instructions not available, use the location in the aircraft of known load carrying capacities. Baggage compartments, cabin or cockpit floors are good mounting platforms providing the floor attachments meet the proper strength requirements.

Determine that the location of radio equipment provides sufficient air accumulation to avoid overheating, that it has proper clearance between high temperature areas, and that there is a safe distance and/or protection from flammable materials or hazardous fluids or fumes. Be sure that the location gives protection against damage from baggage or seat deflectors and that the equipment cannot come in contact with adjacent parts of the structure or appliances.

Aircraft can only be returned to service if the radio installation has been made according to approved installation data, has been properly inspected, and operation checks have been made to satisfy all regulations.

Where radio installation or removals have created changes in weight of the aircraft, proper weight and balance computations must be made to determine that the aircraft's gross weight, forward and rearward center of gravity limits, etc., have not been exceeded. This current weight and balance data must be made a part of the aircraft's permanent records.

F. AIRCRAFT FUEL SYSTEMS

A well designed fuel system for aircraft must provide for storage of the required amount of fuel in spaces available within the aircraft structure. In addition, external fuel tanks (such as wing tip tanks) may be utilized when originally incorporated by the manufacturer of the aircraft or later installed according to approved specifications. The fuel system, in addition to storage purposes, must expedite the flow of fuel from the tanks to the fuel metering devices. The system must meter the fuel to these devices at the proper rate of flow and pressure. The fuel system must be relatively simple in its operation and be positive and reliable under all normal conditions of ground and flight operation. There must be incorporated in the fuel system a means of shutting off all fuel flow to the engine/s. The system must be free from potential fire hazards.

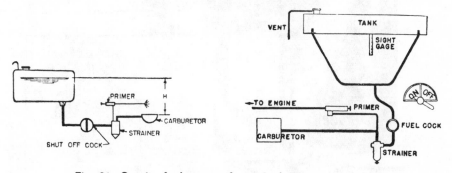

Fig. 91. Gravity fuel system for a single-engine airplane.

GRAVITY FLOW FUEL SYSTEM. The gravity flow fuel system, used in some small aircraft, is an elementary system that is very basic in its design and operation. The gravity system generally consists of a fuel tank (or tanks) from which the fuel can flow by gravity to the fuel metering device (carburetor, etc.). To be operable, the fuel metering device must be at a lower level than the fuel tank outlet. The fuel metering device receives the liquid fuel, vaporizes it, mixes the vaporized fuel with a proper amount of air, and then sends the mixture through the induction system to the respective engine cylinders.

FUEL LINES AND FITTINGS. Fuel lines from the fuel tank/s must be routed, whenever possible, away from any areas of excessive heat. The lines must always be installed below electrical wires or conduits or bundles of wires and must be free of bends, kinks, etc., that would impede fuel flow and lead to fuel line breakdown. In addition, upward loops in fuel lines must be avoided and lines should be installed with the least number of bends and changes in direction. These faults will lend to retarding fuel flow and may create areas of fuel flow lag, with subsequent problems of "vapor lock."

Vapor lock in fuel lines may occur under varying conditions but is most likely when fuel lines become hot. The heat vaporizes the fuel within the line creating a vapor pressure that may be greater than the fuel pressure. As a result, a "lock" is created in the line with a subsequent stoppage of fuel flow. When it is necessary to route fuel lines near a source of heat, the lines should be "lagged." Lagging fuel lines is accomplished by wrapping or otherwise insulating the lines so that the nearby heat will not penetrate the lines and vaporize the fuel.

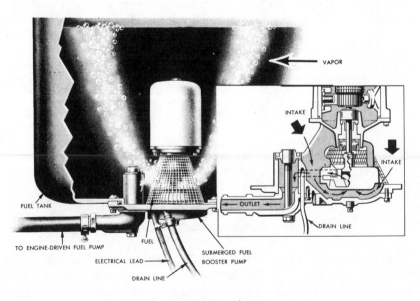

Fig. 92. Centrifugal-type booster pump.

PRESSURE FUEL SYSTEMS. The pressure type of fuel system incorporates an engine-driven vane type fuel pump that assures fuel from the tank/s to the fuel metering devices regardless of the level of the fuel tank outlet/s in relation to the fuel metering devices. Thus, fuel tank outlets may be installed *below* the level of the fuel metering devices with assurance of fuel flow from the tank/s to the carburetor, etc. in all anticipated flight attitudes.

In large aircraft installations there are multiple fuel tanks so arranged with fuel manifolds that engines may receive fuel from one or more of the fuel tanks. Also in large aircraft installations, electrical centrifugal boost pumps are generally incorporated within the fuel tanks to insure a positive flow of fuel from the tanks to the engine-driven fuel pump/s. These centrifugal boost pumps are especially important in large aircraft where the fuel tanks are located some distance from the fuel metering devices on the engines.

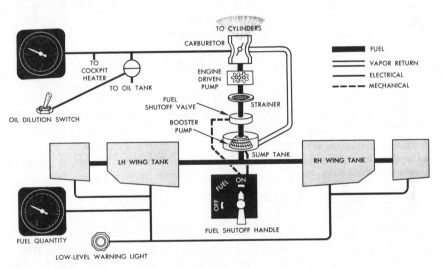

Fig. 93. Fuel quantity gage and fuel pressure gage operation.

FUEL SYSTEM COMPONENTS. Fuel systems, depending on the type of aircraft, will incorporate such components as fuel tanks, boost pumps where required, strainers and/or filters, fuel lines and fittings, shutoff valves, fuel selector valves, fuel pumps, pressure relief valves, vapor eliminators, fuel warning devices, fuel tank quantity indicators, fuel quantity gages, pressure gages, primer systems, fuel flow indicators, etc.

Mechanics should, prior to doing maintenance or repair on fuel systems on which they have not previously worked, familiarize themselves with the particular system by use of manufacturer's diagrams, schematics, specifications, etc.

FUEL SYSTEM MAINTENANCE. Maintenance, servicing, repair, and adjustment of aircraft fuel systems should be done according to the applicable manufacturer's instructions.

FUEL LINES AND FITTINGS. Compatibility of fittings. All fittings are to be compatible with their mating parts. Even though various types of fittings appear interchangeable, in many cases they will have a different thread pitch or minor design difference which prevents proper mating and may cause the joint to leak or fail.

Routing. Make sure that a fuel line does not chafe against control cables, airframe structures, etc., or comes in contact with electrical wiring or conduit. Where physical separation of the fuel lines from electrical wiring or conduit is impracticable, locate the fuel lines below the wiring and clamp it securely to the airframe structure. Electrical wiring or conduit *must never* be supported by, or clamped to, fuel lines.

Alignment. Locate fuel line bends accurately so that the tubing is aligned with all support clamps and end-fittings and is not drawn, pulled, or otherwise forced into place by the support clamps. Never install a straight length of tubing between two rigidly mounted fittings. Always incorporate at least one bend between such fittings to absorb stress caused by vibration and temperature changes.

Bonding. Bond metallic fuel lines at each point where they are clamped to the structure. Integrally bonded and cushioned line support clamps are preferred to other clamping and bonding methods.

Support of line units. All fittings heavy enough to cause the line to sag should be supported by means other than the tubing. Locate clamps and brackets as close to bends as possible to reduce overhang.

FUEL TANKS AND CELLS. Aircraft welded or riveted fuel tanks that are made of commercially pure aluminum (3003, 5052, or similar alloys) may be repaired by welding. Tanks made from heat-treatable aluminum alloys are generally assembled by riveting. In case it is necessary to rivet a new piece in a tank, use the same material as used in the tank undergoing repair and seal the seams with a compound that is insoluble in gasoline. Special sealing compounds are available and should be used in the repair of fuel tanks. Inspect fuel tanks and cells for general condition, security of attachment, and evidence of leakage. Examine fuel tank or cell vent lines, fuel lines, and sump attachment fittings very closely. Purge defueled tanks of explosive fuel/air mixtures in accordance with the manufacturer's service instructions. In the absence of such instructions, utilize an inert gas such as CO_2 as a purgative to assure the total deletion of fuel/air mixtures.

Integral tanks. Examine the interior surfaces and seams for sealant deterioration and corrosion (especially in the sump area). Follow the manufacturer's instructions for repair and cleaning procedures.

Internal metal tanks. Check the exterior for corrosion and chafing. Dents or other distortion, such as a partially collapsed tank caused by an obstructed fuel tank vent, can adversely affect fuel quantity gage accuracy and tank capacity. Check the interior surfaces for corrosion. Pay particular attention to the sump area, especially those which are made of cast material.

Removal of flux after welding. It is especially important, after repair by welding, to completely remove all flux in order to avoid possible corrosion. Promptly upon completion of welding, wash the inside and outside of the tank with liberal quantities of hot water and then drain. Next, immerse the tank in either a five percent nitric solution or five percent sulfuric acid solution. If the tank cannot be immersed, fill the tank with either solution and wash the outside with the same solution. Permit the acid to remain in contact with the weld about one hour and then rinse thoroughly with clean water. Test the efficiency of the cleaning operation by applying some acidified five percent silver nitrate solution to a small quantity of the rinse water used last to wash the tank. If a heavy precipitate is formed, the cleaning is insufficient and the washing should be repeated.

Inspect the interior for checking, cracking, porosity, or other signs of deterioration. Make sure the cell retaining fasteners are properly positioned. If repair or further inspection is required, follow the manufacturer's instructions for cell removal, repair, and installation. *Do not allow flexible fuel cells to dry out.* Preserve them in accordance with the manufacturer's instructions.

FUEL TANK CAPS, VENTS, AND OVERFLOW LINES. Inspect the fuel tank caps to determine that they are the correct type and size for the particular installation.

When substituted for unvented caps, *vented caps* may cause loss of fuel or fuel starvation. Similarly, an improperly installed cap that has a special venting arrangement can also cause malfunctions.

When substituted for vented caps, *unvented caps* will cause fuel starvation and possible collapse of the fuel tank or cell. Malfunctioning of this type occurs when the pressure within the tank decreases as the fuel is withdrawn. Eventually, a point is reached where the fuel will no longer flow, and/or the outside atmospheric pressure collapses the tank. Thus, the effects will occur sooner with a full fuel tank than with one partially filled.

Check *tank vents and overflow lines* thoroughly for condition, obstructions, and correct installation. Also check for proper operation of any check valve and ice protection units. Pay particular attention to the location of the tank vents when such information is provided in the manufacturer's service instructions. Inspect for cracked or deteriorated filler opening recess drains. These may allow spilled fuel to accumulate within the wing or fuselage creating a potential fire hazard. One method of inspection is to plug the fuel line at the outlet and observe fuel placed in the filler opening recess. If drainage takes place, investigate the condition of the line and purge any excess fuel from the wing.

Assure that all *filler opening markings* are stated according to the applicable airworthiness requirements and that they are complete and legible.

FUEL CROSSFEED, FIREWALL SHUTOFF, AND TANK SELECTOR VALVES. These valves must be carefully inspected for leakage and proper operation.

Internal leakage can be checked by placing the appropriate valve in the off position, draining the fuel strainer bowl, and observing if fuel continues to flow into it. Check all valves located downstream of boost pumps with the pump/s operating. Do not operate the pump/s longer than necessary.

External leakage from these valves can be a severe fire hazard, especially if the unit is located under the cabin floor or within a similarly confined area. Be sure and correct the cause of any fuel stains associated with fuel leakage.

Selector handles. Check the operation of each handle or control to see that it indicates the actual position of the selector valve. Assure that stops and detents have positive action and feel. Worn or missing stops and detents can cause unreliable positioning of the fuel selector valve.

Inaccurate positioning of fuel selector valves can also be caused by *worn mechanical linkage* between the selector handle and the valve unit.

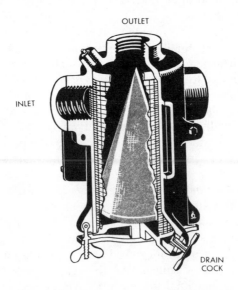

OUTLET

INLET

DRAIN COCK

Fig. 94. Typical main-line strainer.

An improper fuel valve position setting can seriously reduce engine power by restricting the available fuel flow. Check universal joints, pins, gears, splines, cams, levers, etc., for wear and excessive clearance which prevent the valve from positioning accurately or from obtaining fully off and on positions.

Regarding tank selector valves, etc., assure that required placards are complete and legible.

FUEL PUMPS. Inspect, repair, and overhaul boost pumps, emergency pumps, auxiliary pumps, and engine-driven pumps in accordance with the appropriate manufacturer's instructions.

FUEL FILTERS, STRAINER, AND DRAINS. Check each strainer and filter element for contamination. Determine and correct the source of any contaminants found. Replace throwaway filter elements with the recommended type. Examine fuel strainer bowls to see that they are properly installed according to direction of fuel flow. Check the operation of all drain devices to see that they operate properly and have positive shutoff action.

FUEL INDICATOR SYSTEMS. Inspect, service, and adjust the fuel indicator systems according to the manufacturer's instructions. Determine that the required placards and instrument markings are complete and legible.

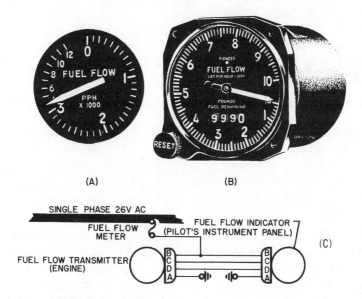

Fig. 95. (A) fuel flow indicator; (B) fuel flow totalizing indicator; (C) single fuel flow indicating system.

TURBINE FUEL SYSTEMS. The use of turbine fuels in aircraft has resulted in two problem areas not normally associated with aviation gasolines. They are (1) entrained water (microscopic particles of free water suspended in the fuel) and (2) microbial contaminants.

Entrained water will remain suspended in aviation turbine fuels for a considerable length of time. Unless suitable measures are taken, the fine filters used in turbine fuel systems will clog with ice cyrstals when the temperature of the fuel drops below the freezing temperature of the entrained water. Some fuel systems employ heated fuel filters or fuel heaters to eliminate this problem. Others rely upon anti-icing additives.

Microbial contamination is a relatively recent problem associated with the operation and maintenance of turbine fuel systems. The effects of these micro-organisms are far reaching. They can cause powerplant failure due to clogging of filters, lines, fuel controls, etc., and the corrosive acids which they produce can lead to structural failure of integral tanks. Microbial contamination is generally associated with fuel containing free water introduced by condensation or other extraneous sources.

Turbine fuel system maintenance. Maintain turbine engine fuel systems and use anti-icing and antibacterial additives in accordance with the manufacturer's recommendations.

FUEL JETTISON (DUMP) SYSTEMS. In general, when the gross takeoff weight (GTOW) is 105% greater than the landing weight (LW) of an aircraft, a fuel dump system must be incorporated. This permits, in an emergency, the dumping of fuel in flight to bring the aircraft weight down to a safe (or satisfactory) landing value.

Generally there are two dump chutes incorporated, one on each side of the fuselage that can be extended from under the wing. The arrangement permits fuel to be dumped from the main tanks quickly, efficiently, and without fire hazard.

FUEL USE AND MANAGEMENT. The arrangement of fuel system controls, indicators, and warning lights are such that the flight crew members have all controls, etc., within easy reach and that the indicators and warning lights are visually and quickly discernible.

Fuel use and fuel transfer is an important flight crew function. It must be accomplished properly in order to maintain correct aircraft balance, good aircraft stability, and to keep the center of gravity within manufacturer's limits.

Fueling of aircraft. There are several precautions that must be taken during fueling operations of *all aircraft*. They are (1) ascertain that the correct fuel octane and/or type of fuel is being supplied, (2) take all possible precautions to prevent any fire hazard, (3) prevent any possible contamination of the fuel, and (4) be sure the aircraft is properly grounded to prevent static electrical discharges from any possible source.

If there is any question as to whether the fuel is free from contamination, the fuel should be strained through acceptable materials such as chamois skins, etc. Straining of fuels is especially necessary when the fuel has been stored in drums, barrels, or similar type containers.

The type of fuel to be used in any particular aircraft should be noted on or near the filler cap/s. If information on the type of fuel is not easily available the manufacturer's specifications should be consulted.

FUEL PRESSURE CONTROLS. Engine-driven fuel pumps are generally of the sliding vane type. Fuel pressure is controlled by a pressure relief valve incorporated in the fuel pump housing. This relief valve can be adjusted to meet the needs of the particular fuel system. In the conventional type of relief valve, clockwise adjustment will increase fuel pressure and counter-clockwise adjustment will decrease fuel pressure. The relief valve "holding" nut should be "cracked" or loosened only slightly when making a fuel pressure adjustment. The holding nut should be well secured after the adjustment has been made.

USE OF POTASSIUM DICHROMATE. In order to prevent fuel tank corrosion, it is very common practice to utilize potassium dichromate as a fuel tank corrosion inhibitor.

FUEL TANK PROPERTIES. Fuel tanks must withstand a pressure of at least 3½ PSI and be so designed to have at least a 2% expansion space. The 2% expansion space in the tank allows for fuel expansion during high temperature conditions.

Fuel tank sumps should be removed and cleaned at specified intervals to keep the fuel lines, etc., free from contamination.

When it is necessary to repair a fuel tank by welding, the work should be done in an open area. The tank should be "grounded" to protect against static discharges and possible fire hazard. The drain/s must be open and the tank empty. With the filler and drain openings fully clear, the tank is then purged of all vapor according to the manufacturer's bulletins. Generally speaking, fuel tanks can be purged of fuel vapors by the use of live steam, the circulation of water, or by the use of an inert gas such as CO_2. Proper circulation can be accomplished by having the tank sump and filler openings clear at all times during the purging procedures.

TYPICAL LARGE TURBOJET AIRCRAFT FUEL SYSTEM. All fuel is carried in the interspar area of the wing structure in four main tanks, two reserve tanks, and a center tank. Each of the four main tanks supplies fuel directly to its respective engine through engine fuel shutoff valves. A fuel manifold enables the center tank or any of the four main tanks to deliver fuel to any or all engines subject to control through fuel manifold valves and engine fuel shutoff valves. Fuel from the reserve tanks can

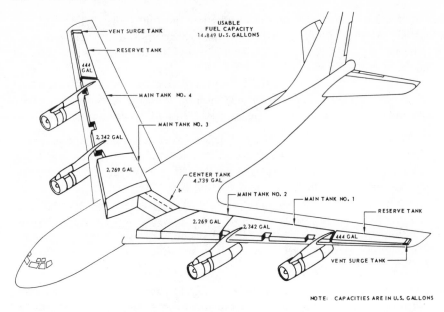

Fig. 96. Fuel tank arrangement on the Boeing 720.

be used only after it has been transferred by gravity through reserve tank transfer valves to adjacent outboard main tanks.

Each fuel tank, except the two reserve tanks, is equipped with two electrically powered boost pumps. Dip sticks are installed to measure the fuel level in each tank. Fuel system controls and indicators are located on the pilots' center panel and the flight engineer's upper and lower panels. An underwing pressure fueling system is used, incorporating automatic shutoff valves. A fueling preset system is incorporated in the quantity indicating system. Provisions are also made for overwing fueling through standard filler receptacles.

The fuel tanks are equipped with a vent system and an engine fuel heater is provided.

FUEL QUANTITY INDICATING SYSTEM. The electronic capacitance type fuel quantity indicating system consists of an indicator, tank units as required, and a compensator tank unit. Individual quantity indicators located on the systems lower panel show the fuel contained in each fuel tank. A total quantity indicator shows the total amount of fuel aboard the aircraft. The indicating system measures the effect of fuel volume and density on the tank units and presents a continuous reading of fuel available in pounds. A fueling preset system incorporated in the fuel quantity indicating system allows a predetermined amount of fuel to be loaded aboard the airplane without having to monitor the indicators on the flight engineer's lower panel. The fuel quantity indicating system is calibrated for a normal in-flight attitude.

A single push-to-test switch located on the flight engineer's lower panel, when depressed, causes all indicator needles to move toward zero, indicating that the system is operable. When the switch is released, the needles will return to their original reading.

DEFUELING. Defueling is accomplished through a manually operated valve in the inboard dry bay in each wing. Operation of the fuel boost pumps and opening of the fuel manifold valves delivers fuel to the defueling valve. The defueling rate is approximately 50 gallons per minute (GPM) for each tank. The reserve tanks are defueled by transferring fuel through the adjacent main tanks.

FUEL VENT SYSTEM. The fuel tanks are vented through a vent surge tank to a single outlet at each wing tip lower surface. The outlet is provided with a non-icing flush ram scoop to insure positive ram air pressure within the tanks at all times.

FUEL DUMP SYSTEM. The fuel dump system provides the means by which fuel can be jettisoned in flight by gravity flow directly from the main and center tanks through retractable dump chutes that are located in the trailing edge of the wing forward of the inboard flaps. The reserve tank fuel is transferred to and dumped with the adjacent outboard main tank fuel. The dump lines are installed in the tanks in such a manner as to retain a predetermined amount of undumpable fuel in the main and center fuel tanks. Fuel is dumped at an average rate of approximately 1,135 kilograms (2,500 pounds) per minute when dumping from all fuel tanks simultaneously. The exact flow rate obtained will depend upon airspeed, airplane center of gravity, and specific fuel load in individual tanks. For each of the six dump valves left closed when dumping fuel or if the tank supplying that valve has reached the standpipe level, the jettisoning flow will be reduced by one sixth. The fuel dump system controls and indicator lights, located on the flight engineer's upper auxiliary panel in a covered panel box, consist of: (1) two fuel dump chute switches; (2) two dump chute position lights; and (3) six dump valve switches with their respective position indicator lights. Do not exceed 239 knots IAS while extending or retracting the dump chutes. Extend dump chutes with wing flaps up. Do not exceed 274 knots IAS with dump chutes extended. Damage to the dump chute structure may result. Flap setting should be maintained in 0°–30° range while dumping fuel. Normal operation of electronic equipment may be continued while dumping fuel. Do not release flares while dumping fuel. Placard speed should be observed if an amber dump chute position light is illuminated. The fuel dump chutes should not be retracted until two minutes after the fuel dump valves have been closed. Chute retraction before closing the fuel dump valves may result in residual fuel, in the chute, spilling into the wing trailing area.

G. AIRCRAFT ELECTRICAL SYSTEMS

The following text on aircraft electrical systems will cover three general areas: (1) general electrical theory, electrical components, and electrical circuits; (2) typical aircraft electrical systems and components; and (3) general electrical system maintenance.

The aircraft mechanic, if he is to assume all-around maintenance responsibilities, must be well acquainted with electricity, electrical components, electrical trouble-shooting, repair, and inspection.

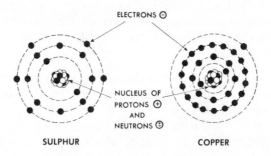

Fig. 97. Sulphur and copper atoms.

FLOW OF ELECTRICITY. When you turn on a switch and an electrical unit or appliance works, it is the result of the flow of tiny negatively charged particles (called *electrons*) through the wires and through the unit or appliance. These electrons, together with *protons* (positively charged particles) and *neutrons* (neutral particles), make up atoms, of which all substances are composed. The protons and neutrons are in the nucleus (center) of the atom and generally cannot move about within a substance. The remainder of the atom is composed of electrons which are in constant motion about the nucleus. Some of these electrons, however, are only loosely held by the nucleus and move freely when an electrical pressure is applied. The flow of these electrons through wires and appliances is called electric current.

The number of free electrons varies with the substance. Some substances have a comparatively large number of free electrons, and current can flow through them with comparative ease. Such substances are called *conductors.* In general, metals and their alloys are good conductors. Copper, for example, is the common conductor used in airplanes to conduct electricity to such devices as navigation, landing, and instrument lights, motors operating landing gear mechanism, flaps, etc..

Other substances contain only a small proportion of free electrons and offer considerable resistance to the transfer of electrons. These substances are called *insulators* or non-conductors. They are used to insulate conductors and keep the electric current from flowing in undesired paths. Some common insulators are mica, glass, rubber, bakelite, plastics, and spun glass. Specially treated cambric and nylon are materials commonly used to insulate electrical conductors in airplanes.

STATIC ELECTRICITY. The problem of static electricity is particularly serious with airplanes. Electrons may accumulate on an airplane while it is in flight, and a heavy static charge may remain on the airplane after it has landed. A discharge of this static electricity during refueling might result in an explosion. For this reason, airplanes are equipped with a ground wire fastened to the metallic part of the airplane. When the airplane lands, this wire drags on the ground and discharges the static electricity. On some airplanes, wheel tires which will conduct the electricity are used instead of a ground wire.

Bonding the metallic parts of the airplane reduces considerably the possibility of fire from static electric charges within the airplane.

ELECTRICAL CIRCUITS. The electricity that is used to operate the various devices in an airplane is not static electricity (which does not move unless there is a discharge) but dynamic electricity. This is better known as current electricity which flows steadily in a circuit when the switch is *closed.*

DIRECT AND ALTERNATING CURRENT. Electricity which flows steadily in the same direction, such as the electricity produced by a battery, is called direct current electricity (DC). The electricity used in most appliances in your home, such as irons, lights, or refrigerators, does not flow steadily in the same direction, but flows first in one direction, and then in the opposite direction, at regular intervals. This kind of electricity is called alternating current electricity (AC).

A complete electrical circuit is a path through which electrons can flow. It consists of a source of electrical pressure (battery or generator), conductors which carry the current, a control switch to start and stop the flow of electricity, and an electrical device (such as a lamp or a motor) in which the electrical energy is used.

ELECTROMOTIVE FORCE. The pressure that forces free electrons through the conductors and electrical devices is called electromotive force, voltage, or difference of potential. On airplanes, electromotive force is supplied by generators or alternators. These convert mechanical energy into electrical energy. The unit of measurement of electromotive force is called a *volt.* Most airplanes have 24 volt electrical systems.

Electromotive force (voltage) is measured with a voltmeter. When you use a voltmeter, connect the two leads (wires) of the voltmeter to the terminals of the unit across which the voltage is to be measured.

CURRENT. The flow of electrons is called electrical current. The unit of measurement of current is called an *ampere,* and the instrument by which current is measured is called an ammeter. When you use an ammeter, connect it into the path of current.

Before the discovery of the electron, it was assumed that the flow of current was from positive to negative. This concept became known as the conventional direction of current flow. Scientists now agree that the electrons in motion are the current, and that current flows from negative to positive. Either the conventional direction of current flow or the electron theory of current flow may be used in tracing electrical circuits.

RESISTANCE. The opposition of the flow of current offered by the conductors and electrical devices within a circuit is called *resistance.* The amount of resistance offered by a conductor depends on the material of which the conductor is made and on the length, cross-sectional area, and temperature of the conductor. The unit of measurement of resistance is called an *ohm,* and the instrument by which resistance is measured is called an ohmmeter.

Resistors may be of the fixed type (incapable of being varied) or of the variable type (adjustable to any amount of resistance within their capacity). Variable resistors are called potentiometers or rheostats.

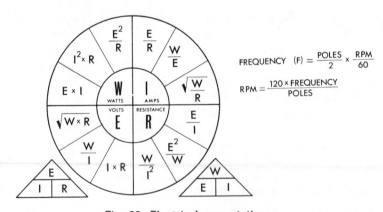

$$\text{FREQUENCY} \ (F) = \frac{\text{POLES}}{2} \times \frac{\text{RPM}}{60}$$

$$\text{RPM} = \frac{120 \times \text{FREQUENCY}}{\text{POLES}}$$

Fig. 98. Electrical computations.

OHM'S LAW. There is a definite relationship between the voltage, current, and resistance of any circuit or part of a circuit. If the voltage is increased, the current increases proportionately, and if the resistance is increased,

SYSTEMS AND COMPONENTS 201

the current decreases proportionately. This relationship is known as Ohm's law (so called in honor of George Ohm who first discovered it), and is generally stated as follows: *The current in a circuit is equal to the voltage divided by the resistance.* Mathematically it is written as: $I = E \div R$. In this equation, I stands for current in amperes, E for voltage in volts, and R for resistance in ohms. Thus, if the source of voltage in a circuit is a 6-volt battery and the electrical device is a bulb having 3 ohms of resistance, the current will be $6 \div 3$ or 2 amperes.

The equation for Ohm's law ($I = E \div R$) can be converted mathematically to read as $E = I \times R$. By use of this equation you can find the voltage across a component of a circuit if you know its resistance and the current through it. Thus, if you know the current through a lamp is 2 amperes and the resistance of the lamp is 3 ohms, you know that the voltage across it must be 3×2 or 6 volts.

The equation for Ohm's law can be converted mathematically in another way to read: $R = E \div I$. With this equation you can determine the resistance of any circuit component if you know the voltage across it and the current through it. Suppose you know that the voltage across a lamp is 6 volts and the current through it is 2 amperes. You can find its resistance by substituting in the equation $6 \div 2 = 3$ ohms.

These three equations enable you to find any one of three quantities—current, voltage, or resistance—if you know the other two.

FUSES. A fuse is a strip of metal with a very low melting point, which is so connected that all the current in the circuit flows through it. An alloy of tin and bismuth is used in most fuses. Others, made of copper, are called current limiters and are used in sectionalizing an airplane circuit. A fuse will melt and break the circuit whenever the current becomes excessive. Whereas a current limiter will stand considerable overload for a short time. Two types of fuses most used are the "plug-in" and the "clip" types. Since a fuse is a protective device, it is very important to use one that fits the needs of the circuit in which it is to be used. When you replace a fuse, always consult proper sources for the correct type and capacity.

CIRCUIT BREAKERS. A circuit breaker is a device that breaks the circuit when the current reaches a predetermined value. It is often used in place of a fuse and may sometimes eliminate the need for a switch. The feature which distinguishes a circuit breaker from a fuse is the fact that a circuit breaker can be reset, while a fuse must be replaced. Several types of circuit breakers are used. One is a magnetic type which operates by the pull of an electromagnet on a small armature which trips the breaker. Another type is the thermal-overload breaker or switch, a bi-metallic strip which, when it becomes heated, bends away from a catch on the switch lever and permits the switch to trip open.

Some circuit breakers have to be reset by hand while others are automatic. When a "manual reset" type circuit breaker trips to the off position, you must move it back to the on position to put the circuit back in operation. An automatic type circuit breaker will reset itself. If the overload is still present, the circuit breaker will again trip without damage to the circuit.

CIRCUIT PROTECTORS. A circuit protector is a device which automatically opens the circuit whenever the temperature of the associated unit becomes excessively high. It has two positions, automatic off and automatic on, and is most often used with motors. If an inoperative part (a locked rotor, for example) caused the temperature of the motor to become excessively high, the circuit protector breaks the circuit intermittently. This may cause the rotor to break loose and allow the motor to operate normally. The operation of the circuit protector depends on the bi-metal disk or strip which bends and breaks the circuit when it is heated. On cooling, the bi-metal disk assumes its original position and closes the circuit.

TYPES OF ELECTRICAL CIRCUITS. Electrical circuits can be divided into three general classifications: (1) series; (2) parallel; and (3) series parallel.

SERIES CIRCUITS. A series circuit is one in which there is only one path through which the voltage can force the current. In the circuit illustrated, three resistances and a battery are connected to form a series circuit. Since there is but one path for the current, all the current is forced through each resistance, and the current is the same throughout the circuit. The total voltage drop in the circuit is equal to the sum of the voltages (voltage drops) across each of the resistances. The total resistance is equal to the sum of the resistances of each of the units. If one device (motor or lamp) in a series circuit burns out, there is no longer a complete path for the current, and the other devices in the circuit will not operate.

Here is how you use Ohm's law in solving a series circuit problem. *Problem:*

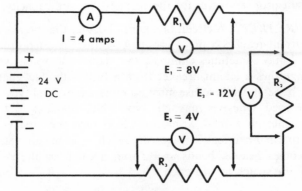

Fig. 99. Series circuit.

In the series circuit diagram shown, 3 resistances are connected in series across a 24-volt power source—the voltages and current were measured and found to be as indicated. Find:

1. The total voltage drop.
2. The total current.
3. The resistance of each unit and the total resistance.

Solution:

1. The total voltage drop $E_T = 8+12+4 = 24$ volts.
2. The current in a series circuit is the same in all parts of the circuit and is equal to 4 amperes.
3. The resistance of each unit is equal to the voltage across that unit divided by the current which flows through the unit:

$$R_1 = \frac{E_1}{I} = \frac{8}{4} = 2 \text{ ohms}$$

$$R_2 = \frac{E_2}{I} = \frac{12}{4} = 3 \text{ ohms}$$

$$R_3 = \frac{E_3}{I} = \frac{4}{4} = 1 \text{ ohm}$$

The total resistance:

$$R_T = R_1 + R_2 + R_3 = 2+3+1 = 6 \text{ ohms}$$

$$Check: R_T = \frac{E_T}{I} = \frac{24}{4} = 6 \text{ ohms}$$

PARALLEL CIRCUITS. Most of the electrical devices in airplane electrical systems are connected in parallel. In a parallel circuit, two or more electrical devices provide independent paths through which the current may flow. The voltage across each device in parallel is the same. The total current in the circuit is equal to the sum of the currents flowing through all the devices. Thus, the *total* amount of current is greater than the current in any individual part, and the total resistance; that is, the resistance of the circuit as a whole, is less than the smallest resistance in it. (By Ohm's law—the current is greater; therefore, the resistance must be less.) The more electrical devices or resistors connected in parallel, the greater will be the total current and hence the smaller the resistance of the complete circuit.

Electrical devices are connected in parallel to decrease the total resistance and to allow them to be operated independently. If one device in a parallel circuit burns out, the others may still be operated. (One path is broken but the others are still complete.)

To find the total resistance of a parallel circuit, use the following equation and solve for the resistance of only two paths at a time:

$$R_T = \frac{\text{Resistance of 1st unit} \times \text{resistance of 2nd unit}}{\text{Resistance of 1st unit} + \text{resistance of 2nd unit}}$$

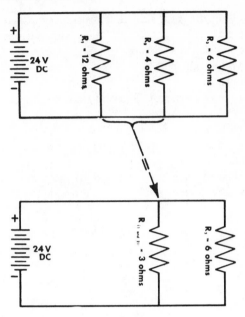

Fig. 100. Parallel circuits.

Example:

Three load units are connected in parallel as given below. Find the total resistance.

1. For the first two paths,

$$R \text{ (}_1 \text{ and }_2\text{)} = \frac{R_1 \times R_2}{R_1 + R_2} = \frac{12 \times 4}{12 + 4} = \frac{48}{16} = 3 \text{ ohms}$$

2. Since 3 ohms is the total resistance of the first two paths, you may substitute a resistor for them, and adding the 6-ohm resistor of the third path, redraw the circuit as shown.

3. Then combining R (₁ and ₂) and R₃,

$$R_T = \frac{R \text{ (}_1 \text{ and }_2\text{)} \times R_3}{R \text{ (}_1 \text{ and }_2\text{)} + R_3} = \frac{3 \times 6}{3 + 6} = \frac{18}{9} = 2 \text{ ohms, the total resistance.}$$

When the load units in parallel have equal resistance, the equation for total resistance may be simplified to the following:

$$R_T = \frac{\text{Resistance of one unit}}{\text{The number of units}}$$

For example, the total resistance of six 5-ohm units connected in parallel is 5/6 ohm.

You may now use Ohm's law with load units connected in parallel as in the following problem:

In the circuit shown, 3 lamps connected in parallel across a 24-volt

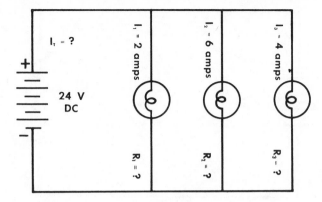

Fig. 101. Parallel circuit.

source of power—the voltage and current were measured and found to be as indicated. Find:

1. The total voltage drop across each lamp.
2. The total current.
3. The resistance of each lamp.
4. The total resistance.

Solution:

1. The voltage in a parallel circuit is the same across each unit. In this circuit, it is 24 volts.
2. The total current (I_T) in a parallel circuit is equal to the sum of the currents in each path.
   $I_T = I_1 + I_2 + I_3 = 2 + 6 + 4 = 12$ amps.
3. Find the resistance of each lamp by dividing the voltage across the lamp by the current which flows through the lamp.

   $R_1 = \dfrac{E_1}{I_1} = \dfrac{24}{2} = 12$ ohms

   $R_2 = \dfrac{E_2}{I_2} = \dfrac{24}{6} = 4$ ohms

   $R_3 = \dfrac{E_3}{I_3} = \dfrac{24}{4} = 6$ ohms

4. The total resistance (R_T) in a parallel circuit equals the voltage divided by the total current.

   $R_T = \dfrac{E}{I_T} = \dfrac{24}{12} = 2$ ohms

Another way to find the total resistance would be the method of figuring two paths at one time.

1. Solve for the first two paths:

   $R \text{ (}_1 \text{ and }_2\text{)} = \dfrac{R_1 \times R_2}{R_1 + R_2} = \dfrac{12 \times 4}{12 + 4} \times \dfrac{48}{16} = 3$ ohms

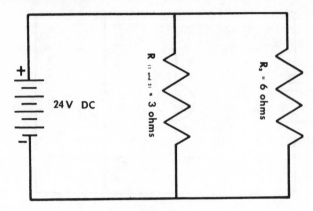

Fig. 102. First equivalent circuit.

2. Now you can assume the circuit to be equivalent to the circuit in the accompanying illustration.
3. Then combining R ($_1$ and $_2$) and R$_3$,

$$R_T = \frac{R\ (_1\ and\ _2) \times R_3}{R\ (_1\ and\ _2) + R_3} = \frac{3 \times 6}{3 + 6} = 2 \text{ ohms}$$

Note: the total resistance of the circuit is only 2 ohms—less than the resistance of any of its individual paths.

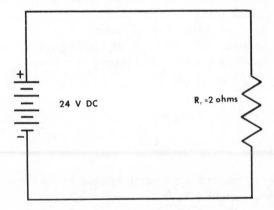

Fig. 103. Final equivalent circuit.

BATTERIES AND BATTERY CONTAINERS. The drain and venting provisions for the battery containers should be checked frequently and if found corroded, the compartment and surrounding structure should be washed with a solution of soda and water to neutralize the battery acid.

Major adjustments of items and equipment such as regulators, generators, contactors, control devices, inverters, and relays should be accomplished outside the airplane on the test bench where all necessary instruments and test equipment are at hand. The adjustment procedure outlined by the equipment manufacturer should be followed.

The criteria upon which the selection of electric cable size should be based, when considering an alteration, are current carrying capacity and voltage drop.

The selected cable should not carry current continuously or intermittently in excess of the proper ampere values as shown by a cable chart.

The voltage drop in the main power cables from the generation sources, or the battery to the bus should not exceed 2% of the regulated voltage, when the generator is carrying rated current or the battery is being discharged at the 5-minute rate.

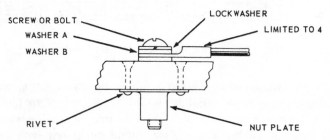

Fig. 104. Plate nut bonding or grounding to flat surface.

Cable terminals are attached to the ends of electric cable to facilitate connection of the cable to junction boxes, terminal strips, or items of equipment. The tensile strength of the cable terminal joint should be at least equivalent to the tensile strength of the cable itself, and its resistance should be negligible relative to the normal resistance of the cable run. Terminals are available which are specifically designed for use with the standard sizes of aircraft cable. Haphazard choice of terminals may lead to overheated joints, vibration failures, and corrosion difficulties.

Cable runs are usually joined at terminal strips. The terminal strip should be fitted with barriers to prevent terminals or adjacent studs from coming in contact with each other. The studs should be anchored against rotation and be long enough to accommodate a maximum of four terminals. When more than four terminals are to be connected together, two or more adjacent studs should be used, and a small strip-metal bus mounted across the studs. In all cases, the current should be carried by the terminal contact surfaces, and not by the stud itself.

Terminal strip stud sizes. If the stud size is too small, it is easily sheared during servicing by applying too much torque on the nut. After a few failures of this sort, the electrician will become overcautious and not tighten the nut sufficiently. A loose connection is hazardous. Consequently, it is good practice to limit stud sizes to No. 10 or larger.

Connector assemblies. Connectors (plugs and receptacles) are used to facilitate maintenance when frequent disconnection is required in service. Since the cable is soldered to the connector inserts, the joints should be individually insulated and the cable bundle firmly supported to avoid damage by vibration. Connectors have been particularly vulnerable to corrosion in the past, due to condensation of moisture within the shell. Special connectors with waterproofing features have been developed, and a chemically inert water-proof jelly is sometimes packed in the connector. This combats the corrosion difficulty.

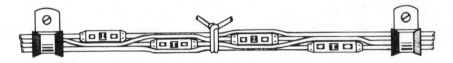

Fig. 105. Staggered splices in wire bundles.

To simplify maintenance and to minimize the damage that may result from a single fault, cable bundles should be limited as to the number of wires in the run. Shielded cable, ignition cable, and cable which is not protected by a circuit breaker or fuse should be routed separately. The bending radius should not be less than 10 times the outer diameter of the bundle. This avoids excessive stresses on the cable insulation.

Soft insulating tubing (spaghetti) cannot be considered as mechanical protection against external abrasion of cable, since at best it provides only a delaying action. Conduit or ducting should be employed in such cases.

Clamps, preferably of non-metallic material, should be used to support the cable bundle along the run. Lacing may be used between clamps, but should not be considered as a substitute for adequate clamping. Adhesive tapes are subject to age deterioration and therefore should not be used as a means of clamping.

Where cables pass through bulkheads or other structural members, a grommet or suitable clamping should be provided to prevent abrasions.

Cable separation from flammable fluids. An arcing fault between an electric cable and a metallic flammable fluid line may puncture the line and result in a fire. Consequently, every effort should be made to avoid this hazard by physical separation of the cables from lines or equipment containing oil, fuel, hydraulic fluid, or alcohol. When separation is impractical, the electric cable must be placed above the flammable fluid line and securely clamped to the structure. In no case should the cable be clamped to the flammable fluid line.

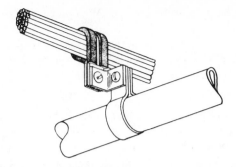

Fig. 106. Separation of wires from plumbing lines.

Conduit installation. Conduit is available in metallic and non-metallic materials and in both rigid and flexible forms. Primarily its purpose is for mechanical protection of the cable, although some radio interference shielding may be provided.

Junction boxes should be made from a fire resistant, non-absorbent material, such as aluminum alloy or an acceptable plastic material. Where fireproofing is necessary, a stainless steel junction box is recommended. A rigid construction will prevent "oil-canning" of the box sides, which may result in internal short circuits. In all cases, drain holes should be provided.

In installing junction boxes, it is desirable to mount them with their open side facing downward. Loose metallic objects such as washers, bolts, etc., will tend to fall out of the junction box rather than wedge between terminals.

The original layout of the junction box should take into consideration the necessity for adequate wiring space, and possible future additions. Electric cable inside the box should be laced or clamped in such a manner that terminals are not hidden, relay armatures are not fouled, and motion relative to any equipment is prevented. Entrance openings for cable should be protected against chafing by grommets or other means.

Bonding metallic parts of the airplane. Bonding is defined as the process of electrically connecting the various metallic parts of the airplane to achieve one or more of the following:

(a) A low resistance ground path for electrical equipment, thereby eliminating ground wires.

(b) A reduction in radio interference.

(c) Less probability of lightning damage to such airplane elements as control hinges.

(d) Prevention of the buildup of static charges between parts of the airplane, which may be a fire hazard.

Bonding jumpers should be as short as practicable. They should be installed in such manner that the resistance of each connection does not exceed .003 ohm.

Reasonable access for maintenance should be provided. The jumper should not interfere with the operation of moveable aircraft elements such as surface controls lest normal movement of these elements result in damage to the bonding jumper.

TYPICAL AIRCRAFT ELECTRICAL SYSTEM AND COMPONENTS. In general, there are two sources of electrical energy created and utilized during the operation of aircraft. They are (1) DC generator/s that supply DC to the storage battery/s and which incorporate inverters to convert the battery DC to AC for the operation of AC circuits and (2) alternators that supply AC direct to AC circuits and which incorporate recti-fier-transformers to convert the alternator AC to DC for the purpose of battery charging.

DC generators and AC alternators are the regular source of electrical energy and the batteries (lead-acid or nickel-cadmium types) are the auxil-iary source of electrical energy. During normal engine operation, electrical energy for operating the aircraft electrical equipment is derived from the engine-driven generators or alternators. The storage batteries are actually being depleted of electrical energy when the engine/s are not operating or when the generators are malfunctioning.

The number of generators or alternators is determined by the number of engines and by the electrical power requirements of the particular aircraft.

The generator system of an aircraft consists of one or more generators, controls, and indicating devices. The controls usually consist of voltage regulators, reverse current relays (RCR), overvoltage and field control relays, a battery disconnect switch (master switch), circuit breakers, and generator switches.

The indicating devices usually consist of voltmeters, ammeters (load-meter), and when several generators are incorporated, switches whereby individual readings of each generator can be made. Also, if several genera-tors are utilized in an aircraft, provisions are made so that the output of all generators can be adjusted to the same value. This is known as "paralleling generators" and is an important function that is accomplished prior to departure of the aircraft.

The generator converts mechanical energy to electrical energy. It re-stores to the battery the current used in starting the engine and also supplies current to carry the electrical load of the lights and electrical units up to the limit of the generator capacity.

The generator consists essentially of an armature, a field frame, field coils, and a commutator with brushes. It may employ collector rings with brushes. The brushes establish electrical contact with the rotating element of the generator. The magnetic field of the generator is usually produced

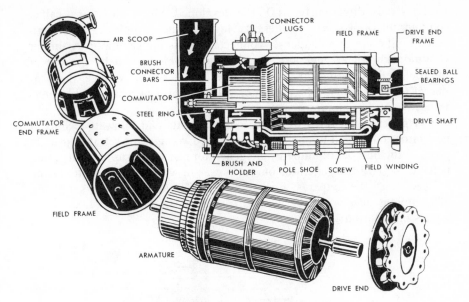

Fig. 107. Typical 24-volt airplane generator.

by electromagnets or poles magnetized by current flowing through the field coils. Soft iron pole pieces are contained in the field frame that forms the magnetic circuit between the poles. Two, four, six, and eight-pole frames are commonly used. However, generators may have any number of even poles.

The armature assembly consists of a steel shaft on which are located a soft iron core, armature windings, and a commutator. It serves as the magnetic circuit and as the mechanical support for the windings. The ends of each winding are brought out to the commutator. The commutator is made up of a number of individual segments insulated from each other and the armature shaft by mica and insulated end-rings. A splined drive shaft couples the armature shaft to the engine.

The field frame is the cylinder in which the generator field is located. Field coils are wound on poles and make up the electromagnet. The pole pieces are wound so that similar poles are opposite each other or so that north and south poles are next to each other. The supports at each end of the field frame are the end frames. They contain bearings on which the armature shaft rotates. The commutator end frame supports the brushes and brush holders, while the opposite end frame contains flanges for mounting the generator on the accessory section of the engine.

An aircraft generator contains many armature windings. The greater the number of windings the greater the generator voltage and the more constant the output. By appropriate spacing of the armature windings, an almost steady direct current (DC) is developed. Slight pulsations, called commutator ripples may be present. However, these ripples can be eliminated by the use of a condenser connected across the generator. A similar condition known as generator reaction can be eliminated by the use of interpoles. By building up the low values of the output voltage, the condenser tends to smooth the generator voltage to a point where radio interference is at a minimum.

A generator is rated in power output (watts and kilowatts). Since a generator is designed to operate at a specified voltage, the rating is usually given as the number of amperes the generator can safely supply at its rated voltage. To find the generator output power in watts, simply multiply the voltage times the amperes. For example, a 28-volt generator supplying 300 amperes develops 8,400 watts of power or 8.4 kilowatts (1,000 watts equals one kilowatt).

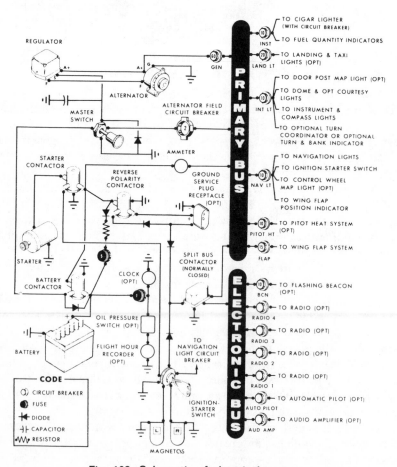

Fig. 108. Schematic of electrical system.

Four-engine aircraft are generally equipped with four generators (one for each engine). In addition, an auxiliary powerplant equipped with another generator can be incorporated. This generator is usually of lesser capacity than one of the regular engine-driven generators.

Aircraft equipped with two or more generators require the process of *paralleling* or of checking to see if the generators are in parallel. Paralleling is accomplished to be sure that each generator is adjusted so that it carries its share of the load. This check must be made on the ground before takeoff. If it is found that the generators are not in parallel, they must be paralleled at this time.

Alternators are, in a sense, similar in operation and purpose as the generator. They are driven mechanically either directly by the engine or through a linkage such as the CSD (constant-speed-drive). The alternators develop AC which is commonly used in large aircraft both jet and reciprocating type. Engine-driven alternators provide alternating current to the aircraft's electrical system. However, a rectifier-transformer is utilized to convert the alternator AC to DC to charge and maintain the storage batteries for electrical systems that require DC. Where storage batteries supply circuits that require AC, inverters are then utilized. Inverters do just the opposite of rectifiers; that is, inverters convert DC to AC.

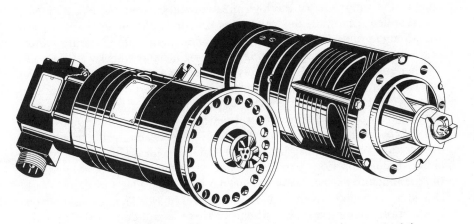

Fig. 109. Combination starter (left); direct-cranking starter (right).

A *starter* is a device or mechanism used to develop a considerable amount of mechanical energy that can be applied to the crankshaft of a reciprocating type engine to cause the engine to start. The starter consists primarily of an electric motor, a flywheel, and a gear reduction unit. Some starters operate on the principle of a transfer of energy in a rotating flywheel to the crankshaft. These are called inertia starters. Others apply energy directly to the crankshaft without the use of a flywheel. These are called direct-cranking starters.

Starter motors are series motors with small resistance in the windings. When the starter switch is closed, a starter motor draws an extremely high current and develops a powerful starting torque. However, as the motor gains speed, the amount of current drawn decreases. Since a starter motor will "race" when the load is removed, do not operate a starter motor at full voltage with no load on it.

Starter motors are designed for intermittent operation. Therefore, operate the starter for short periods only. If the engine fails to start, after normal starting procedures have been applied, disengage the starter and allow the starter motor to cool before making another attempt to start the engine.

Typical jet engine starting system. The starting system consists of an air turbine starter geared to the N_2 (high pressure) compressor shaft of each engine, start switches, start levers, and two compressed air bottles. The air required to drive the turbine starters may be supplied under low pressure to any engine or high pressure to the inboard engines. The low pressure air is supplied to the starters through a pneumatic manifold from an operating turbocompressor or a low pressure external ground cart. The cart connects to the manifold through an external pneumatic ground service connection located on the lower right side of the fuselage, near the leading edge of the wing. Air for the high pressure starts of engine No. 2 or No. 3 is supplied from individual high pressure air bottles mounted in the aft wing-body fairings. High pressure air from an external high pressure source can be connected through a fitting in the right hand wheel well area for starting engine No. 3 only. Controls for operating the starting systems are accessible to both pilots. A three-position ground start selector switch and four engine start control switches are located on the overhead panel. Engine start levers are on the center control pedestal.

Electrical specifications, etc. The aircraft mechanic, prior to working on electrical systems, must first become familiar with the particular system either through previous experience and/or through a thorough study of the manufacturer's specifications, schematics, plans, blueprints, etc. Work should never be attempted on a "trial and error" basis. The help and advice of superiors should be sought whenever there is any doubt existing as to the procedure and work to be done.

A *"growler"* is used to check for shorts in generator fields and electrical motor armatures. The growler is an easy means of finding breakdowns in fields and armatures which have already or soon would cause shorts and subsequent damage to the unit. Most important, however, checking with growlers can prevent the loss of essential electrical energy while the aircraft is in flight.

The *commutator* is the circular end of the armature that has a combination of conductors and insulators that make possible the conversion of field alternating current to direct current. The direct current is "picked off" the

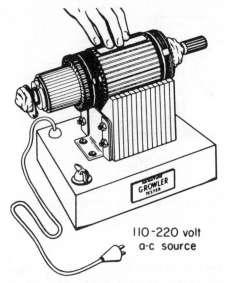

110-220 volt
a-c source

Fig. 110. Checking armature for short circuits.

armature by brushes and sent to charge batteries (DC) and other DC circuits.

Commutator cleaning. Sometimes referred to as commutator "dressing," it is important that this be done using fine sandpaper to remove surface dirt, etc. Metallic cleaning devices such as emery cloth, etc., should not be used for commutator cleaning. The fine metal particles of emery cloth, etc., can work their way into the commutator grooves that are normally a means of insulation between the commutator conductor elements. Obviously, the metal particles would nullify the benefit of the insulator grooves and cause shorts across the commutator conductors.

Residual voltage. This is a small amount of voltage that remains in the generator when the generator comes to rest after being in operation. On the next subsequent start of the generator the residual voltage permits immediate resumption of generator voltage output.

Flashing a generator field. If, for any reason, the generator becomes "motorized," it will become necessary to "flash" the generator field. Motorizing is a condition whereby the generator flow has become reversed and flashing the field is a procedure that places the generator flow back to normal. A major cause of generator motorizing is due to sticking or fusing together of the reverse-current relay points.

Generator brushes. Generally made of carbon material, the brushes conduct the generator output from the armature (commutator section) to the storage battery or to some other unit that has need for direct current. Brushes should be installed, dressed, or properly seated by the use of fine sandpaper.

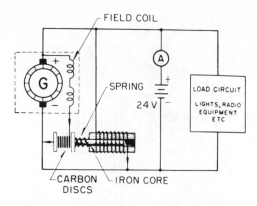

Fig. 111. Carbon pile voltage regulator.

Sometimes referred to as "generator ripple," *armature reaction* is the uneven flow or rippling of the generator field action. This condition, though not serious, can be reduced or completely eliminated by the use of interpoles (a type of condenser) in the generator circuit.

Carbon-pile voltage regulators. This type of voltage regulator consists of a "pile" or set of carbon discs (conductors) that are so arranged and operated that their distance between each other can be varied automatically. The distance between them permits an intervening air gap that acts as a resistor. Air is normally a resistor to electrical flow. Thus, the smaller the air gap between the carbons the less the resistance and the greater the air gap, the greater will be the resistance.

During generator operation, when the generator output is too low, the carbon discs (through an electromagnet and spring arrangement) are moved closer together. This decreases the air gap between them with a subsequent lower resistance. This lower resistance permits an increase in generator output to meet the particular needs of the aircraft electrical units. Just the opposite takes place when the generator output is too high. In this case, the carbon discs are spread apart creating a larger air gap between them. The larger air gap creates a greater resistance with a corresponding decrease in generator output.

Generator speeds. The speed of generators in both RPM and CPS (cycles per second, also known as frequency and the letter "F") can easily be found by the use of formulas. They are:

(1) $RPM = \dfrac{120 \times CPS}{\text{No. of poles}}$

(2) $F = \dfrac{\text{No. of poles}}{2} \times \dfrac{RPM}{60}$

Generator torque relationship. The series-wound direct current motor has the ability to create a high starting torque at relatively low **RPM** and high amperage. This capability makes the DC motor very effective for engine starting purposes. The parallel wound type of motor has the opposite results in its operation. It can create very high RPM with relatively low torque values. It is best used where a high RPM is required and the torque value is of little importance.

Generator voltage regulation. Generators are rated for a given power output. Accurate regulation of generator output is necessary to prevent it from delivering electrical energy beyond its capacity. Electrical equipment and electrical units throughout the aircraft depend on the generator for its operating power. These units are of a given capacity and, if overloaded by generator output, may overheat causing damage and/or complete failure.

The carbon-pile voltage regulator is in common use. Another type of voltage regulator that gives good service is described as a vibrating type of regulator.

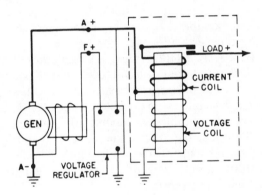

Fig. 112. Reverse current relay.

Reverse-current cutout relays. This type of relay is necessary to protect the generator being "motorized" when the battery output is greater than generator output especially at low generator operating speeds (low engine RPM or idle). When this unbalanced condition occurs (battery output greater than generator output) the relay points open and prevent battery flow back to the generator. If the relay points fail to open under these circumstances, the greater battery flow into the generator field tends to create a flow reversal in the generator with a subsequent motorizing effect. This condition causes a loss of generator efficiency and, if not corrected, will result in complete generator failure.

To correct a generator problem of this type, it becomes necessary to "flash" the generator field. This correction will redirect the generator flow back to the battery in a normal manner. Obviously, the cause of the relay points sticking must be found and corrected or the generator will again become motorized.

Magnetic clutches and brakes on electric motors. These are used on motors that operate items or units that move through an angular or arc pattern where the unit must be stopped quickly in a given position. An example of this type of unit is the landing gear lights that may be extended or retracted through an angle of varying degrees. The exact desired angle of the lights will vary depending on whether the aircraft is taxiing, taking off, landing, etc. The pilot may, through the landing light controls in the cockpit, easily stop the angular movement of the lights at any desired position. The stopping action is caused by the effect of the magnetic brake on motor rotation.

Reversible motors. Double field windings make possible the operation of an electrical motor in either direction (clockwise or counter-clockwise). This type of motor is so designed that it may be operating in one direction and be completely reversed. The use of reversible electric motors in aircraft is quite common in the operation of units that must be moved or rotated in either direction. Examples are landing lights, flaps, landing gear, electrically operated propeller blades, etc.

American Wire Gage (AWG). This is a standardized system for the determination of electric wire or electric cable size depending on the required wire length, voltage drop in the particular circuit, ampere (current) load, and the use to which the wire or cable is to be placed; i.e. intermittent rating, continuous rating, cables in conduit and bundles, etc. This wire standard is also known as the "Brown and Sharpe (B and S) Wire Gage." The diameters form a geometrical series in which No. 0000 is 0.46 inches and No. 36 is 0.005 inches. Note that the smaller the size number the larger the cable diameter in inches.

Solenoid switch chatter. This can occur when the voltage input to the solenoid unit is too low or there is some obstruction in the circuit itself that causes an increased resistance to the flow of electricity to the solenoid unit.

Static wicks and dischargers. Bonding of aircraft structures is a means of creating an extremely low path of resistance so that static electricity can easily flow to a common discharge point (static wick or dischargers). A common location for static dischargers is at the outboard trailing edge of the wing.

The ideal inter-connection of the metal structure is by the use of bonding strips of similar metal. This prevents dissimilar metal corrosion between the structure member and the bonding strip. In general practice, however, aluminum strips are used when bonding aluminum parts; copper strips are used when bonding steel parts of the structure. Bonding strips are clamped when connecting structural members and they must not have a resistance greater than .003 ohm.

Anti-collision lights are of high intensity requiring high amperage and relatively heavy duty electrical wiring. For this reason, installation should be as close to the source of electrical power as possible.

Position lights are used for the purpose of night aircraft recognition. They are also an added safety device for night operations. Their location and standard colors indicate to other planes in the general vicinity the relative position and speed of the particular aircraft being observed.

The standard colors and specific light locations are: (1) a red light on the left (port) wing tip, (2) a green light on the right (starboard) wing tip, and (3) a white light located at the aft section of the tail area. These lights are installed in parallel with each other (if one burns out the others remain on) and in series with the navigation light switch in the cockpit.

A *rectifier-transformer* is an electrical unit that changes alternating current to direct current and, at the same time, steps up or steps down the voltage. When alternators are used in aircraft in lieu of generators it is necessary to incorporate a rectifier-transformer between the alternator (AC) and the battery (DC). Ground power units that utilize hangar electrical power (AC) must also incorporate a rectifier-transformer to convert the hangar AC to DC that plugs into the aircraft ground-use receptacle.

When it is necessary to change direct current to alternating current, an *inverter* is incorporated in the electrical system. A common location for an inverter installation is in the storage battery (DC) output circuit. Some aircraft instruments utilize AC. Thus the inverter converts the battery DC to AC to operate the AC instruments.

Transformers. An electrical device that is used for stepping up or stepping down voltage values in an electrical circuit.

Induction coils. A coil with a primary winding of a few turns of heavy wire and a secondary coil of many turns of fine wire. This type of coil is used to obtain intermittent high voltage in the secondary winding by supplying an intermittent current in the primary coil. The primary current is broken by an *interrupter* (breaker points that act as a switch). The faster or sharper the interruption the greater the rate of change of the primary current. This greater rate of change in the primary results in a greater secondary voltage produced. A common use of induction coils is in the engine magneto (ignition) circuit. The primary voltage is relatively small. However, the action of the induction coil increases the primary voltage to approximately 20,000 volts. This high voltage is sufficient to "jump" the air gap between the spark plug electrodes and create a "hot" spark to ignite the fuel-air mixture that has been compressed at the top of the engine cylinder.

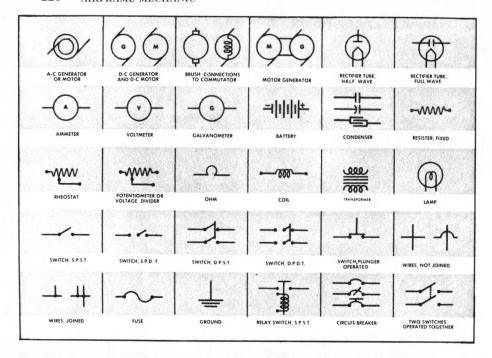

Fig. 113. Conventional electric symbols (top), and wiring diagram symbols below.

Alternating current in aircraft. The use of alternators that are engine-driven and that produce alternating current have a number of advantages over generators that produce direct current. It is generally assumed that alternating current used in aircraft is more efficient than DC and is easier to install and maintain.

Alternators produce AC that can be sent directly into the AC aircraft circuits. Where DC is required, such as in the charging of storage batteries, rectifiers in conjunction with transformers efficiently convert AC to DC.

General electrical maintenance data. All electrical equipment, electrical assemblies, and wiring should be inspected frequently for damage, general condition, and proper functioning. This will assure continued satisfactory operation of the many electrical units and components.

The adjusting, repair, and testing of all electrical equipment and systems should be accomplished in accordance with recommendations and procedures set forth in Maintenance Instructions or approved manuals published by the aircraft and equipment manufacturers.

Output and checks of electrical systems operation are generally made with approved meters and check instruments. Some of these are known as voltmeters, ammeters, ohmmeters, meggers, growlers, continuity lights, etc.

Standard wire gages should be referred to when replacing electrical wiring, especially when wires are to be placed in conduit or bundles, so that each particular wire can safely carry its load and not overheat or burn out.

Soapstone should be used when stringing wires through conduits or bundles. Soapstone makes an excellent non-conducting lubricant.

Frequent cleaning of electrical equipment to remove dust, dirt, grime, etc., is highly recommended. Fine emery cloth may be used to clean terminals and mating surfaces when they appear corroded or dirty. Very fine sandpaper should be used to clean and polish commutators and slip rings.

Electrical fires should be extinguished with dry chemicals. Never attempt to extinguish an electrical fire with any material that may conduct electricity.

H. POSITION AND WARNING SYSTEMS

Maintenance personnel should familiarize themselves with position and warning systems and be sure they have a thorough knowledge of their purposes and functions.

Skid detectors. This is an anti-skid device incorporated in the hydraulic brake system of large aircraft to prevent excessive skidding during heavy braking on the ground.

On a typical large aircraft, a skid in any main gear wheel will cause a flywheel detector to energize an anti-skid valve. This action releases brake

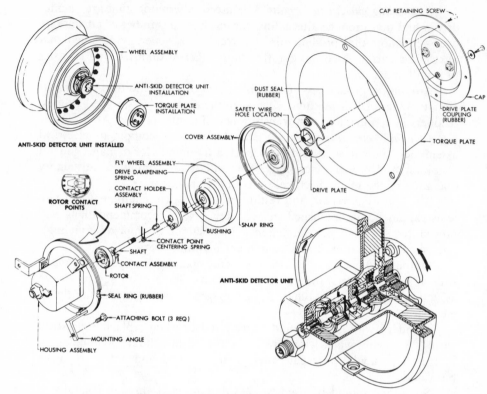

Fig. 114. Brake anti-skid detector details.

pressure on the affected tandem pair of wheels. This action is indicated by a "kickback" felt through the rudder pedals and by one of four brakes-released indicators on the overhead panels. Protection is provided for the possibility of a locked wheel by time delay circuits which extend the brake released time by approximately 0.7 second. These circuits, energized when the landing gear is extended (as on an approach), prevent the brakes from being applied until 0.7 second after touchdown.

Landing gear warning signals. In general, the following light indications are used in landing gear operation: (1) when the gear is up and locked, there will be no light indication, (2) if gear is in transit, there will be a red light indication, and (3) when the gear is down and locked, there will be a green light indication.

There are various additional systems of safety and warning lights as well as horn signals incorporated in the retractable landing gear system of various types of aircraft. Aircraft mechanics must familiarize themselves with the particular system being worked on in order to do a complete and satisfactory job of trouble-shooting and repair.

I. ICE AND RAIN PROTECTION

It is important to understand the difference between the term "de-icing" and the term "anti-icing." De-icing equipment provides a means to eliminate ice formation after the ice has formed on the leading edges of the wing, vertical stabilizer, etc., and where de-icing equipment, such as de-icer boots, are installed. Anti-icing is a means by which ice is prevented from forming on surfaces by the use of heat, etc. Anti-icing is turned on whenever the flight is entering an area that has potential ice forming atmospheric conditions. Potential aircraft icing conditions are present whenever a flight is entering an area of near or below freezing temperature (approximately 2°C. to − 15°C.) and there is visible moisture (clouds, rain, fog, etc.) present in the area.

Typical large turbojet icing system. In a typical large turbojet aircraft, a de-icing system is provided for the empennage (tail section) and anit-icing systems for the wing, engine nose cowl, engine inlet guide vanes, pitot tube, and Q-inlets.

Anti-icing and anti-fogging are provided for the control cabin; passenger cabin windows are anti-fogged. Windshield wipers are provided for the pilot's and copilot's forward windows. The wings are thermally anti-iced by low pressure compressor bleed air which is ducted along the wing leading edges and exhausted overboard. The leading edges are not heated in the vicinity of the engine strut-wing intersections, nor inboard of the inboard engines.

De-icer boot operation and installation. De-icer boots are installed on the leading edges of wings, stabilizers, etc., to break away ice after the ice has formed to a given thickness. De-icer boots are made of a rubberized material with several internal air passageways running parallel to their length. These passageways are, pneumatically charged (air charge) alternately when ice has formed on the boot area. The air charge is created by vane-type pumps. The alternate inflation and deflation of the internal passageways causes a "knuckling" action along the boot which cracks the ice formation. The ice formation, now broken up, is carried away by the airstream.

De-icer boots are generally installed with rivnuts. However, in some installations glue is used as a means of fastening the boot to leading edges of aircraft.

The surfaces of leading edges, prior to installation of de-icer boots, should be thoroughly cleaned and, if painted, the paint should be removed and then the surfaces cleaned. Talcum powder or a like substance is applied to the boot and surfaces on which the boot is to be installed as a part of the installation procedures.

Propeller de-icing. Propeller blades are de-iced by electrically heated elements imbedded in boots that are installed on the blade leading edges. These elements are controlled by switches located in the flight deck for either automatic or manual operation.

The use of anti-ice for propeller blades is accomplished by the action of "slinger" rings, so installed that an anti-icing fluid of alcohol and glycerine flows on the propeller blades by centrifugal force. This fluid coating on the blades acts as a very excellent means of preventing ice formation.

Windshield anti-icing. In large aircraft, thermal anti-icing for the windshield has proven most satisfactory. In addition, isopropyl alcohol windshield anti-icing is utilized by spraying the alcohol over the windshield external surfaces. A supply tank and pump furnish alcohol for the windshield. This same tank and pump also furnishes alcohol for the carburetor anti-icing system. In either system, alcohol is a deterrent to the formation of ice. The rate of alcohol flow is controlled from the flight deck by a conveniently located valve mechanism.

Normally windshield anti-icing is accomplished by hot air supplied to the windshield from a cabin heater through an insulated duct system. The hot air is routed up the center post of the windshield and forced between the inner and outer panes. In addition to supplying heat for anti-icing, the system also supplies varying degrees of heat to maintain the vinyl layers of the windshield in a satisfactory state to retain its impact-resistance qualities. This is a most important consideration in the case of bird strikes, etc.

J. FIRE PROTECTION SYSTEMS

Smoke and fire detectors are installed in the various designated fire zones and compartments of large aircraft. In addition, fire extinguisher systems are installed in zones and/or compartments of large aircraft where possible fires may be most critical to the safety of the aircraft and its occupants.

Automatic detectors and manually selected electrically operated extinguishers are provided in large aircraft so that a fire can be detected quickly and proper action taken by the flight crew.

Smoke indicator systems are installed to indicate the presence of smoke in designated zones, compartments, electronic racks, etc. Air from each of the six areas (on a typical aircraft) of the main compartment and one from the electronic rack is routed through lines to the smoke detector indicator and then overboard through vents. The indicator, located on the systems operations center panel, consists of seven windows, a shrouded bulb, a detect light, and a PRESS TO TEST button. When the button is depressed, the shrouded bulb illuminates but is not normally visible in the window. However, if any smoke is present, the smoke particles will reflect the light and the appropriate window will appear to illuminate.

The detect light is an indication that the shrouded bulb is servicable. The air is continually routed through the indicator and only a momentary depression of the button is necessary to detect the presence of smoke.

Fire extinguishing equipment. There are three classes of fires: (1) Class A—trash and combustibles; (2) Class B—liquids, greases, etc.; and (3) Class C—electrical. Water extinguishers are used on Class A fires. Dry and/or smothering chemicals are used on Class B and C fires. Chemicals used to extinguish electrical fires must be of the non-conducting type.

On a typical large turbojet aircraft, there are two CO_2 and two water portable extinguishers provided in the airplane. One CO_2 fire extinguisher is located in the emergency equipment panel in the control cabin and one in the aft coat closet. The CO_2 fire extinguishers are used primarily for extinguishing electrical fires. A trigger mechanism, lockwired and sealed until used, operates the extinguisher. Water fire extinguishers are provided in two locations—one on the aft side of the control cabin bulkhead adjacent to the forward main entry door, and one in the aft coat closet. The water extinguishers are used primarily for non-electrical fires. To retain service-ability at low temperatures, an anti-freeze agent is added. The handle, when rotated, punctures a CO_2 cartridge which will force the water to discharge when the thumb operated trigger is depressed. The handle is lockwired and sealed until used.

Typical engine-nacelle fire extinguishing systems of a turbopropeller aircraft consists of two fire extinguisher bottles installed in each outboard nacelle providing fire protection for both nacelles on each wing. Each bottle has two discharge bonnets and contains 15.5 pounds of Freon 13 B1 pressurized with nitrogen to 625 PSI at 70°F. The four discharge bonnets of the two bottles on either side of the aircraft permit discharge of one shot to each nacelle or two successive shots to either at the pilot's discretion. The two bottles in nacelle No. 1 can only be discharged into either of the port nacelles, and the two bottles in nacelle No. 4 can only be discharged into the starboard nacelles. Thus, when both bottles on one side of the aircraft have been used, there is no further chemical fire protection available for that side. A yellow discharge indicator disc and a red thermal discharge indicator are mounted adjacent to the bottles on the skin of the outer nacelles. If either of these discs is damaged or missing, it indicates that the bottles are empty and must be replaced before flight. The system is armed by moving the condition lever of the affected engine to the FEATHER position and lifting it to arm the *firex* system.

Ground fire extinguishing. In a typical large turbojet aircraft, the cowling for each engine is equipped with two push-in panels, one on each side. In case of fire on the ground the panel is to be pushed in by the nozzle of the ground fire extinguisher and its charge released.

General aviation light aircraft mostly use CO_2 and/or water-type portable fire extinguishers. The number and location of these extinguishers depends on the particular aircraft and whether it is in commercial operation or is privately owned and operated.

The amount of fire protection material in a fire bottle or extinguisher can be determined by weighing the container.

Fire warning indicators are installed in aircraft locations deemed to be a potential fire zone. The complete system consists of fire detectors, smoke detectors, detector circuits, test circuits, fire warning bell, fire warning lights, and controls in the flight deck for selection and discharge of fire extinguishing agents.

TYPICAL WRITTEN EXAMINATIONS FOR THE AIRFRAME MECHANIC RATING

All applicants for the FAA mechanic certificate with an airframe rating must pass written examinations in the following categories: (1) general, (2) airframe structures, and (3) airframe systems and components.

In the pages that follow, the questions are grouped together by subject matter in the order noted above. The questions in the FAA exams are presented in the same approximate order, with some variation in each category. The number of questions in each subject of a category may vary according to the particular issue of the exam you are given by the FAA.

The typical examinations here are similar to, but not copies of, the actual airframe mechanic examinations given by the FAA. The questions are designed to test your knowledge of the same subjects covered by the FAA. If you can answer more than 70% of them correctly, chances are that you will do at least as well on the FAA exams. If you score *less* than 70% correct, further study of the text of *Airframe Mechanics Manual* is obviously called for.

Note: To save time and effort, be sure you're familiar with the text material in this manual and the FAR's inside the back cover before you try to answer these questions.

1. THE GENERAL EXAMINATION

The general written examination includes subject matter applicable to both *airframes and powerplants*. Identified as Section 1 of the FAA examinations, it need be passed only once, even by applicants for *both* airframe and powerplant ratings.

An applicant is not required to take the general section of an airframe or powerplant test if he can show that he has previously passed it. Proof of passing may be in the form of a mechanic certificate with the alternate rating or an Airman Written Examination Report for the alternate rating that shows a passing grade on the general section. In the latter case, the passing credit must have been earned within the preceding 24 months.

Even if you already have a powerplant rating or have passed the powerplant general exam, it is suggested that you take the following general test as an aid in preparing for the FAA airframe oral-practical test that you must pass after completing the FAA written tests.

Subjects covered in the general exam include: (1) basic electricity, (2) aircraft drawings, (3) weight and balance, (4) fluid lines and fittings, (5) materials and processes, (6) ground operation and servicing, (7) cleaning and corrosion control, (8) mathematics, (9) maintenance forms and records, (10) basic physics, (11) maintenance publications, and (12) mechanic privileges and limitations. The questions that follow, like those in the FAA tests, cover these subjects but not necessarily in the same order.

The answers to the questions in Section 1 are on page 367.

1. In electrical systems incorporating a generator and battery, a reverse current relay is utilized. The purpose of the reverse current relay is to:
   (1) prevent the generator from motorizing the battery
   (2) prevent the generator from building up hysteresis
   (3) prevent the battery from motorizing the generator

(4) permit sufficient battery charging even though the engine is at idle speed

2. Alternating current is used in aircraft electrical systems because:
   (1) weight of the installation is lighter than DC
   (2) AC is easier to regulate through the use of transformers
   (3) AC is more efficient than DC
   (4) all three of the above answers are correct

3. Current flow in the electrical system is shunted around (or by) the:
   (1) ammeter
   (2) voltmeter
   (3) rectifier
   (4) inverter

4. Which of the following aircraft circuits is considered to be intermittent in use?
   (1) cabin and cockpit lights
   (2) instrument lights
   (3) inverters
   (4) engine starters

5. Regarding electrical terms, which of the following is correct?
   (1) voltmeters are placed in parallel and ammeters in series when installed in electrical circuits
   (2) the henry is a measure of inductance
   (3) the farad, micro-farad, or micro-micro-farad is a measurement of capacitance
   (4) all three of the above answers are correct

6. Which of the following is an electrical device that uses mutual induction to change (step up or down) voltage of an AC current?
   (1) rectifier
   (2) transformer
   (3) inverter
   (4) alternator

7. The voltage regulator, in relation to the generator, is in series with the:
   (1) generator field
   (2) reverse current relay
   (3) generator switch and ammeter relay
   (4) equalizer bus

8. A shunt is incorporated with which of the following?
   (1) the feeder lines
   (2) the reverse current relay
   (3) an ammeter
   (4) a voltmeter

9. A total of 27,500 watts is applied to the 30-volt electrical system of a four-engine aircraft in flight. The total electrical load in amperes is:
   (1) 875
   (2) 975
   (3) 897
   (4) 917

10. The drain and venting provisions of the battery or battery container should be checked at frequent intervals. Acid (electrolyte) storage batteries create fumes that are extemely harmful and of an explosive nature; therefore, any lack of ventilation can be hazardous. Acid-type battery containers and/or battery areas should be:
    (1) washed thoroughly with warm water, rinsed, and, after drying, coated with zinc chromate to prevent corrosion
    (2) washed thoroughly with a suitable solvent, rinsed with water, and, after drying, coated with acid paint, asphalt paint, or bituminous paint
    (3) washed with a solution of soda and water, rinsed with water, and, after drying, coated with acid paint, asphalt paint, bituminous paint, or other approved battery corrosion inhibitor
    (4) washed thoroughly with a solution of boric acid, rinsed with a suitable solvent and, after drying, coated with acid paint, asphalt paint, and bituminous paint

11. Thermocouples are excellent temperature measuring units. They are commonly used to measure the cylinder head temperatures (CHT) of reciprocating engines and the exhaust gas temperature (EGT) of turbojet engines. They operate on the principle of:
    (1) dissimilar metal contact
    (2) standard alternating current (AC)
    (3) small electrical currents induced when metals become heated
    (4) electromagnetic induction

12. A rheostat is a variable resistor used to limit the amount of current through a circuit, or to any particular unit of a circuit. One of the most common uses of the rheostat is to vary the:
    (1) resistance in the magneto primary circuit
    (2) intensity of lights throughout an airplane

(3) operation of the inverters

(4) action of various solenoids throughout the electrical system

13. A solenoid consists of a coil and a soft iron core mounted so that the core can move inside the coil. Solenoids in an electrical circuit are used to move a part, arm, or switch of a remote unit. This is accomplished without a mechanical connection between the unit and the operator. When the coil of the solenoid is energized by electricity, it sets up a magnetic field and:

(1) draws the core into the coil

(2) moves the core well out of the coil

(3) causes the core to rotate

(4) sets the core in its original fixed position

14. When testing for continuity and/or grounded electrical circuits, which of the following is appropriate?

(1) ammeter

(2) ohmmeter

(3) inverter

(4) rectifier

15. The Wheatstone bridge is a device used for remote indications of temperature on aircraft instruments. A sensitive galvanometer is incorporated with the Wheatstone bridge. The galvanometer is a simple voltmeter used to indicate:

(1) ohms and millivolts

(2) farads and micro-farads

(3) capacitance and henrys

(4) current and voltage

16. The potentiometer or voltage divider is used principally in the aircraft electrical system to indicate:

(1) system voltage

(2) system amperes

(3) fuel level

(4) reluctance

17. All wires in an aircraft electrical system should be easily identified by means of an attached number or by means of different colored wires. In general, when replacing a defective wire, splicing is permitted:

(1) on low voltage circuits only

(2) where a circuit consists of auxiliary devices not necessary to engine operation

(3) where a circuit consists of auxiliary devices not necessary to aircraft flight operation

(4) under no circumstances

18. The reverse current relay (RCR) is, in principle, an automatic switch that prevents the battery from discharging through the generator whenever the:
   (1) engine speed reaches maximum RPM
   (2) generator output is greater than battery output
   (3) generator output is lower than battery output such as occurs at low engine RPM or idle RPM
   (4) generator reaches a certain operating temperature

19. When insufficient charge is given a storage battery during charging procedures, the battery can become discharged relatively easy. However, if high and excessive voltage is used during charging:
   (1) the battery will accept only its maximum voltage, with no damage
   (2) the water evaporation will be retarded or decreased
   (3) no damage will occur to the battery as the battery will only accept a minimum stated charge at any time
   (4) violent gassing, excessive heating, and possible damage to the battery will occur, creating conditions that may cause a battery explosion

20. When aircraft batteries are to remain idle for some time, they should be removed from the aircraft and placed in a battery room for storage and necessary servicing. It is important that storage batteries of adequate ampere-hour rating be used in order to have a strong battery even though it may be subjected to high loads over a period of time, especially during engine starting procedures. A battery rated at 100 ampere-hours indicates that the battery (during a no-charge or static condition) will:
   (1) deliver a 5 ampere current for 20 hours
   (2) deliver a 10 ampere current for 10 hours
   (3) deliver a 2 ampere load for 50 hours
   (4) all three of the above answers are correct

21. Impedance is an apparent resistance met by an alternating current passing through a conductor. This is due to self-induction which sets up an electromotive force (EMF) counter to the impressed EMF and will persist as long as the current flows. Impedance can also be defined as:
   (1) the combined capacitance forces in an alternating current electrical circuit
   (2) the combined resistance forces in a direct current electrical circuit
   (3) the combined resistance forces in an alternating current electrical circuit
   (4) any combination of forces that includes farads, joules, capacitance and inductance

22. The difference between a conductor of electricity and an insulator is:
    (1) the conductor has no free-moving electrons, the insulator does
    (2) the conductor has free-moving electrons, the insulator does not
    (3) the conductor has more impedance than an insulator
    (4) any conductor's strength is measured by reluctance and any insulator's strength is measured by henries

23. Copper electrical cable has greater conductivity than aluminum electrical cable of the same size and length, i.e., aluminum cable, size for size, has greater resistance. According to this conclusion, which of the following is correct?
    (1) copper electrical cable, for a given load and length, will be smaller in area (circular mils) than the required aluminum cable
    (2) if the length of a given copper electrical cable is doubled, the resistance will be doubled
    (3) if the diameter of a given copper cable is reduced to one-third its original diameter (or replaced by a cable one-third the original diameter) the resistance of the smaller cable will be nine times greater
    (4) all three of the above answers are correct

24. A cabin heater draws 20 amperes at 110 volts. How many amperes are required if the voltage is reduced to 90 volts?
    (1) 16.4
    (2) 10
    (3) 15
    (4) 19

25. Which of the following best describes the principle of electromagnetic induction?
    (1) a condition whereby electromotive force (EMF) is created by a combination of meggers, farads, and joules
    (2) the production of an EMF by a change of magnetic flux linking a circuit
    (3) an EMF created by a combination of reluctance, impedance, and micros
    (4) an EMF created by a combination of inverters, rectifiers, and transformers

26. The characteristics of magnets and magnetic lines of force are:
    (1) a permanent magnet, such as steel, will create an invisible field of lines of force within its immediate vicinity
    (2) a temporary magnet, such as soft iron, will create an invisible field of lines of force within its immediate vicinity only when it is exposed to a magnetic influence

(3) a magnet will create a field of invisible lines of force that, when cut or collapsed through a conductor, will induce an EMF

(4) all three of the above answers are correct

27. An electrical alternating current circuit contains resistance and inductance. Which of the following is correct?
    (1) inductive reactance is inversely proportional to the inductance
    (2) inductive reactance will increase as the frequency is increased
    (3) the actual effects of capacitance and inductance will be the same
    (4) inductive reactance will decrease as the frequency is increased

28. Which of the following occurs when capacitive reactance is equal to inductive reactance?
    (1) reluctance
    (2) impedance
    (3) resonance
    (4) electromagnetic induction

29. The capacity of a condenser is known as capacitance and is measured in *farads*. The opposition which capacitance offers to the flow of current is known as *capacitive reactance* and is measured in *ohms*. In alternating current circuits, where the current is always changing from instant to instant, the effect of induction is a measurable property and is called *inductance*. The effect of inductance in an AC circuit is to produce opposition to the flow of the current and tend to make the current *lag behind the voltage in time or phase*. The opposition to the flow of alternating current due to inductance is called inductive reactance and is measured in:
    (1) ohms
    (2) volts
    (3) amperes
    (4) watts

30. Direct current (DC) as produced by a generator, moves in one direction but is pulsating, that is, the current strength rises and falls. Alternating current (AC) reverses its direction of flow at fixed intervals. Generally, 60 cycles per second (CPS) is the most common AC. The voltage and amperage of AC can be changed by means of a:
    (1) megger
    (2) transformer
    (3) commutator
    (4) growler

31. Nickel-cadmium batteries have many advantages over the lead-acid storage battery. Some of these advantages are: (a) more durable and longer lasting under heavy load requirements, (b) a much lower freezing

point, (c) less maintenance (they do not require continual addition of water, etc.), (d) they can be completely discharged and return to normalcy with proper charging, and (e) the battery strength can be checked with a voltmeter. Regarding nickel-cadmium batteries:

(1) the battery area can be cleaned up with boric acid solution and then, to prevent corrosion, painted with acid paint, asphalt paint, or bituminous paint

(2) they are relatively heavier than lead-acid batteries and more expensive

(3) each cell develops approximately 1.2 volts

(4) all three of the above answers are correct

32. A thermocouple instrument that indicates cylinder head temperatures:
   (1) has no external power source
   (2) is installed with a DC current hookup
   (3) is installed with an AC current hookup
   (4) operates on a voltage power source from the battery

33. Unless otherwise specified in alternating current, any values given for current or voltage are assumed to be:
   (1) mean values
   (2) average values
   (3) instantaneous values
   (4) effective values

34. The effective value of alternating current is the same as the value of a direct current which can produce:
   (1) an equal heating value
   (2) an equal amount of flux lines
   (3) a multiple average of heating
   (4) an unequal amount of heating values

35. A thick electrical cable is replaced by a thin electrical cable of the same length and heat capacity. This will:
   (1) decrease the resistance of the circuit
   (2) increase the resistance of the circuit
   (3) not change the resistance of the circuit
   (4) have no effect on circuit resistance since all cables of the same length have the same resistance

36. Which of the following takes place when capacitive reactance occurs?
   (1) the voltage leads the current
   (2) an alternation of the voltage leading and then lagging the current
   (3) the voltage lags the current
   (4) none of the above answers is correct

37. In an AC circuit the factor which causes the impedance to be larger than the resistance is called the reactance. This reactance is due to the presence of inductance and is called:
    (1) inductive reactance
    (2) reactive inductance
    (3) capacitance
    (4) residual voltage

38. Capacitance is measured by the electrical units *farad* and *microfarads*. The electrical unit *henry* measures:
    (1) ohms
    (2) watts
    (3) residual voltage
    (4) inductance

39. What is used (a) to increase or decrease AC voltage, and (b) to determine capacitor values?
    (1) transformer and impedance
    (2) inverter and farad
    (3) transformer and farad
    (4) rectifier and amperes

40. Capacitance is measured by which of the following:
    (1) volts
    (2) amperes
    (3) ohms
    (4) microfarads

41. A particular aircraft has a 28-volt electrical system and incorporates an 80-ampere generator. At a time when the engine has been shut down, all of the electrical units are turned on and the ammeter indicates a discharge of 62 amperes. Based on this ammeter reading:
    (1) the electrical drain is 77.5% of the generator capacity
    (2) if another 22.5% load is added to the electrical system the electrical drain will equal the generator capacity
    (3) the electrical drain on the generator capacity is under 80% of the generator capacity
    (4) all three of the above answers are correct

42. A 28-volt generator is driving a 28-volt motor. The motor output is 1/5 horsepower and it is 75% efficient. In addition, the generator is supporting three 20-watt light bulbs, an electrical unit requiring 3 amperes and one unit requiring 5 amperes. Under these circumstances the generator draw (total generator output) is:
    (1) .646 horsepower
    (2) 17.23 amperes

(3) 482.5 watts

(4) all three of the above answers are correct

Solution: Motor output is 1/5 horsepower (149.2 watts), which is 75% of the motor input (motor draw from the generator). The motor input must be 100%, thus the output must be increased by 25% or 1/3 more than the output. Thus 1/3 of 149.2 watts is 49.73 watts. Then 149.2 watts equals 198.5 watts for the motor input or motor draw from the generator. The total generator draw for all electrical units will be: motor 198.5 watts, 3 20-watt bulbs equal 60 watts, the 5 ampere unit 140 watts, and the 3 ampere unit 84 watts for a grand total of 482.5 watts. Thus 482.5 watts equals 17.23 amperes (amperes equals watts divided by volts) total draw from the generator. The total generator draw in horsepower equals .646 horsepower (482.5 watts divided by 746 watts).

43. A certain 24-volt flap motor needs a current value of two horsepower. This combination will require how many amperes?

(1) 31.6

(2) 62.2

(3) 42.5

(4) 56.6

44. A 24-volt generator is driving a 24-volt motor. The motor output is one horsepower and it is 80% efficient. Under these circumstances the generator draw (generator output to the motor) is:

(1) 932.5 watts

(2) 38.85 amperes

(3) 1.25 horsepower

(4) all three of the above answers are correct

Solution: Motor output is one horsepower (746 watts), which is 80% of what the motor is drawing from the generator (100%). The generator draw must be 20% more than the 80% output, or, in terms of fractions, the generator draw must be 1/4 more than the output of the motor. 1/4 (20% more than 80%) is found by taking the 746-watt motor output plus 1/4 more of the 746 watts (186.5) and adding them, with a result of 932.5 watts. This is being drawn from the generator and supplied to the motor in order to have the motor output of 746 watts based on the 80% efficiency. This can be proved by taking 80% of the 932.5 watts input to the motor and finding the 80% to be 746 watts, the output of the motor. To solve the amperes draw, use Ohm's Law: amperes equals watts divided by volts; thus 932.5 watts divided by 24 volts equals 38.85 amperes. To solve for the horsepower draw on the generator, divide 932.5 watts by 746 watts, which equals 1.25 horsepower.

45. Which of the following indicators utilize the principle of the galvan-
    ometer?
    (1) ohmmeter and ammeter
    (2) ohmmeter and voltmeter
    (3) voltmeter and ammeter
    (4) oscilloscope and ignition analyzer

46. What is the proper method when connecting voltmeters and ammeters
    in an airplane electrical circuit?
    (1) voltmeters in series and ammeters in parallel
    (2) ammeters in series and voltmeters in parallel
    (3) voltmeters in shunt and ammeters in shunt
    (4) voltmeters in shunt-parallel and ammeters in shunt-parallel

47. The ammeter is used to measure flow through a conductor and is placed
    in series with the circuit. A device called a *shunt resistor* is used when?
    (1) the current to be measured is greater than that which the moving
        coil can safely pass or otherwise it would be burned out
    (2) the current to be measured is less than that which the moving coil
        can safely pass or otherwise it would be burned out
    (3) it can be placed in the direct series part of a circuit with no possibility
        of the ammeter being burned out
    (4) it has nothing whatsoever to do with a parallel relationship in the
        AC circuit

48. Which of the following statements are correct regarding electrical terms
    and their proper relationship?
    (1) farad, micro-farad, micromicro-farad measure capacitance
    (2) a kilowatt (KW) is equal to 1,000 watts
    (3) 1,000 milliamperes is equal to 1 ampere
    (4) all three of the above answers are correct

49. A typical solenoid switch relay used in 24-volt circuits to connect the
    battery to the airplane's electrical system is essentially a remotely
    controlled:
    (1) single-pole, single-throw switch
    (2) single-pole, double-throw switch
    (3) double-pole, single-throw switch
    (4) double-pole, double-throw switch

50. In air carrier transport planes spare fuses must be provided equal to
    at least 50% of the number of each rating required for complete circuit
    protection. General aviation planes (non-air-carrier) must be provided
    with:
    (1) one spare set of fuses
    (2) three spare fuses of each magnitude

(3) one spare set or three spare fuses of each magnitude

(4) one spare fuse for each intermittent circuit

51. A voltmeter, when used in testing a circuit, is connected in:
    (1) series
    (2) series-shunt
    (3) series-parallel
    (4) parallel

52. There are several facts to be considered when determining the resistance of an electrical conductor: (a) copper cable has less resistance than aluminum; (b) the smaller the conductor area the greater the resistance; and (c) resistance varies proportionately with the length of a conductor. During an electrical installation job you replace an electrical cable with one that has half the area of the original cable and is twice as long. The resistance of the replacement cable as compared to the original is:
    (1) the same
    (2) twice as much
    (3) four times as much
    (4) eight times as much

53. A 24-volt series circuit has three resistors of eight ohms each. This circuit has how many amperes?
    (1) 24
    (2) 8
    (3) 16
    (4) 1

54. An electrical circuit of 28 volts has three series resistors only. Their resistance is respectively 10 ohms, 15 ohms, and 20 ohms. In this case the ammeter, which is also placed in a series, should read:
    (1) 6.22
    (2) 1.260
    (3) .62
    (4) 62

55. You are working out an electrical circuit of 24 volts that has three equal parallel resistors only. The ammeter that is placed in the series portion of this circuit reads 9 amperes. According to this data, the resistance of the middle (or center) resistor is:
    (1) 24 ohms
    (2) 8 ohms
    (3) 12 ohms
    (4) 4 ohms

56. You are working on a 30-volt circuit that has three 10-ohm resistors in parallel. What is the voltage drop across the middle 10-ohm resistor?
    (1) 10
    (2) 20
    (3) 30
    (4) 40

57. You are working with a circuit of 24 volts and it is producing 1,000 watts. The amperes and resistance of this circuit are respectively:
    (1) .575 ampere and 42.67 ohms
    (2) 35.72 amperes and .928 ohms
    (3) 41.67 amperes and .575 ohms
    (4) .575 ampere and 41.67 ohms

58. A 24-volt series circuit has three resistors of eight ohms each. This circuit has 1 ampere. The voltage drop across each of the eight-ohm resistors is:
    (1) 24
    (2) 16
    (3) 8
    (4) 1

59. In parallel circuits the voltage drop is:
    (1) proportional to the number of parallel circuits involved
    (2) split evenly between the parallel circuits and all series circuits in the particular system
    (3) the same as system voltage in each segment of the parallel circuit
    (4) not measurable unless there are also series circuits involved

60. Which of the following is considered to be one AC cycle?
    (1) zero to peak to zero
    (2) zero to peak to zero to peak to zero
    (3) zero to peak to peak to zero
    (4) zero to peak to zero to peak to zero to peak

61. Which of the following is correct regarding the installation or removal of aircraft storage batteries?
    (1) during installation, connect and tighten the positive cable first and then connect and tighten the negative cable
    (2) during removal, disconnect the negative cable first or remove and then disconnect the positive cable
    (3) when installing or removing storage batteries, first remove the cable that will prevent possible arcing
    (4) all three of the above answers are correct

62. When putting in a storage battery you should install:
   (1) the negative or ground lead first
   (2) the positive lead first
   (3) the negative and positive leads at exactly the same time
   (4) either the negative or positive lead first

63. When *removing* a storage battery you should remove:
   (1) the negative or ground lead first
   (2) the positive lead first
   (3) the positive and negative leads at exactly the same time
   (4) either the ground or positive lead first

64. The negative terminals of two batteries are joined together and then connected directly to the metal structure to provide a ground return (sometimes a large switch is placed in the ground lead). The positive terminals of the batteries are connected:
   (1) directly to the battery bus bar
   (2) to the battery bus bar via a fuse or other protective device
   (3) directly to the solenoid switch relays through fuses or other protective devices, then to the battery bus bar
   (4) directly to the battery bus via an SPDT

65. Aircraft storage battery installation should be:
   (1) in a location compatible with the use of light electrical cable and low amperage current
   (2) near the engine starter due to the low amperage required for starter operation
   (3) near the starter because of the high amperage load required and the necessity of using heavy cable
   (4) near the starter because of the high amperage load required and the necessity of using light electrical cable

66. A six-volt battery output passes through a three-ohm resistance. The current available is:
   (1) two EMF (electromotive force)
   (2) three EMF
   (3) two amperes
   (4) three amperes

67. Impedance of an AC circuit is:
   (1) the resistance of any part of the complete circuit
   (2) the total opposition that the circuit offers to the flow of current
   (3) the reluctance to flow of any part of the complete circuit
   (4) a certain voltage value in relation to a constant wattage output

68. Which of the following is correct regarding the symbol for a battery in an electrical circuit?
    (1) the longer line represents the negative side of the battery
    (2) the shorter line represents the positive side of the battery
    (3) either line may represent the positive or negative side of the battery since the length of the symbol lines have no significance whatsoever.
    (4) in all cases the long line represents the positive terminal side and the short line represents the negative terminal side of the battery

69. You have four 12-volt and six 24-volt batteries that are to be connected to a constant current battery charger. In order to charge all of them at the same time you should connect:
    (1) the four 12-volt batteries in parallel and the six 24-volt batteries in series
    (2) the 24-volt batteries in parallel and the 12-volt batteries in two parallel banks, each bank to have two 12-volt batteries in series
    (3) the four 12-volt batteries in series and the six 24-volt batteries in parallel
    (4) all of the batteries in series with the 24-volt batteries closer to the charger

70. The hydrometer reading of a storage battery in near or fully charged condition is 1.275 to 1.300. To correct battery hydrometer readings:
    (1) add one point for each 3°F. above 80°F.
    (2) subtract one point for each 3°F. below 80°F.
    (3) add one point for each 3°F. above 80°F. and subtract one point for each 3°F. below 80°F.
    (4) all three of the above answers are correct

71. What is the preferable first step when you begin to check a discharged battery?
    (1) remove the caps from the vent holes
    (2) cover up the vent holes and make sure the covering is secure
    (3) remove all cables leading to the battery
    (4) blow out the vent holes and cover with a wet cloth

72. When two or more batteries are utilized in an aircraft they are usually connected in parallel. The net voltage will be the same as when a single battery is used. The capacity in ampere hours will:
    (1) also be the same as for a single battery
    (2) be less than for a single battery
    (3) be the sum of the capacities of each separate battery utilized
    (4) remain constant regardless of how many batteries are utilized in parallel

73. Which of the following design factors affect storage battery voltage and capacity?
   (1) number of cells
   (2) quality of the plate material
   (3) area in the battery plates
   (4) all three of the above answers are correct

74. When an airplane storage battery is being discharged, lead sulphate is formed on both the positive and negative plates. What corrective procedure should you take if the battery plates are found to be excessively sulphated?
   (1) discharge the battery at a·high rate for a short period and then charge the battery fully at a slow charging rate
   (2) charge the battery in a normal manner as this procedure will remove the sulphation from the plates
   (3) discard the battery as the plates will be shorted out as a result of the sulphation
   (4) return the battery to the manufacturer as the battery will require a complete overhaul

75. A given battery is rated at 35 ampere hours. If it is in good condition and fully charged it will furnish which of the following electrical loads?
   (1) seven amperes for five hours
   (2) 35 amperes for one hour
   (3) 10 amperes for 3½ hours
   (4) all three of the above answers are correct
   Note: Theoretically all three answers are correct. However, answer (1) is best because the efficiency of a battery (or battery capacity) is lowered under a heavy discharge load, as indicated by answers (2) and (3).

76. Storage battery containers are usually made of hard rubber or some type of material that is able to withstand mechanical shocks and extremes of high and low temperatures, and is resistant to the action of the acid electrolyte placed within the cells. A separate compartment is provided for each battery cell. In a fully charged battery:
   (1) there are positive plates made of lead peroxide
   (2) the negative plates are made of a spongy type of stainless steel
   (3) the specific gravity of the electrolyte should be at least 1.240 as measured with a hydrometer
   (4) the cell vents should normally be closed

77. When working around or checking storage batteries, extreme caution must be taken because of the explosive nature of battery acid and its corrosive effects on clothing, metal, etc. The freezing of fully charged batteries is greatly reduced owing to:

(1) the action of the acid bubbling and becoming turbulent between the plates of the battery cells
(2) the chemical action of the battery acid that creates sufficient heat to prevent freezing
(3) the very low freezing point of the acid solution of the charged battery
(4) the compact pressure maintained in the battery cells due to the closed battery cell vents

78. Wet-cell storage batteries give off the most gases when:
(1) on a low charge rate
(2) being discharged at a low rate
(3) being charged excessively
(4) on a fairly high discharge rate

79. Drawings and schematic diagrams make up the graphic language used in the industrial world (aviation in particular) to express and record ideas and information necessary for the building of machines, structures, and components. The drawing may represent a new item, a modified item, or the means by which an item can be repaired. Blueprints or other copy methods are utilized so that any number of persons may have a copy of the drawing or schematic for the purpose of checking, shop work, etc. The aircraft mechanic, when working with drawings, schematics, blueprints, etc., must:
(1) be able to understand the "language" of drawings since the drafts-man uses standard lines, views, and descriptive data in his work and makes no attempt to give a complete picture of the item
(2) be able to make drawings and schematics to describe work that he contemplates doing himself or to supervise work done by others
(3) understand how pattern draftsmen produce a drawing or schematic and how these can be used to make templates that are passed to shop personnel for production of the actual part, unit, etc.
(4) all three of the above answers are correct

80. In the column on the left below (labeled A through I) are some terms applied to basic drawing and measuring instruments. In the column on the right (but not in the same order) are the definitions or explana-tions of these terms. Select that one of the four answers that correctly matches all nine terms with the proper definitions:

A. Ruler.   1. Assures that lines are perpendicular.
B. Protractor.   2. Assures that lines are straight.
C. Micrometer.   3. Measures distances and divides lines.
D. T-Square.   4. For drawing circles.
E. French curve.   5. For plotting varying curved lines.
F. Compass.   6. Assures horizontal lines are parallel.
G. Dividers.   7. Measures thousandths of an inch or smaller.

H. Straight edge.   8. Measures degrees of an angle.
I. Square.   9. Measures inches and fractions of an inch.

(1) A-9, B-8, C-7, D-6, E-4, F-5, G-3, H-2, and I-1
(2) A-9, B-8, C-7, D-6, E-5, F-4, G-2, H-3, and I-1
(3) I-1, H-2, G-3, F-4, E-5, D-6, C-7, B-9, and A-8
(4) I-1, H-2, G-3, F-4, E-5, D-6, C-7, B-8, and A-9

81. Title blocks on a drawing or blueprint are usually located in the lower
    right-hand corner of the drawing or blueprint. The information gener-
    ally contained in the title block is:
    (1) name of company and its location, name of the unit (or machine)
        being represented, and name of part (if a detail drawing)
    (2) drawing number, part number (if a detail drawing), and scale of
        the drawing
    (3) names or initials of draftsman, tracer, checker, and approving au-
        thority, each with appropriate date
    (4) all three of the above answers are correct

82. Following are four lines used in aircraft drawings:

    A. |◄————2"————►| 1. cutting-plane line
    B. — ·· ———— ·· ——   2. center line
    C. ——— · ——— · —   3. adjacent part line (phantom line)
    D. A|____ ·· —— ·· _|A   4. dimension line

    Choose the answer below that correctly matches each drawn line with
    its proper identification:
    (1) A-3, B-2, C-1, D-4
    (2) D-1, C-2, B-4, A-3
    (3) A-4, B-3, C-2, D-1
    (4) A-2, B-3, C-1, D-4

83. Which one of the following statements about aircraft drawings is correct:
    (1) to indicate short breaks in the drawing of an item there would be
        a line of equal dashes, each dash about ¼ inch long (— — — —)
    (2) a long break in the drawing of an item is indicated by a very thin
        straight line (——————)
    (3) hidden lines in the drawing of an item is indicated by a heavy
        solid straight line (——————)
    (4) a short break is shown by a freehand jagged line and long breaks
        are indicated by a series of thin straight lines broken at intervals
        by a brief up and down line ( ) ( )

84. For the purpose of weighing aircraft and determining the empty weight
    (EW) and the empty weight center of gravity (EWCG) you should
    refer to the:

    (1) aircraft log books
    (2) aircraft specifications
    (3) airworthiness certificate
    (4) airworthiness directives (AD's)

85. Which of the following is the point of reference for all weight and balance measurements?
    (1) any point so long as it does not exceed the forward or aft center of gravity (CG) limits
    (2) the lateral axis that runs through the wing from tip to tip
    (3) the datum
    (4) the firewall of the engine/s

86. Where would you locate the information that should be known to compute weight and balance data for a particular aircraft?
    (1) the owner's manual
    (2) aircraft registration certificate
    (3) aircraft specification data sheets
    (4) glove compartment or the inside door compartment

87. To determine the moment of an item of equipment in an aircraft you would:
    (1) multiply the weight of the item times the distance the item is located from the datum (in inches)
    (2) multiply the weight of the item times the distance (in inches) the item is located from the empty weight center of gravity
    (3) refer to either the airworthiness or the registration certificate of the particular aircraft
    (4) weigh the item in pounds and then determine the distance the item will be from the leading edge of the wing

88. Information on which to base the record of weight and balance *changes* to the aircraft may be obtained from:
    (1) the pertinent aircraft specification
    (2) the prescribed aircraft operating limitations or approved airplane flight manual
    (3) the pertinent aircraft weight and balance report
    (4) all three of the above answers are correct

89. Regarding rotorcraft weight and balance with respect to control with changes in center of gravity positions:
    (1) rotorcraft are less critical than fixed-wing aircraft
    (2) rotorcraft are more stable than fixed-wing aircraft due to the longer permissible travel of the center of gravity between the forward and rearward limits

(3) rotorcraft, in general, are more critical with respect to changes in CG positions and, in general, the CG limit of travel is very limited as compared to a fixed-wing aircraft

(4) there is no difference between the rotorcraft and fixed-wing aircraft in regards to center of gravity changes and controllability

90. The maximum weight is the maximum authorized weight of the aircraft and its contents as listed in the aircraft specifications. The gross weight of an aircraft is the total weight for a given flight at takeoff, however, it cannot exceed the maximum authorized weight. Which of the following weight and balance combinations is correct?

(1) EW + UL = the GW

(2) GW − the UL = the EW

(3) EW + PL + NPL = GW

(4) all three of the above answers are correct

Note: EW = empty weight, UL = useful load, GW = gross weight, PL = payload, and NPL = non-payload.

91. An airplane has a GW of 4,000 pounds (maximum authorized weight) and an EW of 2,300 pounds. A proposed flight will carry a pilot and five passengers, 60 gallons of fuel, and four gallons of oil. On the basis that fuel weighs six pounds per gallon, oil weighs seven and one-half pounds per gallon, and each occupant is assumed to weigh 170 pounds each, how many pounds of baggage and/or cargo can be carried on this flight?

(1) 325 pounds

(2) 290 pounds

(3) 265 pounds

(4) 340 pounds

92. The basic rules regarding weighing an aircraft to establish its empty weight center of gravity (EWCG) are: (a) place aircraft in a closed hangar; (b) place each wheel on a suitable scale; and (c) using blocks, jacks, etc., place the aircraft in level flight attitude. For proper weight and balance procedures, which of the following is most appropriate:

(1) the weight of the items used to level the aircraft (blocks, jacks, etc.) are considered to be "tare weight" and must be subtracted from the scale readings

(2) when a forward weight and balance check (to establish the most possible forward CG position) is made use the maximum weight of items that are located forward of the most forward CG allowable limit

(3) all items are measured in inches forward or aft of the datum (zero reference line) and minimum fuel (in pounds) for a weight and

balance check is found by the formula: $\dfrac{\text{METO} \times .55}{2}$

(4) all three of the above answers are correct

Note: Former formula for answer 3 was METO ÷ 2

93. You are in charge of weighing an airplane. It has been placed in a closed hangar to eliminate any effect of air movement on the scale readings. The airplane has been set up in level flight condition. The right main gear scale reads 3,250 LBS. net, the left main gear scale 3,250 LBS. net, and the nose gear scale 1,950 LBS. net. The main gear is at an arm of 121 inches and the nose gear at an arm of 10.5 inches. The center of gravity is:
    (1) 112.72 inches aft of the datum
    (2) 15.87 inches forward of the datum
    (3) 95.50 inches aft of the datum
    (4) 67.24 inches forward of the datum

94. In order to obtain accurate and useful weight and balance data, the airplane:
    (1) should be in level flight condition during the weighing process
    (2) should be in a closed hangar protected from the wind
    (3) should have all fixed equipment secured in its proper and permanent location
    (4) all three of the above answers are correct

95. You are making an aft weight and balance extreme check. The items of useful load must be computed at their minimum weights if they are located forward of the aft CG limit and at their maximum weights if they are located aft of the aft CG limit. In other words, you are computing the weights and arms to make the aircraft as tail-heavy as possible which in turn places the CG as far back as is possible to determine if the CG is still within the allowable rear limit. To make a forward weight and balance extreme check:
    (1) follow the same rules as for an extreme aft check
    (2) the items of useful load must be computed at their minimum weights if they are located aft of the forward CG limit and at their maximum weights if they are located aft of the forward CG limit
    (3) the items of useful load must be computed either at their minimum or maximum weights depending on their distance, in inches, from the datum or reference line
    (4) the items of useful load must be computed at their minimum weights if they are located aft of the forward CG limit and at their maximum weights if they are located forward of the forward CG limit

96. A particular aircraft engine has 450 METO horsepower. The minimum fuel capacity is 67.5 gallons (METO × .15 gallon). The minimum oil capacity is six gallons (one gallon per each 75 METO horsepower). The minimum fuel required for a weight and balance check is:
    (1) 44.5 gallons
    (2) 20.6 gallons
    (3) 34.5 gallons
    (4) 32.5 gallons
    Current solution: $\dfrac{\text{METO} \times .55}{2} = \dfrac{450 \times .55}{2} = \dfrac{247.50 \text{ lbs.}}{2} = 123.75$

    lbs. or 20.63 gallons.

    Former solution: $\dfrac{\text{METO}}{12} = 37.5$ gallons or 225.0 lbs.

97. You are making a weight and balance check. The airplane is equipped with a 450 METO horsepower engine. What is the minimum fuel load required when making a forward or rearward CG limit computation? (Use *former* solution.)
    (1) 30.5 gallons or 183 pounds
    (2) 37.5 gallons or 225 pounds
    (3) 45.5 gallons or 273 pounds
    (4) 27.5 gallons or 165 pounds

98. Which of the following weight and balance computations is correct?
    (1) $+W \times +A = +M$
    (1) $+W \times -A = +M$
    (3) $-W \times +A = +M$
    (4) $-W \times -A = -M$

99. Two weights are added to a bar and the bar balances properly. This would indicate that the:
    (1) weights are equal
    (2) arms or distance to the datum are equal
    (3) moments are equal
    (4) bar is mounted on its fulcrum

100. The mechanical efficiency of the engine will not contribute to high oil consumption, and low wear values is not a property of aircraft engine lubricating oils. When computing the empty weight of an aircraft, lubricating oil in the oil supply tank should be considered as:
    (1) a part of the empty weight of the aircraft
    (2) a portion of the aircraft payload
    (3) always creating a plus moment on the aircraft in establishing the center of gravity location
    (4) a part of the useful load of the aircraft

101. Approved radio equipment has been installed in an aircraft forward of the datum. As the mechanic in charge, you should:
    (1) re-compute the weight and balance and complete a new Form 337
    (2) make only a note of this work in the aircraft log since this is a very minor alteration
    (3) make a good estimate of the change in weight and balance and so indicate on the operations and limitations record
    (4) place a special memorandum on the airworthiness certificate in the space reserved for this type of alteration

102. When additional radio equipment is installed in an aircraft:
    (1) the empty weight of the aircraft is increased
    (2) the useful load of the aircraft is reduced
    (3) the maximum allowable gross weight remains the same
    (4) all three of the above answers are correct

103. A beading machine is used to turn beads on pipes, tubing, etc., for stiffening, gripping, and for ornamental purposes. Beads may also be placed on sheet stock that is to be welded. In the latter case:
    (1) place the bead near the weld
    (2) beading prevents buckling of the sheet stock
    (3) beading will make the joint stronger
    (4) all three of the above answers are correct

104. During the installation of fluid lines along the same route as electrical cables and/or electrical bundles, you should:
    (1) install the fluid lines (by proper clamping, etc.) above the electrical cables and/or bundles
    (2) install the electrical cables or bundles above the fluid lines
    (3) under these conditions, install the fluid lines and electrical bundles parallel to each other
    (4) install in such a manner that there is no contact between them and that they are properly tied to the structure

105. When inspection shows a metal line to be damaged or defective, replace the entire line or, if the damaged section is localized, a repair may be inserted. In replacing lines, always use tubing of the same size and material as the original line. You can use the old tubing as a template in bending the new line unless it is too greatly damaged, in which case a template can be made from soft iron wire. Soft aluminum tubing (1100, 3003, or 5052) under ¼″ outside diameter may be bent by hand. For all other tubing use an acceptable hand or power tube bending tool. A small amount of flattening in bends is acceptable, but do not exceed an amount such that the small diameter of the flattened portion is:

(1) less than 75% of the original outside diameter (OD)
(2) less than 75% of the original inside diameter (ID)
(3) 100% down to 90% of the original OD
(4) 95% down to 85% of the original ID

106. The minimum bend radii (in inches) when bending aluminum alloy tubing is approximately:
(1) 2 to 3 times the tube ID
(2) 2 to 3 times the tube OD
(3) 4 to 6 times the tube ID
(4) 4 to 6 times the tube OD

107. Many tubing connections are made using flared tube ends (37° flare) and standard connection fittings: AN-818 and AN-819 sleeve. In forming flares, cut the tube ends square, file smooth, remove all burrs and sharp edges, and thoroughly clean. The tubing is then flared using the correct forming tools for the size of tubing and type of fitting. Regarding tube connections you should also:
(1) use a single flare on tubing and a double flare on soft aluminum tubing ⅜ inch OD and under
(2) use hydraulic fluid as a lubricant and then tighten
(3) do not overtighten due to possible damage to the tube or fitting and do not undertighten as this may cause leakage which could result in a system failure
(4) all three of the above answers are correct

108. Minor dents and scratches in tubing may be repaired. Scratches or nicks no deeper than 10% of the wall thickness in aluminum alloy tubing, that are not in the heel of a bend, may be repaired by burnishing with hand tools. Replace lines with severe die marks, seam, or splits in the tube. Any crack or deformity in a flare is also unacceptable and cause for rejection. A dent less than 20% of the tube diameter is not objectionable unless it is in the heel of the bend. Dents may be removed from tubing by:
(1) tapping lightly with a ball peen hammer
(2) forcing a proper size solid steel rod through the tubing
(3) drawing a bullet of proper size through the tube by means of a length of cable
(4) none of the above answers is correct

109. The condition of a completed or finished weld is: (a) a smooth seam and uniform in thickness, (b) the weld metal is tapered smoothly into the base metal, (c) no oxide has formed on the base metal at a distance of more than ½ inch from the weld, (d) the base metal shows no signs of pitting, burning, cracking, or distortion, (e) the depth of penetration

insures fusion of base metal and filler rod, and (f) welding scale is removed by wire brushing (of similar metal) or sandblasting. The completed weld should also:
(1) be magnafluxed under all circumstances
(2) show no signs of blowholes, porosity, or projecting globules
(3) be properly lubricated with an appropriate lubricant
(4) be exposed to a mild nitriding treatment

110. Metals are annealed (copper tubing, etc.) to relieve internal stresses,
soften the metal, make it more ductile, and to refine the grain structure.
Annealing consists of heating the metal to the proper temperature:
(1) then cooling it back to room temperature immediately
(2) and then cooling it back to room temperature over a period of several hours
(3) holding (or soaking) it for a specified length of time, then cooling
it back to room temperature
(4) then, to produce maximum softness, the metal must be cooled very
quickly

111. Which of the following properly describes the type of flare used in conjunction with hydraulic line fittings?
(1) all hydraulic line tubing requires a double flare
(2) all hydraulic line tubing requires an elliptical type flare
(3) a double flare is used on soft aluminum tubing ⅜ inch OD and
under, and a single flare is used on all other tubing
(4) for tubing ½ inch OD a 45° double flare is used, and a single lap
flare is used on all other tubing

112. A *double* flare is used on soft aluminum tubing ⅜ inch OD and under.
A *single* flare is used:
(1) only on tubing that comes within the classification of hard aluminum
(2) on all other tubing
(3) only on tubing that has been heat-treated
(4) only on tubing that has been properly annealed

113. Fuel lines in a particular engine installation have straight white lines
and a broken red line. This indicates that:
(1) the fuel system should be serviced with at least 100 octane fuel
(2) the system is serviced with aromatic fuels
(3) all types of fuel may be used in this particular fuel system
(4) all fuel system connections are of left hand threads

114. Mechanics should be familiar with the International Color Code for
aircraft piping. Each separate aircraft system (fuel, oil, hydraulic, etc.)
is identified by a color and symbol on a white band. Examples are:
(a) fuel lines—red with a four-pointed star symbol; (b) hydraulic lines—

blue-yellow with a circle symbol. Oil system lines are identified by:
(1) brown and diamond symbol
(2) green and a rectangular symbol
(3) blue and a hash mark symbol
(4) yellow and a square (staggered) symbol

115. The color code for the oil lubrication system lines is:
(1) red
(2) orange
(3) blue
(4) yellow

116. When installing airplane plumbing, threads should be lubricated with:
(1) Parker Threadlube
(2) Parker Greaselube
(3) a good grade of engine lubricating oil
(4) any type of mineral based lubricant

117. The lubricant used when assembling oxygen fittings is:
(1) anti-seize or thread-sealing compound conforming to specifications MIL-T-5542-B or the equivalent
(2) a seize-proof synthetic mineral lubricant
(3) a chromate type of lubricant specifications MIL-T-7710-C
(4) none of the above answers is correct

118. Regarding metal hydraulic lines:
(1) when inspection shows a hydraulic line to be damaged or defective, the entire line should be replaced
(2) if the damaged section is localized, a repair section may be inserted
(3) in replacing hydraulic lines, always use tubing of the same size and material as the original line
(4) all three of the above answers are correct

119. You are about to connect an aircraft engine oil line made of one-half inch copper tubing. The size of the AN fitting should be:
(1) 818-14
(2) 818-22
(3) 818-8
(4) 818-12

120. Replace control cables and wires if injured, distorted, worn, or corroded even though the strands are not broken. Control cables may be spliced when they become worn, distorted, corroded, or otherwise injured. Standard swaged cable terminals develop the full cable strength and

may be substituted for the original terminals whenever practical. Regarding control cables:
(1) flexible cable 7 by 7 has 49 individual wires (7 strands of 7 wires each)
(2) extra-flexible 7 by 19 has 133 individual wires (7 strands of 19 wires each)
(3) carbon steel 19 wire cable is non-flexible
(4) all three of the above answers are correct

121. Most bolts used in aircraft structures are either general purpose AN bolts or NAS (National Aircraft Standard) internal wrenching or close tolerance bolts. Which of the following answers properly describes respectively a bolt with a dash on the head, a bolt with an X in a triangle on the head, bolt grip, and the condition of a bolt when installed?
(1) AN standard steel bolt (corrosion resistant), NAS close tolerance bolt, distance from underside of bolt head to the beginning of the threads, and bolt should be properly oiled prior to installation
(2) NAS close tolerance bolt, AN standard steel bolt (corrosion resistant), equal to the material thickness, and clean, dry, and thoroughly degreased before installation
(3) AN standard steel bolt (corrosion resistant), NAS close tolerance bolt, distance from underside of head to the beginning of the threads, and clean, dry, and thoroughly degreased prior to installation.
(4) AN standard steel bolt (corrosion resistant), special bolt, equal to material thickness, and clean, dry, and thoroughly degreased before installation

122. You should lock or safety all bolts and/or nuts, except self-locking nuts. Do not reuse cotter pins and safety wire. Recommended torque values must be followed when tightening nuts unless other instructions require a specific torque for a specific nut. When tightening castellated nuts on bolts:
(1) the cotter pinholes will always line up with the slots in the nuts if recommended torque values are used
(2) except in cases of highly stressed engine parts, the nut may be overtightened to permit lining up the next slot with the cotter pinhole
(3) under no circumstances can the nut be overtightened
(4) the nut must always be backed off in order for the pertinent slot on the nut to be lined up with the cotter pinhole

123. Refrigerated (ice box) rivets attain 75% of their strength in about an hour after being driven. These types of rivets will acquire the other 25% of their strength through:
(1) immersion in an acid solution
(2) complete immersion in a salt bath

(3) aging for approximately 48 hours
(4) aging for approximately 4 days

124. Do not file welds in an effort to make a smooth-appearing job, as such treatment causes a loss of weld strength. Do not fill welds with solder, brazing, or any other filler. During welding procedures:
   (1) remove all old weld before rewelding if rewelding becomes necessary
   (2) avoid welding over a weld because reheating may cause the material to lose its strength and become brittle
   (3) never weld a joint which has been previously brazed
   (4) all three of the above answers are correct

125. Micrometers and calipers are utilized to make precise measurements, small holes are measured by using a micrometer and a hole gage, and a vernier scale permits measurements to a very close tolerance. To check the alignment of a shaft (crankshaft, for example) you should use a:
   (1) micrometer with a circular variance rule
   (2) dial indicator, flat blocks, and a surface plate
   (3) dial indicator, V-blocks, and a surface plate
   (4) dial indicator, M-blocks, and a curved plate

126. When installing flareless fittings:
   (1) install snugly by hand and then tighten with a wrench ¼ turn
   (2) install snugly by hand and then tighten with a wrench ½ turn
   (3) tighten by wrench to a satisfactory torque value
   (4) tighten by hand only as further tightening with a wrench will cause damage

127. Magnetic particle inspection can be used on magnetic materials such as iron and steel. After this type of inspection, carefully demagnetize and clean the parts and complete the inspection by coating the parts with a suitable preservative. X-ray or radiograph (radiography) inspection may be used on either magnetic or non-magnetic materials to detect subsurface voids such as open cracks, blowholes, etc. Flourescent Penetrant (Zyglo) inspection can be made on metal, plastics, etc. Ultrasonic flow detection can be used to inspect all types of materials. The ultrasonic test instrument requires:
   (1) access to all sides of the material being tested
   (2) no particular ability on the part of the operator
   (3) access to only one surface of the material being tested
   (4) spraying a dye penetrant on the material to be inspected prior to the ultrasonic test

128. When a photographic film or plate is used to record X-ray (in a similar manner to exposing a photographic film), the process is known as radiography. X-ray or radiographic inspection may be used:
    (1) on either magnetic or non-magnetic materials for detecting subsurface voids such as open cracks, blowholes, etc.
    (2) in the same manner as magnetic particle inspection
    (3) as a means to identify properly the type of material being inspected for surface cracks, blowholes, etc.
    (4) to detect subsurface voids such as open cracks, blowholes, etc., on non-magnetic materials only

129. Of the following, which is correct concerning the selection of aircraft aluminum for replacing parts?
    (1) aluminum designated 2017 should not be replaced by aluminum designated 2024
    (2) designation 2017 can be replaced by designation 2024 provided the deficiency in strength is compensated for by an increase in the material thickness
    (3) aluminum 2024 can be replaced by 2017 provided the deficiency in strength will be compensated for by an increase in material thickness
    (4) 2024 must be utilized in place of 2017 because the former is much stronger

130. Heat-treatment of aluminum alloys involves the heating of the material to a specified temperature, holding it there for a specified time then quenching it promptly. After quenching, "age-hardening" (also called precipitation heat-treatment) is necessary for the material to gain its full strength. Regarding heat-treatment of alloys:
    (1) alloys that complete age-hardening (precipitation) at room temperature generally gain about 90% of their strength in the first half-hour
    (2) when a material has gained its full strength on completing the precipitation heat-treatment it is said to be in fully heat-treated condition
    (3) after the alloy is in full heat-treated condition it is designated by the letter T followed by one or more digits
    (4) all three of the above answers are correct

131. The letter *H* indicates a metal is strain hardened. In the new coding, the number that follows the *H* as in *H1* to *H8*, indicates minimum and maximum hardening. In the old coding, for example in *H36*, the number *6* designates the:
    (1) degree of hardness, in this case ¾ H
    (2) classification of the Alclad
    (3) degree of hardness, in this case ½ H

GENERAL257

(4) material has been strain hardened to the 6th degree
Note: The "3" indicates strain-hardened and then stabilized.

132. The metal designation 2024-T3 indicates the:
   (1) method employed in cold working
   (2) alloy and percent of titanium in the metal
   (3) alloy and temper of the metal
   (4) method employed in strain-hardening

133. In 2024-T36 material the alloying agent is expressed by the number:
   (1) 20
   (2) 2
   (3) 24
   (4) 36

134. In the metal designation 2024-T3, the T3 indicates the metal is:
   (1) annealed (cast products only)
   (2) pure aluminum and strain-hardened
   (3) cold-worked aluminum and artificially aged
   (4) solution heat-treated and then cold-worked

135. Which of the following steel designations would have the greatest amount of carbon content?
   (1) 1020
   (2) 1030
   (3) 1040
   (4) 1050

136. The following steels are readily weldable: plain carbon, nickel steels of the SAE 2300 series, chrome-nickel alloys of the SAE 3100 series, chrome molybdenum steels of the 4100 series, and low nickel-molybdenum steels of the SAE 8600 series. Which of the following is considered very difficult to weld or is not usually welded?
   (1) wrought iron
   (2) low-carbon steels, or mild steels
   (3) steels of very high carbon content
   (4) chromium molybdenum steels

137. Which of the following describes a bolt with the symbol X on the bolt head?
   (1) AN standard steel bolt (corrosion resistant)
   (2) NAS close tolerance bolt
   (3) AN standard steel bolt
   (4) a reworked bolt

138. Nuts are classified by number and fit a corresponding size tube (using outside diameter). A No. 8 nut will fit a tube whose OD is:
    (1) ¼ inch
    (2) ½ inch
    (3) ⅜ inch
    (4) ⅝ inch
    Note: The numerical value of the nut classification is equal to 1/16 inch. Other examples are: No. 3 is 3/16 inch; No. 4 is 1/4 inch; No. 5 is 5/16 inch, etc.

139. You are working with one-half inch tubing. You should use AN nut size No.:
    (1) 6
    (2) 8
    (3) 10
    (4) 2

140. Three kinds of friction bearings are used between metal surfaces: *sliding, rolling,* and *ball* bearings. Sliding bearings are used as main and connecting rod bearings, and require more complete lubrication than roller or ball bearings. Of the three kinds:
    (1) thrust bearings that secure the crankshaft on the nose case are of the roller bearing type
    (2) roller bearings require the least amount of lubrication
    (3) thrust bearings require a maximum of pressure lubrication
    (4) ball bearings give the least amount of friction and require the least amount of lubrication

141. Type A rivets have relatively low strength because of the 1100 series material from which they are made. 1100 series material is commercially pure aluminum (99% pure). Type AD rivets:
    (1) consist of 2017 alloy (17ST)
    (2) consist of 2117 alloy (A17ST)
    (3) do not have a dimple on the head for identification
    (4) none of the above answers is correct

142. The firewalls of aircraft should be made of:
    (1) proper alloyed aluminum
    (2) an appropriate registered stainless steel
    (3) an acceptable fire-resistant material
    (4) a properly hard-pressed micarta material

143. Of the following, which would most likely be used for engine firewalls?
    (1) Alclad
    (2) aluminum alloys

(3) copper alloys
(4) titanium

144. Which of the following is satisfactory for firewalls?
   (1) a single sheet of terne plate not less than 0.028-inch thick or a single sheet of stainless steel not less than 0.015-inch thick
   (2) two sheets of aluminum alloy not less than 0.20-inch thick fastened together and having between them an asbestos paper or asbestos fabric sheet at least ⅛-inch thick
   (3) heat- and corrosion-resistant steel 0.015-inch thick, or low carbon steel, suitably protected against corrosion, 0.018-inch thick
   (4) all three of the above answers are correct

145. Aircraft parts or units made from aluminum:
   (1) may easily corrode, leaving a white powder on the external surfaces
   (2) are susceptible to fairly fast deterioration
   (3) will not corrode easily because the exterior surfaces are coated with pure aluminum (Alclad or Pureclad)
   (4) will not deteriorate easily because the aluminum (Alclad or Pureclad) has an alloy exterior surface and a pure aluminum center core

146. When using a micrometer you obtain .9375 as a reading. This is equal to:
   (1) 7/8 inches
   (2) 15/16 inches
   (3) 56/64 inches
   (4) 13/16 inches

147. You do not have a stretch gage. However, you could check a valve for stretch by:
   (1) a micrometer reading of the stem at the head end
   (2) measuring the overall length
   (3) using a telescopic gage
   (4) use of a dial indicator and a V-block

148. Which of the following would be a logical procedure for extinguishing fires in the induction system of a reciprocating engine during starting operations?
   (1) immediately reduce the throttle to idle position
   (2) shut the engine down immediately and fight the fire with CO_2
   (3) keep the engine operating so that the fire in the induction system can be drawn into the cylinders and then out the exhaust system
   (4) none of the above answers is correct

149. Hand signals, by a person in full view of the pilot/s, should be used to direct aircraft movements on the ground. Aircraft should be securely tied down during outside storage in such manner to resist any possible known wind or gusty conditions that may cause damage. Prior to engine starting procedures, the propeller of a reciprocating engine (especially the radial type) should be turned by hand (or by the starter) with ignition switch in OFF position to check:
    (1) the coming-in speed of the right and left magneto
    (2) for liquid lock in the engine cylinders
    (3) that the intake and exhaust valves are seating properly
    (4) that there is an indication of oil pressure prior to actual engine start

150. If, during the daily preflight inspection check, liquid or hydraulic lock has occurred, the oil can be removed by:
    (1) removing the affected cylinder/s and clean them out with a suitable solvent
    (2) alternating engine rotation by hand first to the right and then to the left
    (3) rotating the propeller in the direction opposite to normal rotation
    (4) removing the spark plugs in the lower cylinders and rotating the propeller in the normal direction

151. For ground servicing of aircraft, the type of fuel for the particular aircraft should be noted on or near the filler caps. If not there or the type of fuel has been obliterated, the aircraft specifications should be referred to. Fueling areas should be free of dirt, dust, and/or blowing contaminants. If, at any time, it becomes necessary to "strain" fuel to remove contaminants, which of the following is appropriate?
    (1) very fine wire screens
    (2) chamois skin
    (3) several layers of cheese cloth
    (4) a thin layer of felt

152. When starting an aircraft engine that is equipped with a float-type carburetor:
    (1) place the throttle at one-half open position
    (2) place the fuel-air mixture control at full lean position
    (3) place the fuel-air mixture control at full rich position
    (4) the throttle will be properly positioned while the automatic mixture control (AMC) remains in a static position

153. Reciprocating engines that utilize pressure injection carburetors and mixture controls that have an idle cut-off position should be started on fuel supplied by the primer with the mixture control in:

(1) idle cut-off position
(2) full-rich position
(3) full-lean
(4) automatic position

154. A reciprocating engine is equipped with a two-speed supercharger generally described as a low-blower and high-blower supercharger. This is
an internal engine type that is driven indirectly by the crankshaft.
During engine starting procedures:
(1) the supercharger control should be placed in high-blower position
(2) it makes no difference whether the supercharger control is in low-
blower or high-blower as an automatic device will always switch
it to low-blower if an error has been made on the part of the person
starting the engine
(3) the supercharger control should be placed in low-blower position
(4) high-blower position will give a better engine start and a faster
engine warmup as a general overall procedure

155. For use in reciprocating aircraft engines, gasoline must meet rigid
requirements. It must vaporize readily and be capable of producing
high power without detonation. Tetraethyl lead is added to gasoline
to improve the anti-knock value, but this causes lead deposits to form
in the combustion chamber. In order to break up the lead deposits,
which of the following is added to tetraethyl-type fuel?
(1) toluene fluids
(2) ethlyene dibromide
(3) benzene
(4) cracked kerosene

156. Some gasolines contain such chemical compounds as toluene, xylene,
and benzene. These compounds are known as aromatics and gasoline
to which they have been added are known as aromatic fuels. Which
of the following is correct if an aircraft that has been using gasoline
is to be serviced with aromatic fuels?
(1) be sure that the fuel system hose and seals are made of natural
rubber
(2) there is no need to be concerned about taking precautions to prevent
vapor lock
(3) be sure the fuel system flexible hose and seals are made of aro-
matic-resistant materials
(4) the flexible hose lines must have a lateral white line along their
length for proper identification

157. Volatility of a fuel is a property that enables the liquid to change readily
into a vapor condition. Volatility is important to economy since gasoline

which does not completely vaporize is not only wasted but washes lubricant from the cylinder walls. Too high a volatility is undesirable since it can easily create vapor lock in the fuel lines and components. Even though fuel may be of the correct volatility, vapor lock can be caused by:

(1) fuel lines located too close to heat sources such as the exhaust system
(2) upward loops in fuel lines when fuel line *slack* is to be eliminated
(3) inadequate fuel pressure in the fuel system, especially low fuel pressure between an outboard fuel tank and the engine-driven vane fuel pump
(4) all three of the above answers are correct

158. Turbojet engines have a higher fuel consumption, relative to their output, than reciprocating engines. However, the cost per gallon of jet fuel is substantially less than reciprocating engine gasoline. Comparing jet fuel and gasoline, which of the following is correct?
(1) jet fuel is basically kerosene and can be used in reciprocating engine operation
(2) gasoline is a highly refined petroleum based fuel. However, gasoline cannot be used in jet engine operation
(3) generally speaking, gasoline can be used in jet engine operation even though it is the basic fuel for reciprocating engines
(4) gasoline is unsatisfactory for jet engine operation due to the ease at which condensation (moisture) develops within its mass whereby jet fuel (kerosene) is entirely free of condensation within its mass

159. The type of fuel to be used for a specified aircraft is important since incorrect fuel can be damaging to the engine and cause a forced landing. Gasoline is rated by octane number and jet fuel by "JP" numbers. The type of fuel for a given aircraft should generally be found:
(1) on or near the fuel tank filler opening/s or in the particular aircraft specifications
(2) in the particular aircraft's engine maintenance log book
(3) by referring to AC 43.13-1
(4) on a placard mounted in the flight deck area

160. During aircraft fueling there are three things necessary to produce a fire: fuel, oxygen, and heat. All fire extinguishers are designed to eliminate one or more of these elements. Extinguishers ordinarily provided to fueling crews are either dry chemical, carbon dioxide (CO_2), or foam. No one type of extinguisher is "best" for all classes of fires, but all three types are effective against fires in gasoline, turbine fuels, oil, grease, and flammable liquids when used correctly. Turbojet aircraft fire extinguishers are generally charged with:
(1) carbon monoxide and nitrogen
(2) nitrogen and argon

(3) freon and nitrogen

(4) nitrogen and carbon dioxide

161. Which of the following procedures would be the correct one if you experienced slow acceleration during a jet engine start?
(1) the thrust lever should be moved to the idle position
(2) fuel and ignition should be turned OFF and engine rotation continued
(3) slowly move the thrust lever forward
(4) place the start lever in the shutoff position

162. The most positive indication of a jet engine light-up is:
(1) an oil pressure rise
(2) a fuel flow rise
(3) a rise in engine RPM
(4) a rise in exhaust gas temperature (EGT)

163. Which of the following is used to measure the fuel level while the aircraft is on the ground?
(1) dielectric cells
(2) float mechanism
(3) dip stick
(4) a potassium

164. A turbojet aircraft is being checked with ground power connected. The DC electrical power for the aircraft is furnished by:
(1) an inverter-transformer
(2) an ammeter-motor in shunt
(3) the AC transformer-rectifier unit
(4) none of the above answers is correct

165. Inverters are used to convert DC current into AC current, and rectifiers are used to convert AC current into DC current. The use of current obtained from a hangar electrical system for ground maintenance of aircraft where aircraft lights, etc., are to be used:
(1) is not permissible
(2) is permissible when an inverter is used in the electrical system between the hangar current outlet and the aircraft external plug-in
(3) is permissible when a rectifier is used in the electrical system between the hangar current outlet and the aircraft external plug-in
(4) is not advisable because of the AC current from the hangar electrical system and the DC current needs of the aircraft

166. Proper operation of the primer before starting an aircraft engine is to:
(1) pull out abruptly and push in slowly

(2) pull out slowly and push in abruptly
(3) pull out slowly and push in slowly
(4) pull out abruptly and push in abruptly

167. Priming an engine prior to starting may be accomplished by a hand primer or by electrical means, depending on the type of engine. In a nine-cylinder, radial, single-row, air-cooled engine the primer lines go from the spider to:
(1) the top five cylinders
(2) cylinders number 1, 2, 3, 8, and 9
(3) cylinders 9, 8, 3, 2, and 1
(4) all three of the above answers are correct

168. The various grades of aviation fuel may be recognized by their color. Which of the following is most appropriate regarding the color of aviation fuels?
(1) 80 grade is red and 91/98 grade is blue
(2) 100/130 grade is green and 108/135 grade is brown
(3) 115/145 is purple and turbine fuels are colorless
(4) all three of the above answers are correct

169. A particular engine specification calls for the use of 100/130 fuel. This indicates that:
(1) at rich setting the octane rating is 100
(2) at lean setting the performance number is 130
(3) this engine will develop 130% more power at rich setting
(4) this engine will develop 1.3 times as much power on this fuel under rich mixture conditions (130) as it will on a fuel having a rich performance number of 100

170. When starting an aircraft engine, the carburetor air heat control should be in which of the following positions?
(1) ON
(2) hot position
(3) cold position
(4) midway between hot and cold position

171. The best material for cleaning turbine engine compressor blades is:
(1) a very small sand, almost a dust
(2) a fine sawdust produced from soft woods
(3) a very fine shot of soft lead
(4) walnut shells

172. Fretting corrosion is a surface phenomenon characterized by surface stains, corrosion, pitting, and the generation of oxides. Which of the following is correct regarding fretting corrosion:

(1) it may occur when repeated relative motion of small amplitude is allowed to take place between closely fitting components

(2) certain aircraft parts have been known to fail because of fretting corrosion

(3) where evidence of fretting corrosion is found, the affected parts should be replaced

(4) all of the above answers are correct

173. Avoid damage to aircraft through the use of harmful cleaning, polishing, brightening, or paint-removing materials. Use only those compounds which conform to existing government or established aircraft industry specifications. Always observe the product manufacturer's recommendations concerning use of his agent. The effect of caustic cleaning products or products with "strong" cleaning agents will:

(1) generally have no detrimental effects on aircraft metals providing a mild cleaning pressure is used

(2) tend to destroy the aluminum alloy external surfaces of aluminum clad materials and expose the pure aluminum core to corrosion and deterioration

(3) tend to destroy the pure aluminum coating on the external surfaces of aluminum clad materials and expose the aluminum alloy core to corrosion and deterioration

(4) tend to destroy the pure aluminum coating on the external surfaces of aluminum clad materials. However, the internal aluminum alloy will resist any corrosion tendencies

174. When using chemical cleaners on assembled aircraft the mechanic should:

(1) use them with great care and caution

(2) avoid the danger of entrapping corrosive materials in fraying surfaces and cracks or crevices

(3) use materials which are relatively neutral and easy to remove

(4) all three of the above answers are correct

175. When corrosion is detected, remove it as quickly as possible and protect the surface from further corrosion. Regarding acceptable cleanup procedures and corrosion resistant practices, which of the following is correct?

(1) for steel parts, except for highly stressed steel or stainless steel surfaces, the use of abrasive papers, buffers, hand wire brushing, and steel wool are acceptable cleanup procedures

(2) for aluminum, use steel wool, emery cloth, steel wire brushes, and, in some instances, severe abrasive materials

(3) for tube interiors, coat the interiors by flushing with hot mineral oil, skydrol, or some other suitable corrosion inhibitor

(4) for commutators and generator brushes, use emery cloth or other similar metallic type material to remove harmful surface deposits and/or corrosion

176. Aircraft tire types are referenced in Technical Standard Order TSO-C62. For maximum safety it is essential that tires be inspected frequently for cuts, worn spots, bulges on the sidewalls, and foreign bodies in the treads. Hydraulic fluid on tires should be cleaned off with a mild soap and water. Tires should be stored inside under which of the following conditions?
(1) temperature about 90°F. and relative humidity 85%
(2) relatively moist conditions with a nominal temperature
(3) in a relatively cool, dry area
(4) none of the above answers is correct

Before beginning this part of the examination it may be helpful to review some basic mathematical definitions, formulas, etc.

Feet per second (FPS) = MPH × 1.47 (a constant) or FPS = knots × 1.69 (a constant).

1 horsepower = 746 watts, 33,000 foot pounds of work per minute or 550 foot pounds per second.

1 gallon = 231 cubic inches.

$$\text{Volume of tanks in U.S. gallons} = \frac{\text{length (in.)} \times \text{width (in.)} \times \text{height (in.)}}{231 \text{ cubic inches}}$$

Speed of sound = 1,120 feet per second or 762 MPH or 663 knots at sea level or MSL.

Mach speed = ratio or percentage of the TAS to the speed of sound.

Mach number = a decimal number that represents the TAS relationship to the speed of sound; e.g., if TAS is 82% of the speed of sound, the Mach number is .82 or the aircraft's speed is Mach .82.

Machmeter = instrument that indicates Mach number.

Fuel-air ratio = for a chemically complete combustion in reciprocating engines, the ratio is one unit of fuel to 15 units of air (15 to 1 or .067).

Piston displacement (PD) = 3.1416 (pi) × radius^2 × stroke (of piston).

Engine displacement = (PD) × number of cylinders.

Combustion chamber volume (CCV) = total cylinder volume minus PD.

$$\text{Compression ratio} = \frac{PD + CCV}{CCV}.$$

$$\text{Wing loading} = \frac{\text{gross weight}}{\text{wing area}}.$$

$$\text{Power loading} = \frac{\text{gross weight}}{\text{horsepower}}.$$

Moment = weight (lbs.) × arm (distance from datum in inches).

Converting a fraction to a decimal equivalent: divide the lower (whole) number into the upper number (the part) and carry to four decimals.

To find the cube of a number (example 10^3) = multiply the number by itself twice. (1,000).

To find the square of a number (example 10^2) = multiply the number by itself. (100).

Work accomplished in foot-pounds: multiply weight × distance moved (W = lbs. × distance).

Area of a rectangle = length × width.

Area of a square = length of one side multiplied by itself.

Area of a circle = 3.1416 (pi) × radius².

Volume of a rectangular container = length × width × height.

Volume of a cube container = length of one side times itself twice.

Surface area of an airfoil (wing, etc.) = length × chord (average distance between the leading edge and trailing edge from the root to the tip in the case of a tapered wing).

To find the percent of one number to another, divide the larger number into the smaller number. Carry the answer to four decimals and them move the decimal point two digits to the right and add a percent (%) sign. This will give the answer to the nearest one-hundreth of a percent, a common practice in aircraft calculations.

177. An object is moving at a speed of 22 MPH (statute). What is the object's speed in feet-per-second?
    (1) 35.72
    (2) 32.34
    (3) 30.87
    (4) 29.98

178. An aircraft is cruising at 505 knots. What is the aircraft's speed in feet-per-second?
    (1) 827.75
    (2) 875.21
    (3) 853.45
    (4) 897.22

179. A generator (24 volts) is delivering 3,000 watts to aircraft units and appliances, this generator is developing:
    (1) 120.17 amperes
    (2) 4.75 horsepower
    (3) 131.27 amperes
    (4) 4.02 horsepower

180. A machine has moved a 10,000-pound object through 10 feet of space in one minute. How much horsepower has the machine developed to accomplish this job?
    (1) 3.03
    (2) 4.03
    (3) 2.03
    (4) 5.27

181. The volume of a fuel tank is 6,237 cubic inches. How many gallons of fuel will it hold if there is no expansion space allowed?
    (1) 29.9
    (2) 27.0
    (3) 25.0
    (4) 30.7

182. The volume of a fuel tank that measures 30 inches × 16 inches × 8 inches is:
    (1) 4,040 cubic inches
    (2) 3,725 cubic inches
    (3) 3,840 cubic inches
    (4) 3,792 cubic inches

183. The speed of sound on a particular flight is 663 knots and the Mach speed is .82 at the flight level. Which of the following is correct under these circumstances?
    (1) TAS is 543.66 knots
    (2) the TAS is less than the speed of sound
    (3) the flight of the aircraft is not exceeding the speed of sound
    (4) all three of the above answers are correct

184. Each cylinder of an 18-cylinder twin-row radial air-cooled engine has a bore of six inches, a stroke of six inches, and a total volume (from inside top of cylinder to piston at bottom center) of 197.92 cubic inches. According to this information, the piston displacement (PD) of each cylinder is:
    (1) 159.62 cubic inches
    (2) 179.72 cubic inches
    (3) 169.65 cubic inches
    (4) 153.67 cubic inches

185. In the preceding question, the combustion chamber volume (CCV) is.
    (1) 26.75 cubic inches
    (2) 28.30 cubic inches
    (3) 31.75 cubic inches
    (4) 27.89 cubic inches

186. From the information in the preceding two questions, the compression ratio of each cylinder and the total piston displacement (engine displacement) of the 18 cylinders are respectively:
    (1) 6.99 to 1 and 3053.70 cubic inches
    (2) 7.57 to 1 and 3125.19 cubic inches
    (3) 5.47 to 1 and 2925.27 cubic inches
    (4) 7.97 to 1 and 3217.27 cubic inches

187. An airplane has a gross weight of 10,750 pounds and incorporates two reciprocating engines of 450 horsepower each. The power loading of this airplane is:
    (1) 9.97 pounds per horsepower
    (2) 13.22 pounds per horsepower
    (3) 14.77 pounds per horsepower
    (4) 11.94 pounds per horsepower

188. An airplane has a gross weight of 5,500 pounds and has a wing (cantilever type) that has a span of 36 feet and a chord (average) of 5 feet. The wing loading of this airplane is:
    (1) 30.56 pounds per square foot
    (2) 31.97 pounds per square foot
    (3) 29.27 pounds per square foot
    (4) 27.98 pounds per square foot

189. The square of the numeral 10 (10^2) is:
    (1) 10
    (2) 100
    (3) 1,000
    (4) 10.2

190. The cube of the numeral 10 (10^3) is:
    (1) 10
    (2) 10.4
    (3) 100
    (4) 1,000

191. An object that weighs 320 pounds has been moved a distance of three feet. How much work (in foot-pounds) has been accomplished?
    (1) 860
    (2) 1,060
    (3) 960
    (4) 860

192. A rectangle is measured and found to have a length of 43 inches and a width of 37 inches. The area (in square inches) is:

(1) 1,511
(2) 1,205
(3) 1,591
(4) none of the above answers is correct

193. One side of a square item is measured and found to be 57 inches. The area (in square inches) is:
    (1) 3,249
    (2) 3,149
    (3) 3,349
    (4) 3,049

194. A circle has a diameter of 9 inches. The area of this circle (in square inches) is:
    (1) 59.92
    (2) 63.62
    (3) 65.72
    (4) 57.82

195. A rectangular shipping container is 7'6" long, 3'6" wide, and 1'6" deep. The volume of this container (in square feet) is:
    (1) 36.75
    (2) 41.79
    (3) 39.38
    (4) 43.89

196. A cube shaped container has a measurement of 11'0" along one side (end to end). The volume of this container (in cubic feet) is:
    (1) 1,441
    (2) 1,231
    (3) 1,379
    (4) 1,331

197. The length of a wing (wing tip to wing tip) is 73 feet and the average chord (leading edge to trailing edge) is 6'6". The area of this wing (in square feet) is:
    (1) 457.5
    (2) 482.8
    (3) 474.5
    (4) 463.2

198. The value 17 is what percent of 23?
    (1) 71.25%
    (2) 73.91%
    (3) 75.86%
    (4) 74.11%

199. 1721×15 equals?
    (1) 23,476
    (2) 26,775
    (3) 25,815
    (4) 24,377

200. 1,579÷25×7 equals?
    (1) 442.12
    (2) 463.71
    (3) 435.23
    (4) 45175

201. ⅞×¼×7 equals?
    (1) 1 13/32
    (2) 1 17/32
    (3) 1 23/32
    (4) 1 27/32

202. .278÷.0556 equals?
    (1) 5.1890
    (2) 5.2768
    (3) 5.0000
    (4) 5.3297

203. 1,572−747×9×½ equals?
    (1) 3,712.5
    (2) 3,721.5
    (3) 3,707.5
    (4) 3,732.5

204. ⅞÷⅜×10 equals?
    (1) 25⅓
    (2) 24⅔
    (3) 26⅓
    (4) 23⅓

205. The square and cube of the value 9 are, respectively:
    (1) 81 and 729
    (2) 81 and 6,561
    (3) 729 and 81
    (4) 81 and 1,458

206. An aircraft mechanic should be able to convert decimal equivalents to fractions and vice versa. The decimal equivalent of 13/64 is:
    (1) .2137
    (2) .2264

    (3) .2031
    (4) .1901

207. It is relatively easy to convert fractions to decimals and vice versa without reference to a decimal equivalent chart. Example: 13/64 can be converted to its decimal equivalent by dividing 64 into 13 (.203125). Thus, .203125 can be converted back to a fraction by dividing .203125 by .015625 (the decimal equivalent of 1/64 inch). The constant .015625 is used to convert any decimal to a fraction. After finding the number of 64ths a decimal is equal to, the fraction can then be reduced to the lowest possible fraction value, i.e. 48/64 can be reduced to 24/32 then to 12/16 then to 6/8 then to ¾. What is the decimal equivalent of 7/16 and the fraction equivalent of .828125, respectively?
    (1) .46875 and 57/64
    (2) .37500 and 53/64
    (3) .4375 and 49/64
    (4) .4375 and 53/64

208. You know the following about a particular aircraft: Datum is at the leading edge of the wing; actual measured horizontal distance from the main wheel weight point (center line of the main wheel) to the datum is 37.5 inches; actual measured horizontal distance from the front wheel weighing point (center line of nose wheel) to the main wheel weighing point is 76.5 inches; the right main wheel scale reading is 647 lbs. net (tare 7 lbs.); the left main wheel scale reading is 657 lbs. net (tare 7 lbs.); and the nose wheel scale reading is 501 lbs. net (tare 11 lbs.). The empty weight CG of this nose wheel type airplane is located aft of the datum approximately:
    (1) 16.74 inches
    (2) 16.44 inches
    (3) 19.75 inches
    (4) 20.74 inches

209. Each person performing an annual or 100-hour inspection on an aircraft shall use a checklist while performing the inspection. The checklist may:
    (1) be of a person's own design
    (2) one provided by the manufacturer of the equipment being inspected
    (3) be one obtained from another source than noted in answer 1 and 2
    (4) all three of the above answers are correct

210. The holder of a valid pilot certificate may perform:
    (1) minor repairs and alterations providing the minor repairs and alterations have no effect on the flight characteristics of the aircraft

(2) preventive maintenance on any aircraft owned or operated by him that is used in air carrier service

(3) preventive maintenance on any aircraft owned or operated by him that is not used in air carrier service

(4) preventive maintenance on any aircraft owned or operated by him providing he holds a valid FAA mechanic certificate with appropriate ratings.

211. A person qualified to perform a major repair or alteration, other than a certificated repair station, must:
(1) execute FAA Form 337 at least in duplicate
(2) give a signed copy of executed Form 337 to the aircraft owner
(3) forward a copy of Form 337 to the local FAA District Office within 48 hours of approval to return the subject aircraft or item to service
(4) all three of the above answers are correct

212. With respect to maintenance records "time in service" means:
(1) the time from the moment an aircraft leaves the surface of the earth until it touches it at the next point of landing
(2) the time from the moment an aircraft moves for the purpose of takeoff until it comes to rest at the next point of landing
(3) the time from the moment the aircraft first moves under its own power for the purpose of flight until the moment it comes to rest at the next point of landing (block to block)
(4) time recorded from the moment the engine reaches a given RPM to the time the engine is shut down

213. The primary responsibility for maintaining the airframe and engine logbooks is the:
(1) Authorized Inspector (AI) of the repair and alteration after completion of all FAA Forms, etc.
(2) registered owner for general aviation aircraft and the certificate holder for air carrier aircraft
(3) certificated mechanic that performs the repairs, alterations, or inspections
(4) repair station that performs the repairs, alterations, or inspections

214. After a repair, alteration, or inspection, an aircraft requires a flight test. The results of the flight test are recorded by the:
(1) pilot that conducted the flight test on an FAA Form 337
(2) supervising mechanic on a Form 337
(3) pilot that conducted the flight test in the aircraft records
(4) responsible mechanic in the aircraft records

215. Airworthiness Directives (AD Notes) are issued by the FAA to call attention to repairs, replacements, corrections, etc., on pertinent makes and models of aircraft. The time of compliance is indicated in each AD Note and they are considered to be:
    (1) mandatory and must be complied with by the registered owner of the particular aircraft involved
    (2) not mandatory but advisable and should be complied with by the owner of the particular aircraft involved
    (3) mandatory and must be complied with by at least the next 100 hour inspection of the aircraft involved
    (4) mandatory only if the correction indicated on the AD Note involves a flight characteristic of the particular aircraft involved

216. Aircraft electrical adjustment, overhaul, repair, and/or testing should be done in accordance with:
    (1) FAR's in general
    (2) FCC rules and regulations
    (3) instructions or maintenance manuals that are published for this purpose by the aircraft or equipment manufacturer
    (4) according to standard repair shop procedures

217. Entries in aircraft logbooks should include which of the following?
    (1) color and empty weight of the aircraft involved
    (2) that all outstanding AD Notes have been complied with
    (3) a notation that the center of gravity has been effected by any change in equipment or appliances
    (4) a notation that all FCC rules have been complied with to date

218. Maintenance of aircraft, according to the regulations, means:
    (1) inspection, overhaul, and repair
    (2) preservation and the replacement of parts
    (3) excludes preventive maintenance
    (4) all three of the above answers are correct

219. A major repair, according to the regulations, is a repair:
    (1) that, if improperly done, might appreciably affect weight and balance, structural strength, performance, powerplant operation, flight characteristics, or other qualities affecting airworthiness
    (2) that is not done according to accepted practices or cannot be done by elementary operations
    (3) other than a minor repair
    (4) all three of the above answers are correct

220. A 100-hour inspection on an aircraft has been performed and approved by a certificated mechanic with both the airframe and powerplant ratings. Prior to returning the aircraft to service he must:

   (1) execute two copies of Form 337 and present the original to the aircraft owner and a copy or duplicate to the local FAA office

   (2) obtain an approval from the local Authorized Inspector

   (3) notify the local FAA office of his intention to return the aircraft to service

   (4) make all the proper entries in the appropriate log books

221. Which of the following best describes the requirements for a permanent record of aircraft maintenance?
   (1) pertinent data that is entered in the owners manual of the particular aircraft
   (2) the data found in the aircraft operating limitations record
   (3) the registered owner of an aircraft is responsible for keeping maintenance record information in aircraft (airframe) and powerplant (engine) log books
   (4) the registered owner of an aircraft is responsible for keeping maintenance record information in appropriate aircraft forms and schedules

222. There are 6,080 feet in one nautical mile and 5,280 feet in one statute mile. An object is moving at approximately 13 knots or 15 MPH. How many feet will the object travel in one second or what will its speed in feet-per-second be?
   (1) 22
   (2) 37
   (3) 11
   (4) 67
   Solution: Multiply the speed in knots by the constant 1.69 or multiply the speed in MPH by the constant 1.47.

223. The loss of magnetism from a permanent magnet can occur when:
   (1) the magnet is subject to rough handling
   (2) carelessness permits the magnet to be dropped on a hard surface
   (3) the magnet is subject to excessive heat
   (4) all three of the above answers are correct

224. Reluctance is known as the resistance to flow of magnetism. Generally speaking, which of the following properties does a material have if it offers low resistance to magnetic flux?
   (1) permeance
   (2) impedance
   (3) capacitance
   (4) commutation

225. Methods of heat transfer are:
   (1) conduction

(2) radiation
(3) convection
(4) all three of the above answers are correct

226. The relationship between pressure and velocity in a venturi or over the surface of an airfoil is which of the following?
(1) the relationship of pressure and velocity is proportional
(2) the velocity decreases as the square of the pressure
(3) pressure is inversely proportional to the velocity
(4) no direct relationship can be determined between pressure and velocity in a venturi or over the surface of an airfoil

227. Which of the following represents the three states of matter?
(1) triangular, round, and square
(2) organic, solid, and mineral
(3) cast, extruded, and rolled
(4) solid, liquid, and gas

228. The vaporization rate of fuel is an important factor in the operation of aircraft engines. Which of the following items has very little or no effect on the vaporization rate of aircraft type gasoline?
(1) volatility of the particular gasoline
(2) anti-knock value of the particular gasoline
(3) the pressure of the surrounding air adjacent to the particular gasoline
(4) temperature of the particular gasoline

229. In the study of gases, which of the following best describes Boyle's Law?
(1) the volume of a given mass of gas varies inversely as the pressure, provided the temperature remains constant
(2) the volume of a confined body of gas will vary directly with the pressure
(3) the temperature of a gas will vary inversly with the pressure
(4) when the temperature of a confined gas increases, the pressure will decrease

230. While servicing a bottle that presently has 10 PSI you increase the pressure to 25 PSI with no temperature change. The volume of the bottle:
(1) will be 4 times as much
(2) will remain the same regardless of the increase in PSI
(3) reduces by 2½ times
(4) increases by 2½ times

231. The center of gravity (CG) of an aircraft is:
(1) at the location of the point from which all measurements are taken

(2) a point where a maximum mass weight is concentrated

(3) taken from the point at which the scales are located under the main wheels

(4) a location or point in the aircraft from which, if the aircraft were suspended in mid-air, the aircraft would be perfectly balanced

232. The relationship between temperature and heat energy is best described as:

(1) temperature is the thermal state of an object whereas heat energy is measured by the effect it produces on materials

(2) temperature is an indication of the heat energy of a body whereas heat energy indicates the relative hotness of a material

(3) heat energy is considered to be a rate of temperature

(4) temperature is transferred in the same manner as heat

233. An object is moving in a circular path. After a period of time all external forces are removed from the object. The object will then move in a:

(1) spiraling circular path

(2) descending circular path

(3) straight path

(4) horizontal spiraling path

234. Altimeters operate on the difference of air pressures that vary with altitude. The pressure difference creates a reaction on an internal bellows better known as an aneroid. If the air pressure is relatively high, such as at sea level, the altimeter will read a low altitude indication. When the air pressure is relatively low, such as at the higher altitudes, the altimeter will read a high altitude indication. If not corrected to the existing atmospheric pressure, which of the following will cause the altimeter to read the greatest "too high" reading?

(1) low atmospheric temperature with a high relative humidity

(2) high atmospheric temperature with a high relative humidity

(3) low atmospheric temperature with a low relative humidity

(4) high atmospheric temperature with a low relative humidity

235. When taking off in an airplane, which of the following atmospheric conditions will require the longest runway?

(1) high temperature with a low relative humidity

(2) low temperature with a low relative humidity

(3) high temperature with a high relative humidity

(4) low temperature with a high relative humidity

236. Which of the following is most appropriate to the interpretation of Newton's Laws of Motion?

(1) if a body is in motion, it will remain in motion until acted upon by some force

    (2) if a body is at rest, it will remain at rest until acted upon by some force

    (3) to every action there must be an equal and opposite reaction

    (4) all three of the above answers are correct

237. A table weighing 120 pounds has been raised three feet. The amount of work accomplished (in foot pounds) is:
    (1) 360
    (2) 120
    (3) 3
    (4) none

238. Which of the following combinations would be considered the lightest air?
    (1) 10% wet, 90% dry
    (2) 25% wet, 75% dry
    (3) 40% wet, 60% dry
    (4) 50% wet, 50% dry

239. Of the following combinations, which will be the *least* dense air?
    (1) 90% wet, 10% dry
    (2) 50% wet, 50% dry
    (3) 25% wet, 75% dry
    (4) 10% wet, 90% dry

240. What force will act on a three-square-inch piston if the pressure on an associated one-square-inch piston is 1,000 pounds per square inch (PSI)?
    (1) three times as much on the larger piston as compared to the smaller piston
    (2) 3,000 pounds
    (3) three times as much since the pressure on the small piston will be transferred to the larger piston in direct proportion to the increase in the size of the larger piston
    (4) all of the above answers are correct

241. A hydraulic system utilizes a piston with an area of one square inch. This piston is operating in conjunction with a cylinder that has a piston area of 10 square inches. A force of 20 lbs. has been exerted on the smaller piston. The pressure exerted on the *larger* piston will be:
    (1) 20 PSI
    (2) 200 PSI
    (3) 20 lbs.
    (4) 200 lbs.

242. The overhaul and repair specifications and procedures for different types and models of aircraft engines are quite varied. For this type of information you should refer to:
   (1) current engine maintenance and overhaul manuals
   (2) service bulletins
   (3) instructions published by the manufacturer of the engine to be worked on
   (4) all three of the above answers are correct

243. In order to find placarding data for a certain type of aircraft you should refer to which of the following?
   (1) the aircraft log book for the particular aircraft
   (2) the publication titled AC 43.13–1
   (3) either the aircraft certification or type certificate
   (4) the FAA Aircraft Specifications

244. FAA Advisory Circulars are issued to inform the aviation public, in a systematic way, of items and materials of interest. AC's have a numbering system that corresponds to the numbering system used for Federal Aviation Regulations. Example: AC based on material covered in FAR Part 43 will be identified by AC 43 followed by a pertinent number such as AC 43.13-1, AC 43.13-2, etc. From a compliance standpoint, AC's are:
   (1) mandatory and must be complied with by the date indicated in the particular AC
   (2) nonregulatory and act in the capacity of informing the aviation public of important facts that should be studied and used where appropriate
   (3) mandatory and must be executed in commercial aviation repair and maintenance
   (4) not mandatory in commercial aviation but must be adhered to in general aviation repair and maintenance

245. The location of the datum and the point of leveling an airplane for the purpose of weight and balance computations can generally be found in the:
   (1) aircraft log book of the particular airplane
   (2) Aircraft Circular that is published for the particular airplane
   (3) Type Certificate Data Sheets for the particular airplane
   (4) last issued Form 337 for the particular airplane

246. Color markings on an airspeed indicator are as follows: (a) green arc indicates normal operating speeds, (b) white arc indicates flap operating speeds, (c) yellow arc indicates caution speeds, (d) the red line indicates the never-exceed (V_{NE}) speed, and (e) a white line across the edge of

the instrument case and onto the glass cover in instances where the colored arcs and lines have been painted on the glass; this line will indicate slippage should the glass become loose and rotate from its original position. In general, engine instruments have color markings that indicate which of the following?
(1) red for normal operating and yellow for caution
(2) green for normal operating limits and white for maximum limits
(3) red for maximum and minimum operating limits and orange for normal operating limits
(4) green for normal operating limits and red for maximum and minimum operating limits

247. The FAA requires that certain procedures be followed when applying for a Type Certificate for aircraft, aircraft engines, or propellers. Which of the following is appropriate to the issuance of a Type Certificate?
(1) any person may apply for a Type Certificate
(2) an application for a Type Certificate is made on a form and in a manner prescribed by the Administrator and is submitted to the appropriate FAA regional office
(3) an application for an aircraft Type Certificate must be accompanied by a three-view drawing of that aircraft and available preliminary basic data
(4) all three of the above answers are correct

248. Supplemental Type Certificates are issued by the Administrator under which of the following conditions?
(1) when a major repair has been completed to the satisfaction of the Administrator
(2) when a major alteration has been completed to the satisfaction of the FAA
(3) when the alteration of a product (aircraft, engine, propeller, etc.) introduces a major change in type design that is not great enough to require a new application for a Type Certificate
(4) only after a product that holds a Type Certificate has been completely scrapped and a new like product is being contemplated

249. Engine/propeller speed ratios are determined by which of the following basic reasons?
(1) so that the particular propeller will absorb about 75% of the total engine horsepower
(2) the determination is based on the premise that the propeller tips will slightly exceed the speed of sound at maximum engine RPM
(3) by a strict determination of the weight ratio between the engine and the propeller to be installed on the engine
(4) that the propeller will absorb the required horsepower from the

engine and that the propeller tips at no time will exceed the speed of sound

250. Service manuals for aircraft, engines, propellers, and aircraft appliances are generally available from the:
(1) local FAA district office
(2) Regional offices of the FAA
(3) master files of the Department of Transportation
(4) manufacturer of the particular item

251. Which of the following best describes the checklist that is used when making 100 hour and annual inspections of aircraft?
(1) standard checklist forms are available from the Administrator
(2) since the type of aircraft to be inspected varies considerably in the requirements and type of appliances, units, etc., it is the responsibility of the mechanic and/or inspector to make up his own checklist or obtain suitable checklists from the manufacturer of the aircraft to be inspected
(3) proper checklist forms must be obtained from the local FAA district office as they must all contain precisely the same list of items to be inspected
(4) none of the above answers is correct

252. Which of the following AC's contains methods, techniques, and practices acceptable to the Administrator for inspection, repair, and alterations to civil aircraft?
(1) 61-18A and 61-18B
(2) 63-1A and 63-1B
(3) 43.13-1 and 43.13-2
(4) 65.13-1 and 65.13-2

253. An aircraft part manufactured under a Technical Standard Order (TSO):
(1) indicates that it is guaranteed to be accurate by the Administrator
(2) indicates that it may be used on any type of aircraft
(3) may be used only on aircraft manufactured prior to the date the TSO was issued
(4) assures that the part is manufactured to standard specifications

254. FAA Airworthiness Directives (AD's) are considered to be mandatory until:
(1) the manufacturer involved cancels them
(2) they are cancelled
(3) they appear in the AD summary
(4) they become 6 months old

255. To find information about aircraft of limited production you should check the:
    (1) TSO's
    (2) supplemental type certificates
    (3) AD's
    (4) FAA aircraft listings

256. Which of the following should be referred to for the criteria for determining the classification (major, minor, or preventive maintenance) of airframe repairs and alterations?
    (1) FAR Part 65
    (2) FAR Part 21
    (3) FAR Part 43
    (4) FAR Part 97

257. To be eligible for a mechanic certificate and associated ratings, a person must:
    (1) be at least 21 years of age
    (2) previously hold at least an FAA Repairman Certificate
    (3) be at least 18 years of age
    (4) also be qualified for the issuance of an Inspection Authorization

258. A certificated mechanic may not exercise the privileges of his certificate and rating unless, within the preceding 24 months, the Administrator has found that he is able to do that work or he (the certificated mechanic) has, for at least 6 months:
    (1) served as a mechanic under his certificate and rating or technically supervised other mechanics
    (2) supervised, in an executive capacity, the maintenance or alteration of aircraft
    (3) been engaged in any combination of answers (1) and (2)
    (4) all three of the above answers are correct

259. After any change in his permanent mailing address, the holder of a mechanic certificate shall notify the FAA Airman Certification Branch of his new mailing address, in writing, within:
    (1) 90 days
    (2) 60 days
    (3) 30 days
    (4) 7 days

260. A temporary mechanic certificate and ratings are valid for a period of not more than:
    (1) 3 months and 15 days
    (2) 6 months

(3) 60 days
(4) 90 days

261. A permanent mechanic certificate and ratings is effective until:
(1) an age in years has been reached as designated by the Administrator
(2) it is surrendered, suspended, or revoked
(3) it is surrendered, suspended, or lost
(4) it is surrendered, suspended, or transferred

262. There are specific knowledge, skill, and experience requirements for the certification of persons as licensed airframe and/or powerplant mechanics. In which of the following would these requirements be found?
(1) FAR Part 65
(2) FAR Part 91
(3) FAR Part 39
(4) AC 43.13-1

263. An aircraft registered in the U.S. that is used for hire requires which of the following?
(1) an annual inspection conducted by a certificated mechanic who holds both the airframe and powerplant ratings
(2) an inspection after each 100 hours of "time in service" conducted by a certificated mechanic who holds both airframe and powerplant ratings
(3) an inspection after each 100 hours of "time in service" conducted by a certificated mechanic who holds both airframe and powerplant ratings with a current repairman certificate
(4) both the 100-hour and annual inspection conducted by an approved repair shop in which all maintenance personnel are FAA certificated

264. Publications known as type specifications are now described as:
(1) the aircraft operating manual
(2) AD Notes
(3) AC's
(4) Type Certificate Data Sheets

265. While doing maintenance work on a certain type aircraft you want to check the placarding data. You will find this information in:
(1) the TSO's
(2) the publications titled Inspection Aids
(3) the FAA aircraft specifications
(4) the aircraft's maintenance log books

266. Each person who holds a mechanic certificate shall keep it within the immediate area where he normally exercises the privileges of the certificate and shall present it for inspection upon the request of:
   (1) the Administrator or an authorized representative of the National Transportation Safety Board (NTSB)
   (2) any federal law enforcement officer
   (3) any state or local law enforcement officer
   (4) all three of the above answers are correct

267. In the case of antique aircraft where there are only a few in existence, you would find technical data for these aircraft in the:
   (1) pertinent AD Notes
   (2) FAA aircraft listings
   (3) pertinent Type Certificate Data Sheets
   (4) certification schedules for antique aircraft

268. Which of the following is a regulation that applies to all aircraft registered in the United States?
   (1) a fire detection system must be installed and be of adequate capability
   (2) a fire extinquishing system (built-in) must be incorporated in the aircraft and be of adequate capability
   (3) there must be a fuel shutoff valve or valves that will completely shut off the fuel to the engine or engines
   (4) there must be an FAA approved maintenance manual on board each aircraft

269. A certificated mechanic with both the airframe and powerplant ratings is also FAA certificated as a private pilot and is rated for a particular aircraft that has not had a required inspection for 15 months. The owner of the aircraft has made a request to have the aircraft flown to an airport 57 miles distant so that all necessary inspections, work, etc., can be done and the airplane returned to service. Under these circumstances:
   (1) the certificated mechanic can fly the aircraft to the designated airport providing a special permit (ferry permit) is obtained from the local FAA district office for this flight and no passengers are aboard
   (2) the certificated mechanic cannot fly the aircraft to the designated airport under any circumstances as FAR's do not permit this type of operation
   (3) the certificated mechanic can fly the aircraft to the designated airport under the provisions outlined in AC 43.13-1
   (4) none of the above answers is correct

270. A particular airplane has had an annual inspection and circumstances require that it be test flown prior to its return to service. Under these

circumstances:
(1) a certificated mechanic who holds at least a valid private pilot license and is rated for the particular aircraft can make the test flight
(2) any person who holds a valid private pilot license and is rated for the particular aircraft can make the test flight
(3) the results of the flight test must be entered by the pilot in the pertinent aircraft records
(4) all three of the above answers are correct

271. The performance standards in TSO's are those that the Administrator finds necessary to ensure that the article concerned will operate satisfactorily or will accomplish satisfactorily its intended purpose under specified conditions. Examples of TSO's are: (a) smoke detectors—TSO-C1, (b) airspeed indicator (pitot—static)—TSO-C2, (c) aircraft fabric, grade A—TSO-C15. Performance standards for safety belts can be found under:
(1) TSO-C21
(2) TSO-C23
(3) TSO-C22
(4) TSO-C24

272. A certified inspector (a person holding an FAA Inspection Authorization) has a certificated mechanic working under his supervision. The certificated mechanic, under these circumstances, may not do which of the following?
(1) cover the wing of an aircraft
(2) gas weld a wing strut brace
(3) make a repair on a tubular engine mount by the process of riveting
(4) replace a panel on a fabric wing

273. An aircraft must have a 4-bladed electric propeller removed and a similar type replaced. Who of the following could, according to regulations, remove and replace this type of propeller?
(1) a certificated mechanic who holds an airframe rating
(2) a certificated mechanic who holds a powerplant rating
(3) only a certificated mechanic who also holds an inspection authorization
(4) any mechanic that holds a current and effective repairman certificate

274. The description and technical data of a propeller will be found in the:
(1) appropriate AD's
(2) TSO's
(3) particular propeller manufacturer's specification sheets
(4) an approved owners manual

275. The removal of a control surface and its replacement by one of an identical type is considered to be a:
    (1) function of preventive maintenance
    (2) minor repair
    (3) major repair
    (4) progressive inspection function

276. A major airframe alteration consists of major changes to the basic design or external configuration of any structural component such as wings, tail surfaces, etc. Airframe major repairs are all repairs involving the strengthening, reinforcing, splicing, and manufacturing of primary structural members, or their replacement when replacement is by fabrication such as riveting or welding. Which of the following is correct regarding airframe *minor* repairs?
    (1) they are repairs to non-structural members which may affect the airworthiness of an aircraft
    (2) includes patching and repairing of leaks in non-integral fuel, oil, hydraulic, and de-icer fluid tanks
    (3) includes replacement of components or complete units with parts supplied by the original manufacturer or manufactured in accordance with approved drawings
    (4) all three of the above answers are correct

277. You are a certificated mechanic. According to FAR's you should not exercise the privileges of your certificate unless, within the preceding 24 months you have served as a mechanic under the terms of this certificate for at least:
    (1) 10 months, having been engaged in technical or executive supervision of aircraft maintenance, or any combination of these activities
    (2) 18 months as a line mechanic, engine overhaul specialist, airframe maintenance specialist, or any combination of these mechanical duties
    (3) 6 months, or have been engaged in technical or executive supervision of aircraft maintenance, or any combination thereof
    (4) enough months to assure the FAA you have retained your proficiency as a certified mechanic and that you are current on AD's etc.

2. THE AIRFRAME STRUCTURES EXAMINATION

The subject matter covered in the airframe structures examination is: (1) wood structures, (2) aircraft covering, (3) aircraft finishes, (4) sheet metal structures, (5) welding, (6) assembly and rigging, and (7) airframe inspection. The questions that follow, like those in the Government tests, cover these subjects but not necessarily in the same order.

The answers to the questions in Section 2 are on page 369.

1. Minimizing the shrinking effects of wood can be accomplished by:
    (1) the use of bushings that are slightly short so that when the wood member shrinks the bushings do not protrude and the fittings may be tightened firmly against the member
    (2) the gradual dropping off of plywood face plates by feathering the ends
    (3) the gradual dropping off of plywood face plates by spading the ends
    (4) all three of the above answers are correct

2. Metal ribs are attached to wood spars by:
    (1) small metal screws
    (2) small nails
    (3) butt-type nails
    (4) small fasteners with a series of small pointed teeth

3. Gusset plates are small wood plates used in construction of wood ribs for the wing (form ribs). They are used as reinforcements at each connection of cross-members for the purpose of making the joint stronger and more rigid. They are basically glued on, but also nailed

287

to achieve a holding pressure until the glue sets. Caul blocks are:
(1) used to reinforce the gusset plates
(2) placed between the teeth of a clamp and the wood material being glued to distribute the pressure of the clamp along a wider area
(3) used to lessen the bending tendency of the metal when working with metals close to each other
(4) none of the above answers is correct

4. When repairing stressed plywood skin, the best method is to use a:
(1) scarf patch
(2) plug patch
(3) surface patch
(4) splayed patch

5. Conventional wing ribs which do not have permanent inter-rib bracing should be tied in position by means of cotton tape running parallel to the seams. You would:
(1) apply the tape bracing to both the top and bottom capstrips, maintained parallel to the plane of the cover rather than diagonally between the top and bottom capstrip
(2) apply the tape continuously with one turn around successive capstrips, arranged so that the tape between the ribs is separated from the cover by a distance equal to the depth of the capstrip
(3) tie the turn of tape around each capstrip by means of a short length of lacing cord
(4) accomplish all these steps

6. In a spar splice, one hole is oversize. You should:
(1) do nothing but re-drill the hole to correct size
(2) splice in a new section and re-drill the hole
(3) replace the entire spar to prevent repair weakness
(4) none of the above answers is correct

7. The glue that has preservatives to increase its resistance to moisture is:
(1) casein
(2) synthetic resin
(3) phenol-formaldehyde
(4) resorcinal-formaldehyde
Note: The suitable preservatives are chlorinated phenols and their sodium salts.

8. In the absence of other information, one part glue and two parts water (by weight) may be used in mixing glues for aircraft work. The water should be clean and at a temperature between 60° to 70°F. The glue

powder is dry-mixed and then weighed. Which of the following is also correct?
(1) the glue powder is placed in the mixing bowl, and while it's being stirred the water is added and then both mixed together
(2) both the glue powder and water are added to the mixing bowl at the same time and stirred vigorously
(3) the water is placed in the mixing bowl or cup and the glue powder is stirred in with the mixer running at medium speed, but not faster than 140 RPM.
(4) the water is placed in the mixing bowl or cup and the glue powder is stirred in with the mixer running at an extremely high speed to prevent lumping of the glue mixture

9. Gluing surfaces should be machined true by planes, joiners, or special miter saws. These surfaces should be true and ready for gluing. Under no circumstances should these surfaces be sandpapered. The maximum time that may elapse between final surfacing and gluing is:
(1) 8 hours
(2) 12 hours
(3) 16 hours
(4) 24 hours

10. You would recognize laminated wood by the grain of each layer:
(1) running 90° to each other
(2) being parallel to each other
(3) set at a 45° angle to each other
(4) set at a 25° angle to each other

11. A generally accepted method used to classify hardwoods and softwoods is:
(1) their resistance to crush pressures
(2) the density or weight of the bark
(3) the shape of the leaf of the tree
(4) positive scale readings of a Vickers or Rockwell hardness tester

12. In a wood spar, which of the following conditions would be acceptable?
(1) shakes and checks
(2) splits and minor shakes
(3) mineral lines, providing there is no accompanying decay
(4) compression stress and spike knots

13. The pitch of a saw blade is the number of teeth per linear inch. The set of a saw blade is:
(1) determined by the difference in the distance between each end of the saw blade
(2) determined by the length of the teeth

(3) dependent on the type metal the blade is made of

(4) determined by the manner in which the teeth are bent (or fabricated) outward from each side of the saw blade centerline

14. When working with wood which is to be glued, wetting tests are useful as a means of detecting the presence of wax. Drops of water placed on the surface of wood:
    (1) will soak in if the wood surface is waxed
    (2) will not roll off a curved section of wood if the surface is waxed
    (3) do not spread or wet the wood if there is no wax on the surface
    (4) do not spread or wet the wood if there is wax on the surface

15. The strength of a glued joint cannot be depended upon if the glued joint was assembled and cured at temperatures below:
    (1) 85°F.
    (2) 80°F.
    (3) 75°F.
    (4) 70°F.

16. Why is a certain minimum length required for the slope of a spar splice?
    (1) to remain within the 15 to 1 slope required for wood grain deviation
    (2) to allow for a large splice plate on either side of the splice
    (3) for the purpose of giving sufficient splice surface for gluing
    (4) it is necessary owing to the spar thicknesses that range from ¼ to ½ inch only

17. A plywood knife is used when making emergency repairs on aircraft wood structures. A plywood knife can be made by filing down a short length of a broken hack-saw blade and placing it securely within:
    (1) metal keepers
    (2) wood corners
    (3) a wooden handle
    (4) a soft copper handle

18. The permissible wood substitutes for use in making repairs to wood structures must be:
    (1) of a good weight-strength ratio
    (2) of the same type and quality as the original
    (3) as good or better than the original and approved by the Administrator
    (4) at least 90% as strong as the original and have no knots, streaks, or mineral impregnations

19. Which of the following is correct regarding wood characteristics?
    (1) excessive moisture causes wood to shrink
    (2) excessive dryness causes wood to swell or expand
    (3) kiln drying of wood is not an accepted practice
    (4) excessive moisture causes wood to swell or expand and excessive dryness causes wood to shrink

20. Solid, laminated, and plywood are three classifications of wood generally used in aircraft construction. Which of the following statements are correct regarding aircraft wood?
    (1) the grain of plywood sheets, prior to being glued together, run parallel to each other
    (2) in laminated wood, the grain of each layer run parallel to each other
    (3) the grain of plywood sheets, prior to being glued together, run at 45° to each other
    (4) mineral streaks in wood of aircraft quality are acceptable even though careful inspection reveals decay

21. A bulkhead, made out of laminated wood, has a buckling of the fibers that appear as streaks on the surface of the piece substantially at right angles to (across) the grain. This would indicate:
    (1) a compression failure
    (2) shear failure
    (3) fungus rotting
    (4) excessive decay

22. Oil or grease spots on plywood surfaces should be cleaned off by the use of:
    (1) plain soap and water
    (2) white gasoline
    (3) naptha
    (4) toluene thinner

23. Wood parts or units should be treated with clear varnish:
    (1) for preservation
    (2) to resist moisture penetration into the wood
    (3) so that possible damage to the wood can be easily seen or detected
    (4) all three of the above answers are correct

24. In a wood spar, a longitudinal crack not near a main fitting:
    (1) cannot be repaired
    (2) can be repaired by gluing on reinforcement plates
    (3) can be repaired only after obtaining an authorization from the Administrator
    (4) can be repaired only by splicing in a new spar section

25. Wood propellers are generally made from which of the following types of wood?
(1) spruce laminated layers
(2) white ash layers that have the laminated sheets at 90° to each other
(3) laminated birch
(4) laminated oak

26. On small joints, such as found in wood ribs, the pressure is usually applied by nailing the joint gusset plates in place after:
(1) applying the clamps
(2) spreading the glue
(3) proper surfacing of the entire plate
(4) sandpapering with No. 000 sandpaper

27. Wood used in aircraft construction and repair should not have a grain deviation exceeding 15 to 1. When woods having an unequal moisture content are joined by gluing, the joint will be unduly stressed when the moisture of the woods equalize. Surfaces to be glued must be:
(1) free from grease, paint, dirt, etc.
(2) free from sandpapering, filing, etc., as this closes the wood pores and prevents glue penetration
(3) glued within 7 hours after they have been made ready for glue application
(4) all three of the above answers are correct

28. An airplane is covered with Grade A fabric and has a never-exceed speed of 140 MPH. Before replacement is necessary this fabric may deteriorate to which of the following pounds per square inch in a tension test?
(1) 36
(2) 56
(3) 46
(4) 66
Note: The never-exceed speed of 140 MPH requires intermediate grade fabric. The deterioration can lower to 46 PSI even though Grade A fabric is used.

29. Grade A fabric must be used on airplanes with a never-exceed speed greater than 160 MPH. Mercerized cotton may be used on airplanes with a never-exceed speed of less than 160 MPH. Which of the following is correct regarding the strength and deterioration of these covering cloths?
(1) grade A fabric must test 80 pounds new (undoped) and mercerized cotton must test 65 pounds new (undoped)
(2) grade A fabric must test 56 pounds deteriorated (undoped) for airplanes having a never-exceed speed greater than 160 MPH, and

46 pounds deteriorated (undoped) for airplanes having a never-exceed speed less than 160 MPH

(3) in either case, the allowed deterioration is 30% of the new (undoped) strength for grade A fabric when used on airplanes with a never-exceed speed greater than 160 MPH and 30% of the new (undoped) for mercerized cotton when used on airplanes with a never-exceed speed less than 160 MPH

(4) all three of the above answers are correct

30. A sewed spanwise seam on a metal or wood-covered leading edge should be covered with pink-edged surface tape at least:
(1) 5 inches wide
(2) 4 inches wide
(3) 6 inches wide
(4) 8 inches wide

31. You are to compute the amount of fabric needed to cover a straight rectangular wing. Stretching a flexible tape around one of the form ribs of the wing you get a total distance of 128 inches (all-around measurement of the wing). To this you add 4 inches for a seam and obtain the camber measurement of 132 inches (3⅔ yards). You then measure the overall length of the wing, add one-half the depth of the butt rib and another 6 inches for seams for a total measurement of 150 inches. Using a 36-inch-wide fabric, subtract 1 inch for seam allowance or a net width of 35 inches. The total linear yards of fabric actually needed will be:
(1) 16⅓
(2) 18⅓
(3) 20⅓
(4) 22⅓
Solution: Divide 150″ by 35″ which equals 4 widths plus 10″. This means that 5 widths of 36″-wide fabric is needed. Camber length is 3⅔ yards × 5 widths = 18⅓ linear yards required. Allow and order an extra 5% to 10% for possible mistakes, spoilage, etc.

32. In the preceding question, and on the basis that one gallon of dope covers 117 square feet (one coat), at least how much dope will be needed to apply four coats to the wing fabric?
(1) one 5-gallon can
(2) two 5-gallon cans
(3) three 5-gallon cans
(4) three gallons only

33. An aircraft wing is rectangular in shape and has a length of 19′4″. The width is 6′0″ from leading edge to trailing edge. The least number of widths of 60″-fabric needed to cover this wing is:

    (1) 4
    (2) 5
    (3) 6
    (4) 2

34. In the preceding question, how many 3"-wide 100-yard rolls of surface tape is needed?
    (1) 4
    (2) 3
    (3) 2
    (4) 1

35. In question 33, the wing has 18 ribs. How much reinforcing tape is needed?
    (1) one 36-yard roll
    (2) one 50-yard roll
    (3) three 36-yard rolls
    (4) three 50-yard rolls

36. Which of the following is correct regarding inter-rib bracing?
    (1) conventional wing ribs which do not have permanent inter-rib bracing should be tied in position by means of cotton tape running parallel to the beams
    (2) apply the tape continuously with one turn around successive cap-strips, arranged so that the tape between the ribs is separated from the cover by a distance equal to the depth of the capstrip
    (3) tie the turn of tape around each capstrip by means of a short length of lacing cord
    (4) all three of the above answers are correct

37. Re-covering an airplane with fabric is a:
    (1) major repair
    (2) minor repair
    (3) major alteration
    (4) minor alteration

38. How should the tape for leading edge curvature be cut?
    (1) selvage edges
    (2) on the bias cut
    (3) a vise cut
    (4) pinked edges
    Note: This type of cut will not bunch when stretched around a curve.

39. Which of the following is correct regarding the use of anti-tear strips?
    (1) if the never-exceed speed of an aircraft is in excess of 250 MPH,

anti-tear strips are recommended under reinforcing tape on the upper surface of wings and the bottom surface of that part of the wing in the slipstream

(2) where the anti-tear strip is used on both the top and bottom surfaces, pass it continuously up to and around the leading edges and back to the trailing edge.

(3) where the strip is used only on the top surface, carry it up to and around the leading edge and back on the lower surface as far aft as the front beam

(4) all three of the above answers are correct.

40. Waxed (beeswax) cord or thread is used in aircraft fabric work because:
   (1) of the added strength the wax gives to the cord
   (2) it prevents deterioration of the cord and lessens the tendency of the cord to twist
   (3) otherwise it would easily tangle while working with it
   (4) it leaves a natural finish on the cord, thus eliminating any further doping, etc.

41. Beeswax is applied to cord by:
   (1) melting and pouring the beeswax over the cord
   (2) heating the beeswax and immersing the cord in it
   (3) pulling the cord over a cake of beeswax several times
   (4) slicing the beeswax, melting, and running the cord through the melted beeswax

42. Where would the most likely deterioration occur in airplane fabric?
   (1) upper light-coated surface
   (2) upper dark-coated surface
   (3) lower light colored surface
   (4) lower dark colored surface

43. Before doping fabric, which of the following can be used to remove grease from the fabric?
   (1) acetone
   (2) paint thinner
   (3) toluene
   (4) zinc chromate

44. Which of the following would you *not* do in painting the identification markings (serial numbers) on an aircraft?
   (1) place on both surfaces of a single vertical tail
   (2) place on outer surfaces of a multiple vertical tail
   (3) place on both sides of the fuselage
   (4) paint markings same color as the background

45. You are preparing the layout for the identification number N1683C to be painted on the side of an airplane. If the height of this identification is 12 inches, the width, according to FAR's, must be:
    (1) 52 inches
    (2) 4 feet, 4 inches
    (3) ⅔ of the height for the width of numbers and letters (except the letter *I* and the number *1* which must be 1/6 of the height) and spaces between *all* numbers and letters must be not less than ¼ of the width of the standard characters such as *N* and *8*
    (4) all three of the above answers are correct

46. Which of the following is correct regarding the registration identification painted on an aircraft?
    (1) the registration markings may be any size or height
    (2) the width of the letters and numerals (except the letter *I* and the numeral *1*) must be 3/5 of their height
    (3) the color of the registration markings should be a contrasting color to the color of the aircraft and no ornamental markings may interfere with easy identification of the markings
    (4) color of the registration markings should be of a contrasting color to that of the aircraft and ornamental markings may interfere with identification of the markings to a limited degree

47. Identification markings on aircraft must conform to certain standards. Which of the following answers are most appropriate to these standards?
    (1) the width of each letter *I* or numeral *1* should be 1/6 of its height
    (2) the spacing between each letter or numeral should be 1/6 of the height
    (3) the width of each letter or numeral other than the letter *I* and the numeral *1* should be ⅔ of the height
    (4) all three of the above answers are correct

48. You are using rejuvenator on fabric surfaces to soften the old dope and seal cracks. What is the proper technique used in applying the rejuvenator?
    (1) sand the fabric surface lightly so the rejuvenator will have a greater penetrating effect
    (2) wet-sand the surface heavily to give the rejuvenator a good base to adhere to
    (3) remove the old dope down to the clear dope and then apply the rejuvenator
    (4) spray several coats of rejuvenator over the old dope, the surface of which does not require sanding of any kind

49. Which of the following statements is true?

(1) nitrate lacquer must not be applied over paint, varnish, or enamel as it will soften and blister the underlying surface

(2) zinc chromate primer is applied to metallic surfaces before enamel or lacquer is applied in order to serve as a corrosion-resisting base for the topcoats

(3) spar varnish is used as a transparent protective finish coating for wood, metal, and primers

(4) all three of the above answers are correct

50. During finishing there are two probable causes of bubbles. One is a coat of lacquer which is too heavy and which has been applied over a doped surface. Another probable cause is a doped surface that was not thoroughly dried before the lacquer was applied. The best remedy in either case is:

(1) wash the bubbled surface with a dope thinner until it is smooth, allow the surface to dry for about 12 hours, then sand the surface and follow up by giving the affected area one or more thin coats of lacquer

(2) puncture the air bubbles as quickly as possible to get rid of the trouble

(3) abandon the job completely and start over

(4) check with the local FAA office since this problem may need special engineering data

51. In doping fabric, the total number of coats of dope should not be less than that necessary to result in a taut and well-filled finish. Usually the minimum number of coats is:

(1) 3 coats of clear, 2 coats of aluminum pigmented, and 3 coats of pigment

(2) 1 coat of clear, 4 coats of aluminum pigmented, and 3 coats of pigment

(3) 2 coats of clear, 5 coats of aluminum pigmented, and 1 coat of pigment

(4) 5 coats of clear, 1 coat of aluminum pigmented, and 2 coats of pigment

52. The procedure for installing glass cloth over serviceable fabric with no direct attachment to any structure is considered to be a:

(1) class C application

(2) class B application

(3) class A application

(4) not recommended practice

53. Fungicidal dope is normally used as the first coat for fabrics to prevent rotting. This material is a fine powder and may be already mixed

with purchased dope, or it is feasible to mix the fungicide with the dope prior to doping. If the fine powder is mixed with the dope:
(1) it should be made into a paste (using dope) and then diluted to the proper consistency according to the manufacturer's instructions
(2) it is practical to mix the powder with a large quantity of dope
(3) the first coat should be applied thick in order to saturate both sides of the fabric
(4) none of the above answers is correct

54. Which of the following dope room conditions will minimize blushing when doping fabric?
(1) temperature 90°F., relative humidity 65%
(2) relative humidity 80%, temperature 70° to 75°F.
(3) temperature 70° to 75°F., relative humidity 65%
(4) relative humidity 55%, temperature 80°F.

55. When repairing a damaged fabric wing cover and the damage extends over or is closer than one inch to a rib or other laced member, extend the patch:
(1) 3" beyond the rib or other laced member
(2) 5" beyond the rib and three inches beyond the laced member
(3) 8" beyond the rib or other laced member
(4) none of the above answers is correct

56. In fabric covering work there are three basic seams that can be used. They are: (a) plain overlap, (b) French fell, and (c) folded fell. Generally speaking, which of these seams is considered the stronger?
(1) folded fell
(2) French fell
(3) plain overlap
(4) each of the three seams is of equal strength when properly sewn

57. When working with aircraft fabric you should be familiar with the following terms: (a) filling threads—threads that run crosswise of the woven fabric, (b) warp threads—threads that run lengthwise of the woven fabric, (c) selvage edge—an edge of cloth, tape, or webbing so woven as to prevent raveling. Which of the following is appropriate when working an airplane fabric covering project?
(1) bias means cutting the fabric on an approximate 45° angle
(2) pinking shears should be used when cutting aircraft fabric
(3) bleaching is a chemical process used to whiten textile materials
(4) all three of the above answers are correct

58. Before using fabric rejuvenator products to improve the appearance or condition of doped surfaces, care should be exercised to establish that the fabric strength:

(1) is within 10% of the new undoped strength requirements
(2) has not deteriorated beyond safe limits
(3) is within 20% of the new undoped strength requirements
(4) is within 50% of the new undoped strength requirements

59. Blushing of dopes is very common when doping is accomplished under humid conditions. The condition is caused by the rapid evaporation of thinners and solvents, which lower the temperature on the surface, causing condensation of moisture and producing the white appearance known as blush. When the relative humidity is such that only a small amount of blushing is encountered, the condition may be eliminated by:
(1) decreasing the temperature in the dope room
(2) increasing the relative humidity in the dope room
(3) add blush-retarding thinner to the dope to slow the drying
(4) add blush-retarding thinner to the dope to quicken the drying

60. Carefully washing with naptha is a suitable way to remove oil or grease spots from wood when covering plywood surfaces with fabric. Oil and grease spots can be cleaned from undoped fabric with:
(1) naptha
(2) soap and water or acetone
(3) unleaded gasoline
(4) a suitable paint thinner

61. Using self-tapping screws to attach fabric to the primary rib structure is:
(1) not permitted by regulations
(2) permitted and is a common practice
(3) permitted only on aircraft of a relatively low never-exceed speed
(4) permitted providing special directives are obtained from the FAA

62. If you run out of lacing cord in the middle of a rib, you should tie it with which of the following knots?
(1) splice knot
(2) modified seine knot
(3) square knot
(4) quadrangle knot

63. Aircraft dope is defined as a colloidal solution of cellulose acetate butyrate or cellulose nitrate and generally identified as butyrate or nitrate dope. Butyrate dope is more resistant to fire and can be success-fully applied over nitrate dope. For this reason, butyrate dope is often used as a rejuvenator for finishes that have become badly weathered. Fabric rejuvenation basically improves the appearance of fabric but

must be properly done or the fabric may sag after the rejuvenation process has been completed. Dope is generally:

(1) supplied at a consistency ready for brush coats
(2) thinned according to the particular type of dope using thinner that will be described by the manufacturer on the container label
(3) brushed on for the first two coats and then sprayed (after thinning) for the subsequent coats until the fabric becomes taught, airtight, and of suitable strength
(4) all three of the above answers are correct

64. Acid-type storage battery areas should be cleaned with a solution of bicarbonate of soda (baking soda) and nickel-cadmium battery areas should be cleaned with a solution of boric acid. After cleaning and proper drying, the battery compartments and/or areas should be painted (for corrosion prevention) with:

(1) acid paint, bituminus paint, or asphalt paint
(2) glossy enamel, thin lacquer, or toluene
(3) potassium dichromate, ethylene dibromide, or linseed oil
(4) ethylene glycol, paint retarder, or tetraeythel lead solution

65. Surfaces that come into contact with doped fabric should be treated with a protective coating such as aluminum foil, dopeproof paint, cellulose tape, or a similar protective covering. This procedure protects the surfaces against the action of the solvents in the dope. Clad aluminum and stainless-steel parts:

(1) need not be dopeproofed
(2) should be dopeproofed only if they are a part of the aircraft primary structure
(3) should be dopeproofed only if they are a part of the aircraft secondary structure
(4) should be dopeproofed using a satisfactory grade of aluminum foil

66. Zinc chromate primer is applied to metallic surfaces before enamel or lacquer is applied in order to serve as a corrosion-resisting base for the top coats. However, before the zinc chromate primer is applied:

(1) the metal surfaces must be cleaned of dirt, oil, grease, etc.
(2) the metal surfaces can be cleaned with carbon tetrachloride, naptha, benzol, or other suitable solvent
(3) observe all safety precautions when inflammable, explosive, or noxious solvents are used
(4) all three of the above answers are correct

67. The first two coats of dope are applied to fabric with a brush, using unthinned clear dope. The reason for brushing on the first two coats of dope, rather than spraying, is to:

(1) obtain an uneven surface texture
(2) have enough dope soak through the fabric to a drip consistency
(3) wet the fabric thoroughly and to lay the nap of the fabric
(4) take advantage of the shrinking qualities of raw or new fabric

68. The cause of runs and sags in aircraft finishes is which of the following?
    (1) temperature of the painting area too low
    (2) relative humidity of the area is too high
    (3) coats applied too heavily or because the operator dragged his brush over an edge
    (4) the spray gun nozzle is clogged

69. In almost all paint and dope finishing work on aircraft, the finish is applied with a spray gun. The principle advantage of the spray gun is the speed with which paint or dope may be applied and the smooth finish of the sprayed surface. When working with a spray gun the operator should:
    (1) move the spray gun in an oval pattern
    (2) move the spray gun in a straight line
    (3) obtain a balanced or even spray even though there is a dirty air cap
    (4) be aware of its reliability and not waste time by testing the spray gun against a test surface

70. In general, what is the final step when doping aircraft fabric?
    (1) apply color pigmented dope to the fabric
    (2) spray a thin coat of varnish over the finished dope surface
    (3) spray a moderate coat of enamel over the finished dope surface
    (4) apply dope that has been treated with an approved fungicide

71. The best way to test the serviceability of aircraft fabric is:
    (1) by thumb pressure if the person is well experienced in working with fabric
    (2) by the use of a suitable punch tester
    (3) a laboratory test for tensile strength
    (4) rubbing or scratching the fabric surface with a suitable metal blade

72. When making a fabric recover or repair, beeswax is applied to lacing cord for which of the following reasons?
    (1) to protect the lacing cord from deterioration
    (2) it acts as a lubricant when pulling the lacing cord through stubborn areas
    (3) it prevents the fabric from sticking to the lacing cord
    (4) to prevent the cord from cutting your hand or fingers when pulling it through difficult areas

73. In order to determine the proper fabric to use when recovering an aircraft or portion of an aircraft you must consider the:
   (1) weight of the aircraft and whether it is generally stored outdoors or indoors
   (2) never-exceed speed and the type of inner structure as to whether it is of the monocoque or semi-monocoque type
   (3) wing loading and maneuvering speed of the aircraft
   (4) never-exceed speed and wing loading of the particular aircraft

74. You plan to use some rivets designated AN430A 3-4. Which of the following identifies this rivet?
   (1) they are round head rivets with a dimple identification
   (2) the length is 3/16 inch and the diameter is ¼ inch
   (3) they are round head rivets 3/32 inch in diameter and ¼ inch in length
   (4) the heads are of the universal type and the markings on the head is a cross

75. While working on a sheet-metal repair job you under-buck a high-shear rivet. The most probable result will be:
   (1) the upset head of the rivet will not have proper finished dimensions
   (2) the sheets being riveted may slip and cause the rivet to tip in one or the other direction
   (3) the rivet will not completely fill the hole and may fail under bearing loads
   (4) the collar will not be completely swedged and the rivet will not support full tension loads

76. On properly riveted joints the greatest stress will be:
   (1) tension
   (2) compression
   (3) shear
   (4) torsion

77. Which of the following stresses will rivets withstand best?
   (1) torsion
   (2) tension
   (3) compression
   (4) shear

78. Rivets to be used for aircraft repair may require being kept under refrigeration at freezing temperature (32°F.) in order to:
   (1) prevent loss of the corrosion resistant properties of the rivets
   (2) increase the bearing and shear strength of the rivets
   (3) prevent the rivets from becoming too soft and/or pliable
   (4) retard the aging process of the rivets

79. Safety belts must adhere to the standards set up by the FAA. Weight testing or tensile strength of safety belts is governed by which of the following?
    (1) FAR 43
    (2) AD-175
    (3) TSO-C22
    (4) CAB 320

80. When bolts or rivets are used in the installation of plastics, the holes should be:
    (1) oversize by 1/16 inch
    (2) undersize by 1/16 inch
    (3) oversize by ⅛ inch
    (4) undersize sufficiently to insure a snug fit of the plastic and frame

81. When drilling hard metals, select a twist drill having an included angle of 118° and turn it at *low speeds,* but for soft metals use a drill having an included angle of 90° and turn it at *higher speeds.* Thin sheets of aluminum alloys are drilled with greater accuracy by a drill having an included angle of 118° because:
    (1) the large angle of the drill has more tendency to tear the hole
    (2) the large angle of the drill has less tendency to tear or elongate the hole
    (3) you can then use a greater pressure on the drill without damage to the hole
    (4) all three of the above answers are correct

82. Stringers are used in which of the following types of aircraft fuselage construction?
    (1) monocoque
    (2) semi-monocoque
    (3) Warren-type truss
    (4) Pratt-type truss

83. When drilling high carbon steel or stainless steel you should use:
    (1) high pressure and low RPM
    (2) low pressure and high RPM
    (3) high pressure and high RPM
    (4) low pressure and low RPM

84. You are planning to drill some stainless steel. You should use a:
    (1) 45° included angle drill bit operating at low RPM
    (2) 150° included angle drill bit operating at high RPM
    (3) 118° included angle drill bit operating at high RPM
    (4) 150° included angle drill bit operating at low RPM

85. Which of the following makes the strongest joint?
    (1) countersinking
    (2) dimple and countersinking
    (3) dimpling one side of skin only
    (4) dimpling both sides of the skin

86. You are to repair, by patching, some metal aircraft skin. You know the following: (a) length of the break is 2.0 inches; (b) thickness of original material is .040 inch (24S-T Alclad); (c) skin stress is 75,000 pounds per square inch (PSI); (d) shearing or bearing stress (smaller of the two) is 331 pounds; (e) reinforcement of sheet thickness is .040 inch; and (f) rivet size (.040×3 = 0.120 or ⅛ inch) ⅛ inch diameter A17ST. How many rivets are required for each side of the break?
    (1) 12
    (2) 18
    (3) 24
    (4) 30

87. By computation you need 112 rivets to make a single lap using minimum edge distance, ⅛ inch diameter rivets in triple rows. What is the overlap (in inches) of the splice?
    (1) ½
    (2) ⅞
    (3) 1¼
    (4) 1¾

88. 17S rivets are composed of 2017 alloy, A17S rivets of 2117 alloy, and 24S rivets of:
    (1) 2024 alloy
    (2) 5056 alloy
    (3) 110 alloy
    (4) DD alloy

89. When using a #30 drill for rivet holes to rivet two sheets of metal .032 inch and .064 inch thick, the rivet length should be:
    (1) 7/32 inch
    (2) 5/16 inch
    (3) 11/32 inch
    (4) 17/64 inch

90. Regarding rivets, which of the following is correct?
    (1) the length of a round head or flat head rivet is measured from the end of the rivet to the underside of the rivet head
    (2) the length of a countersunk rivet (flush head) is measured from the end of the rivet to the top of the countersunk head

(3) rivet length is computed by the thickness of the metal sheets to be riveted plus one and one-half times the rivet diameter

(4) all three of the above are correct

91. In the process of riveting you plan to use the following types of rivets: AN 470 (universal head), AN 426 (100% flush head, countersunk), AN 430 (round head), AN 425 (100% flush head, countersunk), and AN 435 (round head). According to these types of rivets, how many rivet sets will be required?

(1) 4
(2) 1
(3) 3
(4) 5

92. About an hour after being driven, an ice box rivet attains about 75% of its strength. The rivet will acquire the other 25% of its strength through:

(1) aging for approximately 4 days
(2) aging for approximately 48 hours
(3) complete immersion in a salt bath
(4) immersion in an acid solution

93. You are to attach a 12″×6″ splice plate using a single-row rivet pattern on all four sides, 4D rivet spacing, and minimum edge distance. This work calls for the use of AN 470AD 4-6 rivets. How many rivets will be required under these conditions?

(1) 68
(2) 74
(3) 62
(4) 56

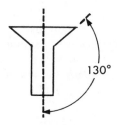

94. The angle of the rivet head, in order to fit a 100° countersink must be:

(1) 130° upward from the centerline of the rivet
(2) 150° upward from the centerline of the rivet
(3) at an angle equal to twice 130° downward from the centerline of the rivet
(4) at an angle approximately 90° from the horizontal

95. Which of the following is correct regarding metal bends?
   (1) the outer part of the bend tends to stretch
   (2) the inner part of the bend tends to compress and the radius of a bend is measured from the inner part of the bend
   (3) the neutral axis of a bend is along the centerline of the bend curvature
   (4) all three of the above answers are correct

96. What is the angle that a flat sheet of metal goes through when bending it to a 15° closed angle?
   (1) 165°
   (2) 105°
   (3) 15°
   (4) 195°

97. Standard rivet edge distance is 2D (2 times the diameter), spacing for single and double row rivets is 4D, and spacing for multiple-row rivets is 3D. The diameter of a rivet, if not otherwise known, can be found by which of the following:
   (1) preferably about three times the thickness of the thicker sheet to be riveted
   (2) preferably about four times the thickness of the thicker sheet to be riveted
   (3) the sum of the thickness of each sheet to be riveted
   (4) none of the above answers is correct

98. Cleaning of transparent plastic should be accomplished by washing with plenty of water and mild soap, using a clean, soft grit-free cloth, sponge, or bare hands. Do not use gasoline, alcohol, benzine, acetone, carbon tetrachloride, fire extinguishing or deicing fluids, lacquer thinners, or window cleaning sprays because they will soften the plastic and cause crazing. Also regarding plastics, which of the following is correct?
   (1) plastics should not be rubbed with a dry cloth since this is likely to cause scratches
   (2) rubbing with a dry cloth will build up an electrostatic charge which attracts dust particles to the plastic surface
   (3) if after removing dirt and grease and no great amount of scratching is visible, finish the plastic with a good grade of commercial wax

in a thin even coat and bring to a high polish by rubbing lightly with a soft cloth

(4) all three of the above answers are correct

99. Never force a plastic panel out of shape to make it fit a frame. If it does not fit easily into the mounting, obtain a new replacement or heat the whole panel and reform. When possible, cut and fit a new panel at ordinary room temperature. In clamping or bolting plastic panels into their mountings, do not place the plastic under excessive stress. Tighten each nut:
    (1) to a firm fit, then back off two full turns
    (2) to a firm fit, then back off ½ turn
    (3) to a firm fit, then back off one full turn
    (4) to a tight fit, then back off ¼ turn

100. There are many types of transparent plastics on the market. Their properties vary greatly, especially in regard to expansion characteristics, brittleness under low temperatures, resistance to discoloration when exposed to sunlight, surface checking, etc. Two types of plastics are commonly used in transparent enclosures of aircraft. They are:
    (1) acrylic and polyester plastics
    (2) acrylic and skydrol plastics
    (3) polyester and toluene plastics
    (4) polyester and neoprene plastics

101. In installations of plastic involving bolts or rivets:
    (1) make the holes through the plastic oversize ⅛ inch diameter
    (2) center the bolts or rivets so that the plastic will not bind or crack at the edge of the holes
    (3) the use of slotted holes is also recommended
    (4) all three of the above answers are correct

102. Replace extensively damaged transparent plastic, rather than repair, whenever possible, since even a carefully patched part is not the equal of a new section either optically or structurally. At the first sign of crack development:
    (1) remove and replace the panel
    (2) drill a small hole at the extreme ends of the cracks
    (3) drill a small hole at the center of the cracks
    (4) drilling small holes to temporarily stop crack development is not permitted

103. Which of the following is correct regarding the use of hand files?
    (1) the forward stroke should be the working stroke and the rear stroke should be made with the file lifted away from the material

    (2) the forward stroke and the rearward stroke should both be working strokes in contact with the material

    (3) either the forward stroke or the rearward stroke can be the working stroke with the file in contact with the material

    (4) none of the above answers is correct

104. If the section of plastic material to be repaired is curved, shape the patch to the same contour by heating it in an oil bath at a temperature of 248° to 302°F., or it may be heated on a hotplate until soft. Boiling water should not be used for heating. Coat the patch evenly with plastic solvent adhesive and place immediately over the hole. Maintain a uniform pressure of from:

    (1) 15 to 25 PSI on the patch for a minimum of 12 hours

    (2) 5 to 10 PSI on the patch for a minimum of 3 hours

    (3) 1 to 5 PSI on the patch for a minimum of 7 hours

    (4) 25 to 50 PSI on the patch for a minimum of 10 hours

105. AN 470AD 4-6 rivets are described as follows: AN (Army-Navy standard); 470 (universal head); AD (2117 material, do not heat treat); 4 (diameter in 1/32"); and 6 (length in 1/16"). How many of this type of rivet will be required to attach a 14"×7" splice plate if you use a single row rivet pattern, 4D spacing, and minimum standard rivet edge distance? (i.e. a single row of rivets along all four sides of the plate)

    (1) 84 rivets

    (2) 88 rivets

    (3) 80 rivets

    (4) 76 rivets

106. You are to use ⅛" diameter rivets on a metal surface patch with a single row pattern. The minimum overlap of the metal will be:

    (1) ½ inch

    (2) ¼ inch

    (3) ¾ inch

    (4) ⅞ inch

107. Which of the following stresses will a properly installed rivet best withstand?

    (1) tension

    (2) compression

    (3) bending

    (4) shear

108. You plan to bend a piece of 6"×3" metal to a right angle of minimum radius. You should bend the metal:

(1) across the grain
(2) at 90° to the grain
(3) perpendicular to the grain
(4) all three of the above answers are correct

109. During installation a rivet was damaged. The rivet should be removed by:
    (1) shearing off the manufacturer's head
    (2) the use of a sharp chisel to shear off the manufacturer's head
    (3) drilling through the manufacturer's head with a drill slightly smaller than the rivet diameter so that the head and shank can be easily removed
    (4) drilling through the manufacturer's head with a drill slightly larger than the rivet diameter so that the head and shank can be easily removed.

110. In order to retain 83% of the material strength, the rivet spacing should be 3D and to retain 75% of the material strength, the rivet spacing should be 4D. Which of the following describes the finished dimensions of a properly bucked rivet head?
    (1) the diameter of the head should be 1½D and the height should be ½D
    (2) the height of the head should be 1½D and the diameter should be ½D
    (3) the diameter of the head should be 1½D and the height should be 1½D
    (4) none of the above answers is correct

111. You plan to rivet two pieces of material using universal head rivets. One sheet is .032 inch thick and the other .040 inch thick. Which of the following rivets would you use?
    (1) AN 425 AD 4-3
    (2) AN 470 AD 4-4
    (3) AN 430 AD 5-5
    (4) AN 470 AD 4-5

112. Which of the following indicates the amount of a rivet that should protrude through the material prior to bucking and that will permit a bucked rivet head of proper dimensions?
    (1) 2D
    (2) 3D
    (3) 1D
    (4) 1½D

113. During installation of high-shear steel rivets in aluminum plates, you should:
    (1) use a zinc chromate dip for the rivets
    (2) stagger the high-shear rivets with heat-treated aluminum rivets
    (3) increase the rivet spacing from the standard minimum
    (4) insure that there is a very close fit

114. To obtain the length of an AN 425 (countersink type) you should measure from the:
    (1) top of the rivet head to the extremeties of the bucked shank
    (2) bottom of the head to the end of the bucked shank
    (3) center point of the head to the end of the rivet
    (4) top of the head to the end of the rivet (overall length)

115. Which of the following is generally used in the construction of firewalls?
    (1) sheet steel of the high carbon category
    (2) stainless steel
    (3) high grade sheet aluminum alloy
    (4) Alclad

116. When preparing to assemble dissimilar metals that will be in contact with each other you should:
    (1) use some type of insulation so that they will not come in direct contact with each other
    (2) use a mild abrasive on the surfaces so that they will make direct contact with each other
    (3) not be concerned whether or not the surfaces come in contact with each other
    (4) permit them to be in contact with each other in order that no electrolytic action will occur

117. In monocoque type aircraft fuselages that utilize bulkheads, stringers, and metal skin:
    (1) the outer metal skin takes most of the stresses in flight
    (2) the bulkheads take most of the stresses in flight
    (3) the stringers take most of the stresses in flight
    (4) the internal truss-type members take most of the stresses in flight

118. The construction characteristics of a cantilever wing is which of the following?
    (1) this type of wing will have both bulkheads and stringers
    (2) standard type external bracing will be installed underneath the wing
    (3) the cantilever type wing has no external supports
    (4) there will be no dihedral, sweepback, or decalage in the cantilever type wing

119. Stop-drilling cracks in sheet metal structures:
    (1) is not approved for aircraft work
    (2) in general, will temporarily stop a crack in the material until proper repair can be done
    (3) adds to the shear and tension strength of the material at the point where the hole is drilled
    (4) none of the above answers is correct

120. When it is essential that bolt holes be of exact size to prevent any play or lost motion and to be assured of a perfect true hole, a reamer is used. When using a reamer:
    (1) the reamer should be turned in the cutting direction only
    (2) great care must be taken in using and storing reamers in order to protect the blades from chipping
    (3) great care should be used to assure even, steady turning, otherwise the reamer will chatter
    (4) all three of the above answers are correct

121. Machine or drill countersinking is accomplished by a suitable cutting tool, machined to the desired angle, which cuts away the edge of the hole so that a countersunk rivet head fits snugly into the recess. When countersinking holes:
    (1) hold the countersink at a slight angle to the material being counter-sunk
    (2) be sure the material through which the countersunk rivet is to be placed is thinner than the depth of the countersunk head
    (3) chattering of the countersink can be caused by improper use or poor condition of the countersink such as being dull
    (4) the countersink can be tipped slightly for the best work

122. Scratches that are small and shallow may be satisfactorily removed from sheet metal by which of the following?
    (1) sanding with OOO sandpaper
    (2) mild etching
    (3) burnishing
    (4) proper polishing

123. Repairs of structural units, such as spars, engine supports, etc., that have been built from sheet metal:
    (1) may be made by the original manufacturer only
    (2) may be made by approved repair shops (stations) only
    (3) should not be repaired if damage has occurred; they should be replaced
    (4) can be made when repaired in an approved manner

124. A fuselage of the semi-monocoque type is best identified by:
    (1) stringers and longerons
    (2) the absence of bulkheads
    (3) spars, ribs, drag and anti-drag wires
    (4) bulkheads and rings only

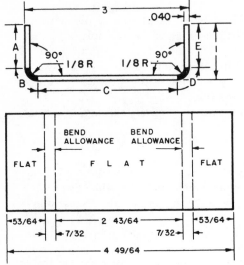

Procedure

1. Determine setback.
   Setback $= K(R+t)$.
   Setback $= 1(0.125 + 0.040)$.
   Setback $= 0.165$ or $^{11}\!/_{64}$.
2. Determine length of all flats.
   Flats A and $E =$ Base measurement $-$ setback.
   Flats A and $E = 1$ inch $- 0.165$.
   Flats A and $E = 0.835$ or $^{53}\!/_{64}$.
   Flat $C =$ Base measurement $- 2 \times$ setback.
   Flat $C = 3$ inches $- 2(0.165)$.
   Flat $C = 3$ inches $- 0.330$.
   Flat $C = 2.670$ or $2^{43}\!/_{64}$.
3. Determine bend allowance for all bends (at B and D).
   B. $A. = (0.01743\ R) + (0.0078\ t) \times N$.
   B. $A. = (0.01743 \times 0.125) + (0.0078 \times 0.040) \times 90°$.
   B. $A. = 0.2242$ or $^{7}\!/_{32}$.
4. Make the layout as shown.
   Total length $= ^{53}\!/_{64} + ^{7}\!/_{32} + 2^{43}\!/_{64} + ^{7}\!/_{32} + ^{53}\!/_{64}$.
   Total length $= 4^{49}\!/_{64}$ or 4.77.

The following four questions refer to the accompanying illustration:

Given: Compute the necessary data to convert a piece of flat metal stock into a channel-type shape that must have the following dimensions—each leg or flange to be 1 inch high, the distance between the outside surfaces of the two flanges to be 3 inches, radius of the two 90° bends is ⅛ inch, and the material is 0.040 inch thick.

125. The setback for each bend (in inches) is:
    (1) 1/8
    (2) 11/64
    (3) 7/32
    (4) 1/2

126. The length (in inches) of flats A and E is:
    (1) 3/16
    (2) 1/2
    (3) 53/64
    (4) 7/8

127. The length (in inches) of flat C is:
    (1) 2 inches
    (3) 2 9/16
    (3) 2 43/64
    (4) 2 3/4

128. The bend allowance (in inches) for bends B and D is:
    (1) 13/64
    (2) 7/32
    (3) 15/64
    (4) 1/4

129. The total length (in inches) of the flat stock required for this metal channel-type shape is:
    (1) 4 43/64
    (2) 4 57/64
    (3) 4 37/64
    (4) 4 49/64

130. In many instances you may have to shape sheet metal to fit into a particular area. Shaping is generally known as *forming* sheet metal and there are many ways in which forming can be accomplished. Which of the following methods of forming is correct?
    (1) shaping or forming malleable metal by hammering or pounding it to form a bump on it is called bumping
    (2) whenever you fold, pleat, or corrugate a piece of sheet metal so that you shorten it, you are crimping the sheet metal
    (3) making bends in sheets, plates, etc., is called folding, hammering a flat piece of metal so that the material will become thinner and cover a greater area is called stretching, and when material is forced or compressed into a smaller area it is called shrinking
    (4) all three of the above answers are correct

131. The rivnut is a hollow rivet counterbored and threaded on the inside. Installation is accomplished with the aid of a special tool, which heads the rivet on the blind side of the work. The rivnut is primarily used:
    (1) in the attachment of deicer boots
    (2) in the primary structure of aircraft
    (3) to attach metal skin to the understructure
    (4) none of the above answers is correct

132. There are many places on the airplane where access to both sides of a riveted structural part is impossible, or where limited space will not permit the use of a bucking bar. For use in such places, rivets are

designed which can be bucked from the front side and are often called blind rivets. These rivets may be described as:
(1) rivnuts
(2) special rivets
(3) cherry or explosive rivets
(4) all three of the above answers are correct

133. The heat treatment of rivets is usually accomplished in a salt bath just as other aluminum alloy parts. The most commonly made rivets are 2117 (A17S) and have a strength value of 30,000 PSI in shear; 2017 (17S) at a strength of 34,000 PSI in shear; and 2024 (24S) at a strength of 44,000 PSI in shear. Under which of the following may 2117 rivets be used to replace 2017 or 2024 rivets?
(1) use more or larger 2117 rivets
(2) use less or smaller 2117 rivets
(3) twice as many 2117 rivets will be required
(4) three times as many 2117 rivets will be required

134. Which of the following is correct when handling and installing rivets that require heat treatment prior to use?
(1) keeping them in a refrigerator at a temperature of 32°F. will delay aging about 24 hours
(2) keeping them on dry ice (solidified carbon dioxide, temperature −150°F.) will delay aging for an extended period of time
(3) in general, remove from the refrigerator only enough rivets that will be used within thirty minutes
(4) all three of the above answers are correct

135. When doing oxy-acetylene welding, several precautions should be exercised by the welder. Which of the following precautions are appropriate?
(1) always keep the torch or working pressure of acetylene below 15 PSI
(2) do not exceed 15 PSI working pressure on the acetylene
(3) do not weld while wearing oily clothes or oily gloves, and be sure there are no leaks in the welding equipment lines
(4) all three of the above answers are correct

136. Welded steel tubing:
(1) may not be spliced under any circumstances
(2) may not be repaired under any circumstances
(3) should not be used in aircraft repairs
(4) may be spliced or repaired at any joint along the length of the tube unless otherwise noted in AC 43.13-1

137. You have joined two ferrous metals by welding. What should be done to relieve the internal stress?

(1) reheat and quench in oil
(2) reheat and quench in cold water
(3) drill several relief holes at each end of the weld
(4) normalize the material

138. The finished weld should have the following characteristics: (a) the seam should be smooth and of uniform thickness; (b) the weld metal should taper off smoothly into the base metal; (c) the base metal should show no signs of pitting, burning, cracking, or distortion; and (d) welding scale should have been removed by wire brushing or sand-blasting. When it is necessary to reweld a joint which was previously welded:
(1) always use a neutral flame
(2) it is satisfactory to weld over a weld
(3) all old-weld material should be thoroughly removed before rewelding
(4) always use an extra large welding rod

139. Welding equipment (gas type) is somewhat standardized and has the following in common: (a) oxygen hose is green, (b) acetylene hose is red, (c) oxygen fittings are of right-hand thread, (d) acetylene fittings are of left-hand thread, and (e) it consists of oxygen and acetylene tanks under pressure, pressure gages and regulators, an ample supply (length) of hose, and a welding torch. The type and size of welding rod to be used is determined by the:
(1) welding flame to be utilized
(2) size of the torch tip being used
(3) quality of flux to be used
(4) type of material to be welded and the thickness of the material

140. When welding aluminum, flux is used to:
(1) prevent oxidation in the weld
(2) prevent the welding rod from overheating
(3) act as a filler agent in the weld
(4) retain the heat at the tip of the welding torch

141. When not made of heat-treated steel, aircraft structural tubing that is slightly bent may be straightened by approved methods. After the straightening process has been completed, the tubing should be:
(1) stronger than before
(2) weaker than before
(3) neither weaker nor stronger than before
(4) re-heat-treated and then strain hardened

142. The interior surfaces of sealed aluminum tubing and steel structures are coated with a preservative to prevent corrosion. A recommended material for this is:
    (1) an acid type paint
    (2) hot linseed oil
    (3) bituminous type paint
    (4) an approved dope-proof paint

143. When preparing to weld, the size of the welding tip is determined by the:
    (1) thickness of the material to be welded
    (2) working tank pressure of the equipment
    (3) type of anticipated flame to be used
    (4) type of material to be welded

144. A soft neutral flame should be used when welding 4130 steel because a:
    (1) soft reducing flame may cause iron oxides and slag to form on the weld
    (2) soft oxidizing flame does not produce sufficient heat to melt the base material
    (3) harsh carbonizing flame will burn the molybdenum out of the weld
    (4) harsh neutral flame will produce cold sluts and laps in the weld

145. You are making a repair and it is necessary to weld and rivet in the same area. You should weld first to:
    (1) prevent tempering of the rivets
    (2) have a more solid area to rivet
    (3) make an easier rivet pattern installation
    (4) prevent a shear stress from being imposed on the rivets

146. In the oxyacetylene welding process, having a 15 PSI or greater pressure on the acetylene could result in which of the following?
    (1) distortion and/or warping of the base metal to be welded
    (2) a highly oxidizing flame
    (3) a possible explosion
    (4) a carbonizing flame of extreme heat value

147. A sharp pointed soldering copper or one that has too fine a point is not considered satisfactory because it:
    (1) will cool too quickly for satisfactory soldering
    (2) may burn completely through the metal being soldered
    (3) becomes too difficult to tin
    (4) will burn or remove the alloy content in the solder

148. The weld may become misaligned due to expansion and contraction. This can be prevented by:
    (1) the use of a soft oxidizing flame
    (2) proper equipment being used correctly, especially when using jigs
    (3) the use of very steady heat during the entire welding process
    (4) correct cleaning of the material prior to welding

149. A new soldering copper cannot be used successfully until after the working faces have been tinned, or coated with solder. This is due to the fact that heat causes copper to oxidize, forming a coating on the surface which will not allow proper heat transference. When preparing to solder:
    (1) the bit may be tinned by heating it to a cherry red and rubbing the working faces over a block of sal-ammoniac
    (2) after the faces have become coated with melted sal-ammoniac, a small quantity of solder should be applied
    (3) if the solder applied to the faces will adhere, the copper will be correctly tinned
    (4) all three of the above answers are correct

150. The condition of flight that a properly rigged fixed-wing aircraft should maintain is:
    (1) slightly nose-heavy to resist any tendency to stall
    (2) slightly tail-heavy to assure the pilot that there will be no sudden down-pitch attitude
    (3) straight and level flight in cruise configuration even though the pilot removes his hands and feet from the controls in the cockpit
    (4) straight and level flight even though the pilot must hold slight pressure on the controls in the cockpit

151. When an airplane is in level flight at a constant or steady speed:
    (1) thrust equals drag and weight (gravity) equals lift
    (2) thrust equals drag and lift is greater than weight
    (3) weight equals lift and thrust is greater than drag
    (4) weight equals lift and drag is greater than thrust

152. In flying a helicopter, a constant heading is accomplished by the use of rudder pedals. Altitude (ascent and descent) is controlled by the collective pitch. Horizontal flight and airspeed are controlled:
    (1) by throttle coordination only
    (2) by cyclic pitch and throttle
    (3) also by collective pitch
    (4) by constant use of the collective and cyclic control

153. The control stick (or wheel) is moved forward. Which of the following control reactions will this create during flight?
    (1) elevator moves down, causing the tail and the nose to move up
    (2) elevator moves down, causing the tail and the nose to move down
    (3) elevator moves down, causing the tail to move up and the nose down
    (4) elevator moves up, causing the tail to move up and the nose down

154. There are three axes associated with an airplane—the longitudinal, lateral, and vertical. Which of the following is correct regarding airplane attitudes?
    (1) the rudder controls the airplane about the longitudinal axis
    (2) the elevator controls the airplane about the horizontal axis
    (3) a combination of the two ailerons controls the airplane about the vertical axis
    (4) yawing about the vertical axis is controlled by the rudder, pitching about the lateral axis is controlled by the elevator, and banking or rolling about the longitudinal axis is controlled by the ailerons

155. Severe vertical vibration in a two-bladed main rotor helicopter indicates that the:
    (1) engine has developed a severe ignition malfunction
    (2) tail rotor pitch is out of adjustment
    (3) main rotor blades are out of track
    (4) center of gravity, due to improper loading, has exceeded the rearmost limit

156. The proper method for tracking helicopter main rotor blades is:
    (1) using a pole-flag arrangement, note exactly where the *swish* of the rotating main rotor blades occurs
    (2) a different colored chalk is placed at the tips of the main rotor blades, a pole-flag unit is held by the person checking the track at the rotor blade tips, and it is then noted how close together the different colored chalks strike the flag part of the checking tool
    (3) the main rotor blade tip, during rotation, is measured from the ground level using an exact pole-flag checking tool
    (4) moisture is placed on the main rotor blade tips, a pole-flag checking tool is raised toward the rotating rotor blades, and the streaks of moisture are checked for closeness

157. Helicopter rotor blades must be in proper track at all times or severe vibrations of the helicopter will occur. Tracking of rotor blades is best accomplished by which of the following?

(1) place the helicopter on a level ground area and measure the distance from ground to rotor blade tip while the blades are at rest (engine not operating)

(2) with engine operating and rotor blades moving at a slow speed, place white chalk on each rotor blade tip, and then with a canvass flag on a pole determine if each rotor blade tip strikes the flag at precisely the same spot

(3) place a chalk of different colors on the rotor blade tips, start engine and operate rotor blades at a given RPM; then use a canvas flag on a pole to determine if each rotor blade tip strikes the flag at the same approximate spot

(4) with colored chalk mark each of the rotor blade tips, start the engine and operate the rotor blades at a particular RPM then hold a canvas flag on a pole toward the rotating blade tips to determine if all colored chalk marks that are transferred to the edge of the canvas flag are within the tracking tolerance for the particular helicopter

158. Recovering of helicopter rotor blades must be done in accordance with:
(1) FAA Manual 24
(2) AC 43.13-1 and/or the manufacturer's maintenance manual
(3) FAA Manual 40
(4) FAA Information Guide

159. A certain airplane has a conventional tail section. The rudder should be rigged:
(1) in prolongation to the fin
(2) at an angular displacement of its leading edge
(3) in neutral position
(4) with greater left rudder pedal travel than right rudder pedal travel
Note: The rudder is aligned or rigged with the vertical fin.

160. On a single-engine jet plane the vertical stabilizer (looking from rear to front of plane) is:
(1) offset to the left of the longitudinal axis
(2) offset to the right of the longitudinal axis
(3) mounted parallel to the longitudinal axis
(4) 25° off the centerline of the longitudinal axis

161. If the trailing edge of a wing is bent downward:
(1) no change on the lift value of the wing will occur
(2) the angle of incidence will be decreased
(3) wing lift for level flight will be decreased
(4) there will be an increase in wing lift in level flight
Note: The trailing edge bent downward increases the angle of incidence and subsequently wing lift.

162. The term *wash-in* describes a rigging procedure whereby the angle of incidence of a wing is increased in order to obtain increased lift. This is the result of an increase in the angle of attack which also increases the wing drag. Accordingly, if a pilot complains to a licensed mechanic that at cruise speed (hands and feet off the controls) his airplane tends to fly with the right wing low, the mechanic knows that:
    (1) the right wing needs wash-in
    (2) the left wing needs wash-out
    (3) the right wing needs wash-in and/or the left wing needs wash-out
    (4) all three of the above answers are correct

163. The propeller rotates in a clockwise direction as viewed from the cockpit. This airplane will have:
    (1) wash-out rigged in the left wing
    (2) wash-out rigged in the right wing
    (3) wash-in rigged in the right wing
    (4) the interplane struts adjusted for more wash-in than wash-out
    Note: Propeller torque creates a left-wing-low and right-wing-high condition that can be corrected by wash-in and wash-out.

164. The principle purpose of flaps, as used in a landing maneuver, is to:
    (1) permit a decrease in the airplane's landing speed yet maintain sufficient lift
    (2) spoil the lift at landing speed thus shortening the landing run
    (3) decrease lift and drag simultaneously
    (4) maintain positive aileron response throughout the landing

165. When attaching wing fittings to the fuselage, caution must be taken not to change which of the following angles?
    (1) angle of attack
    (2) dihedral angles
    (3) sweepback angle
    (4) angle of incidence

166. Helicopter vibrations are classed as low, medium and high frequency. Low frequency is caused by the main rotor, high frequency by the engine, and medium frequency by:
    (1) ground resonance
    (2) ground cushion
    (3) lateral beat of the main rotor
    (4) tail rotor rotation
    Note: High frequency vibrations are caused by tail rotor when its RPM is greater than engine RPM.

167. Swaged terminals are painted to check for:

(1) slippage
(2) proper identification
(3) proper length of cable to be inserted in the barrel of the fitting
(4) correct strength characteristics

168. To check a control cable for fraying:
    (1) run your bare hand over the cable
    (2) use a very light cloth that will easily indicate any fraying of the wires
    (3) using a glove, run your hand along the cable to find any possible fraying
    (4) encircle the cable with a pair of pliers as you move along the cable to pick up any fraying points

169. If control cables are made from tinned steel, the cable requires a coating of rust-preventative oil. Corrosion-resistant steel cable:
    (1) also requires a coating of rust-preventative oil
    (2) is very susceptible to rust formation
    (3) does not require a coating of rust-preventative oil
    (4) none of the above answers is correct

170. Which of the following indicates the correct manner in which a fixed-wing airplane is trimmed during flight?
    (1) the elevator trim tab will move downward to trim for a nose-down condition
    (2) the right wing aileron trim tab will move downward (and the left aileron trim tab upward) to trim for a right-wing-high (left-wing-low) condition
    (3) the rudder trim tab will move to the right to correct for a right-yaw condition
    (4) all three of the above answers are correct

171. If the elevator tab is up, what would the trim control in the cockpit indicate?
    (1) nose up
    (2) nose down
    (3) trim up
    (4) trim down

172. A pulley is worn on one side only. Which of the following should you do?
    (1) repair the pulley by building up the worn side with butyl rubber
    (2) check for pulley vibration and if the virbration is within published limits do nothing
    (3) replace the pulley
    (4) none of the above answers is correct

173. Control surface movement is normally limited by:
    (1) pulley brackets
    (2) swaged terminals painted at intervals
    (3) mechanical stops
    (4) mechanically swaged metal rings encompassing the cables

174. Differential controls are controls which, for the same amount of motion of the stick or wheel, move further in one direction than the other. Regarding differential ailerons:
    (1) the one moving down swings through a relatively small angle such as 15° and the one moving up swings through a relatively large angle such as 30°
    (2) they improve lateral control at stalling speeds
    (3) they tend to eliminate yaw caused by the down aileron
    (4) all three of the above answers are correct

175. Wing stations of an airplane are measured:
    (1) forward or aft of the center of gravity
    (2) outward from the centerline of the longitudinal axis
    (3) outward from the root of the wing
    (4) in the same manner as cargo stations in the fuselage, i.e., along the longitudinal axis

176. An airframe mechanic is making adjustments on a full cantilever wing. The adjustments he is permitted to make are:
    (1) changes in the angle of incidence on some types of planes only
    (2) dihedral on all types of planes
    (3) sweepback on all makes of planes
    (4) all three of the above answers are correct
    Note: A cantilever has no external supports such as struts, wires, etc.

177. When repairing control surfaces, especially on high-performance airplanes, care should be exercised that the repairs do not involve the addition of weight aft of the hinge line. Such procedure may:
    (1) adversely disturb the dynamic balance of the control surface
    (2) adversely disturb the static balance of the control surface
    (3) induce flutter of the control surface
    (4) all three of the above answers are correct

178. Which of the following statements is correct regarding an all-metal airplane?
    (1) during cold weather, control cable tension decreases
    (2) during hot weather, control cable tension decreases
    (3) during cold weather, control cable tension increases
    (4) control cable tension remains the same regardless of weather conditions

179. Prior to rigging an airplane, the mechanic should be aware of the following: (a) the angle of incidence is the angle between the chord of the wing and the longitudinal axis; (b) the angle of attack is the angle between the chord of the wing and the relative wind; (c) dihedral is the upward angle of the wing from the horizontal plane; (d) sweep-back is the rearward angle of the leading edge from wing root to wing tip; and (e) lowering the flaps increases the camber of the airfoil (wing) giving additional lift at lower speeds, increases the climb and glide angle, and, in some instances, flaps may be used as a braking device. The mechanic should also know that:
    (1) control cable tension can be accurately checked by the use of a tensiometer
    (2) moving the wheel (or stick) rearward, moving the wheel to the left, and depressing the left rudder pedal will result in a climbing left turn attitude
    (3) if an airplane is reported to be nose-heavy, a fixed trim tab bent in a downward attitude and mounted on the trailing edge of the elevator could correct the nose-heavy condition
    (4) all three of the above answers are correct

180. A monoplane (airplane with one wing) wing is supported with wing struts. Which of the following is correct regarding this arrangement?
    (1) the angle of attack is adjusted by the rearward wing struts
    (2) the angle of incidence is adjusted by the forward wing struts
    (3) dihedral angle is adjusted by the front or forward struts
    (4) wash-in and wash-out are both adjusted by the front or forward struts

181. In a helicopter equipped with a tail rotor, directional control is maintained by:
    (1) gradually increasing and decreasing rotor-engine RPM
    (2) use of the cyclic control to tilt the main rotor disc in the desired direction
    (3) use of the collective pitch control to tilt the main rotor disc in the desired direction
    (4) changing the pitch or blade angle of the tail rotor blades

182. Static balance of control surfaces can be obtained by adding weight ahead of the hinge line and control surface flutter can be prevented by being sure the hinges and control surface are properly lined up. When rigging an airplane, control surface movement:
    (1) can be found by referring to specifications in the type certificate data sheet or aircraft specifications
    (2) is limited by the use of positive stops
    (3) can be measured in either inches or degrees
    (4) all three of the above answers are correct

183. Control cables are described as 7 × 7 (flexible) and 7 × 19 (extra flexible), the smallest cable used for primary control cables is 3/16 inch diameter, and cables must be cut by mechanical means to prevent fraying. Which of the following is correct regarding running cables through openings and/or changing the direction of control cables?
    (1) fairleads are used to route cables through openings where the cable deflection is not more than 3° and pulleys are used where cable deflection is greater than 3°
    (2) pulleys are used to route cables through openings where the cable deflection is not more than 3° and fairleads are used where cable deflection is greater than 3°
    (3) clevis brackets are used to route cables through openings with a small cable deflection and fairleads are utilized where cable deflection is of a large deflection
    (4) none of the above answers is correct

184. When working with aircraft cables, the Nicopress Process is which of the following?
    (1) a patented process using copper sleeves up to the full rated strength of the cable when the cable is looped around a thimble
    (2) a process that may be used in place of the five-tuck splice on cables up to and including 3/8 inch diameter
    (3) when using a sleeve requiring three compressions, make the center compression first, the compression next to the thimble second, and the one farthest from the thimble third or last
    (4) all three of the above answers are correct

185. In a typical airplane, the angular difference between the angle of incidence of a wing (airfoil) and the angle of incidence of the horizontal stabilizer is described as:
    (1) a tramming angle
    (2) the horizontal dihedral angle
    (3) the positive stagger angles
    (4) decalage

186. During level, unaccelerating flight, thrust equals drag and lift equals weight or gravity. If a change in attitude is made from level unaccelerating flight, which of the following is correct?
    (1) in a climbing attitude lift is less than weight or gravity
    (2) while gliding, lift is greater than weight or gravity
    (3) during a left climbing turn, lift is greater than weight or gravity
    (4) during an accelerating right descending turn, thrust is greater than drag and lift is greater than weight or gravity

187. Which of the following can a universal type propeller protractor be used to measure?

(1) the track of any type of propeller
(2) any positive angle of attack of the wing (airfoil)
(3) wing stations that are graduated in degrees
(4) wing flap movement in degrees

188. When working with swaged type fittings, the junction of the cable and fitting is painted in order to:
(1) prevent the junction area from deteriorating or rusting
(2) later determine if there had been slippage of the cable in the terminal
(3) have a means of identifying the cable
(4) prevent intergranular corrosion in the junction area

189. In order to vary or change the direction of a helicopter flight path, the pilot would:
(1) vary the RPM of the main and tail rotor
(2) tilt the plane of the main rotor blades by use of the collective control
(3) tilt the plane of the main rotor blades by use of the cyclic control
(4) split the needles on the tachometer and use the rudder pedals

190. You are installing a clevis bolt to the fork of a primary control cable. It is important that you:
(1) use a castellated shear nut and proper size cotter pin
(2) use a plain nut and a suitable lock washer
(3) tighten the self-locking nut that is placed on the clevis bolt until the spacer is torqued to a tight fitting
(4) use a plain lock nut with at least two lock washers

191. When recovering a rudder with fabric use caution that surface tape, etc., will not loosen, create a flapping motion, and cause the control surface to flutter or vibrate. After completing the repair work on the rudder, especially on high performance airplanes, you should:
(1) on installation, balance the rudder laterally
(2) on installation, balance the rudder longitudinally
(3) after installation, balance the rudder to the manufacturer's specifications
(4) not balance the rudder since it is installed in a vertical attitude

192. Decalage in a biplane is the difference in the angle of incidence of the upper and lower wing. Positive stagger is a condition where the upper wing is installed ahead of the bottom wing. During the adjustment of the wings of a biplane you should:
(1) adjust the angle of attack by varying the decalage
(2) adjust the angle of incidence with the rear wires
(3) wash out the wings in order to increase the angle of attack
(4) wash-in the wings to decrease the angle of attack

193. You have followed the manufacturer's recommendations in the process of swaging a cable to a terminal. The strength of the swaged fitting should be:
    (1) 90% of the actual rated cable strength
    (2) 80% of the actual rated cable strength
    (3) 75% of the actual rated cable strength
    (4) the same as the rated cable strength

194. Rolling or banking an airplane causes it to move about (or around) the longitudinal axis, yawing causes it to move about the vertical axis, and pitching causes it to move about the lateral axis. After starting a pitching motion during flight, the pitching motion increases. This increase in pitching motion is considered to be:
    (1) good longitudinal stability along the longitudinal axis
    (2) poor longitudinal stability along the longitudinal axis
    (3) good lateral stability along the lateral axis
    (4) poor lateral stability along the lateral axis

195. When a large metal covered airplane is moved from a cold hangar out into warm air, control cable tension will increase. Tensiometers are used to measure control cable tension. Control cable regulators are installed to:
    (1) make ground adjustments of cable tension
    (2) maintain specified or even cable tension
    (3) properly compensate for low atmospheric temperature conditions
    (4) properly compensate for high atmospheric temperature conditions

196. The airplane you are maintaining (as an A. and P. mechanic) is used for hire and was given a periodic inspection on November 15. On this same day you gave a similar inspection to a privately owned airplane which is not for hire. Assuming that each airplane proved to be airworthy and you so certified them in the logbooks, which of the following is correct?
    (1) the airplane for hire must have the next periodic inspection by at least November 15 of the following year
    (2) the privately owned airplane must have the next periodic inspection within the following 100 hours of flight time
    (3) neither of these airplanes is required to have an annual inspection
    (4) the airplane for hire must have the next periodic inspection within the following 100 hours of flight time and the privately owned airplane must have the next inspection within the following 12 calendar months or not later than November 30 of the following year

197. Airplanes must be flight tested for all major repairs, and for some minor repairs such as:
   (1) wing and control surface fairings
   (2) electrical installations of the same general nature
   (3) replacement of wing tips
   (4) cowling adjustment

3. THE AIRFRAME SYSTEMS AND COMPONENTS EXAMINATION

The subject matter covered in this section of the examination includes: (1) aircraft landing gear systems, (2) hydraulic and pneumatic systems, (3) cabin atmosphere control systems, (4) aircraft instrument systems, (5) communications and navigation systems, (6) aircraft fuel systems, (7) aircraft electrical systems, (8) position and warning systems, (9) ice and rain control, and (10) fire protection systems. The questions that follow, like those in the Government tests, cover all these subjects but not necessarily in the same order.

The answers to the questions in Section 3 are on page 371.

1. When aircraft tires are over-inflated they can cause damage to the wheel flange and excessive wear in the center of the tire tread. When under-inflated, excessive wear occurs on the outside of the tire tread. You should obtain proper tire pressure from the manufacturer's specifications. During removal or installation of aircraft tires:
   (1) deflate the tire prior to removal
   (2) inspect the tire for evidence of flex breaks, separation between plies, kinked or broken beads, blisters or heat damage, and cracked, deteriorated, or damaged inner liners of tubeless tires
   (3) reinstalled tires should be inflated, deflated, and again inflated to insure that the innertube is not pinched.
   (4) all three of the above answers are correct

2. Aircraft tires may be retreaded by approved tire repair and retread stations as long as they do not show damage or wear that exceeds

the limitations set forth by the Administrator. Aircraft tires should be stored in closed areas where the atmospheric conditions will be:
(1) cool and dry
(2) warm and moist
(3) cold and dry
(4) cool and relative humidity not over 85%

3. On aircraft with retractable landing gear there must be an emergency means to extend the landing gear. Servicing an oleo (air-oil) landing gear strut must assure that the oil level and air pressure charge do not permit the strut to bottom on landing or taxiing. You can adjust landing gear toe-in by:
   (1) balancing and proper inflation of the tires
   (2) adding, removing, or change the washers on the scissors attachment
   (3) proper adjustment of the anti-skid system
   (4) adjustment of the packing gland seal

4. Power for aircraft hydraulic brakes is derived from the master cylinder. If the master cylinder leaks (worn piston) the brake pedal will be soft or fade. If the return spring is weak or broken, the brakes will not release and cause a dragging action. Regarding aircraft brakes, which of the following is correct?
   (1) brakes should not be applied when the wheel has been removed from the assembly as this can cause damage to the brake system
   (2) air in the brake system will cause a hard brake pedal
   (3) installation of tires on wheels does not require any particular relative positioning between the wheel and tire
   (4) internal leakage in the master cylinder will have no adverse effect on brake operation

5. The debooster in a hydraulic power brake system serves to:
   (1) increase brake pressure to higher requirements of the main hydraulic system
   (2) decrease the hydraulic pressure between the engine-driven hydraulic pump and the nearest actuating cylinder
   (3) decrease main system hydraulic pressure to satisfactory brake pressure
   (4) serves as an emergency means to reduce brake pressure when necessary

6. You observe a raised H on a valve core. This indicates:
   (1) that the valve core is a part of the hydraulic system
   (2) high grade fluids only are to be used in the system
   (3) the system is of high pressure
   (4) the point of a high pressure test location

7. When removing air from an oleo strut, you should first "crack" the valve to release the air slowly. Air in the brake system should be eliminated by bleeding the lines to eliminate the air. When checking an oleo strut, you find the fluid level is up to the filler hole. You should:
   (1) remove and replace the metering pin
   (2) remove and replace the entire strut
   (3) fill with fluid and reinstall the filler plug
   (4) assume the fluid level is satisfactory and return the oleo strut to service

8. The tube of an airplane tire should be lined up with:
   (1) the red dot on the tire
   (2) the line that indicates a possible slippage of the tire on the wheel
   (3) the valve core of the tube at the red dot on the tire
   (4) any point on the tire as the ultimate balance will always be perfect after mounting the wheel

9. Hydraulic fluid has been spilled on an airplane tire. You should clean it off with:
   (1) gasoline
   (2) kerosene
   (3) warm water and mild soap
   (4) paint thinner

10. The location of the brake debooster is:
    (1) between the power brake control valve and the brakes
    (2) in the hydraulic system just ahead of the power brake control valve
    (3) between the relief valve and the power brake control valve
    (4) within the brake control unit since it is an integral part of the brakes and brake shoes

11. How do you know when the landing gear struts are full of oil during filling procedures?
    (1) back off the filler plug until the shock strut is half-compressed, then fill with fluid to the filler level
    (2) release air to 75% compressed level and fill to the filler opening
    (3) back off the filler plug until the shock strut is fully compressed, then fill with fluid to filler level or until the fluid runs out of the filler opening
    (4) none of the above answers is correct

12. If the landing gear strut collapses upon landing but is okay during taxiing, the probable cause is:
    (1) extremely low air charge
    (2) a low oil (fluid) supply

(3) of no concern since this sometimes occurs when the system checks out satisfactorily

(4) the actuating cylinder has worked itself out of line with the housing

13. The brake pedal is spongy after complete and proper bleeding. The cause could be:
    (1) trapped air in the brake lines
    (2) compression of the brake fluid
    (3) worn or deteriorated flexible hose line in the brake system
    (4) loss of fluid through the master cylinder return manifold line

14. After bleeding the brakes a mechanic finds they are still spongy. This is probably caused by:
    (1) extremely rigid flexible hose installed in the brake system
    (2) flexible hose that has deteriorated
    (3) incorrect fluid in the brake system
    (4) servicing with fluid of light density

15. When for any reason the hydraulic system is not functioning on a multi-engine airplane, there is usually incorporated which of the following for braking purposes?
    (1) a coaxial, emergency, foolproof type of brake
    (2) an independently operated brake emergency air system that provides a means of applying the brakes
    (3) a coordinated cable-operated emergency expander type brake
    (4) an emergency system that utilizes the best points of the spot and multiple-disc type of brakes

16. Surface cracks on the friction surface of brake drums occur frequently because of high operating temperatures. These surface cracks may be:
    (1) hazardous to brake operation and should be immediately removed when detected
    (2) disregarded unless they become cracks about an inch long, in which case the brake drum should be replaced
    (3) completely ignored regardless of size since only a small portion of the total brake drum area can possibly be affected by them
    (4) disregarded unless they become cracks about a fourth of an inch long, in which case the brake drum should be replaced.

17. In the air-oil shock struts the landing shock is controlled by:
    (1) compressed air
    (2) a tapered metering pin that gradually restricts the flow of oil
    (3) air pressure regulated by the snubber bleed valves
    (4) an orifice check valve that gradually restricts the compression of the outer oleo cylinder

18. Aircraft brakes are classified as: (a) expanding shoes, (b) spot type, (c) multiple disc, (d) single disc, and (e) expander tube type. Of these types of brakes, which one utilizes a single notched disc splined to and turning with the wheel and which runs between several parallel pairs of round linings called pucks?
    (1) a
    (2) b
    (3) c
    (4) d
    (5) e

19. If the main relief valve in a hydraulic system fails, the pressure pump output is bypassed back to the hydraulic reservoir. The shear section (or shear pin) on the shaft of an engine-driven hydraulic pump:
    (1) strengthens the basic pump housing
    (2) will release pump pressure (or operation) if all pressure relief valves fail to operate properly
    (3) lengthens the effective stroke of the pistons within the pump
    (4) relieves the hydraulic fuse when system surges occur

20. The flap overload valve, incorporated in the flap operating mechanism, prevents the flaps from lowering at excessive aircraft speeds thus preventing damage to the flaps and flap system. A normal hydraulic system can be automatically changed to an emergency system by the use of:
    (1) a shuttle valve
    (2) an orifice check valve
    (3) a power control unit
    (4) an interconnected by-pass valve

21. The basic purpose of an accumulator in a hydraulic system is to act as a surge dampener which decreases the shock of sudden pressure changes. The accumulator also temporarily stores fluid under pressure when the engine-driven hydraulic pump is not operative. The accumulator:
    (1) is generally mounted with the hydraulic fluid side on top and the air side on the bottom
    (2) incorporates a bladder (or flexible separator) between the fluid and air and is generally charged to ⅓ of the particular system pressure
    (3) if not properly charged, can cause too frequent cycling of the hydraulic system
    (4) all three of the above answers are correct

22. When necessary to adjust several valves in a hydraulic system, the units with the highest pressure settings are adjusted first, units with next highest pressure second, etc. The type of hydraulic pressure pump that has four check valves is known as a:

    (1) single-double actuating pump
    (2) standard gerotor pump
    (3) double action pump
    (4) standard sliding vane positive displacement pump

23. You should, prior to tightening the packing gland nut on a landing gear:
    (1) jack up the aircraft
    (2) release all pressure on the hydraulic accumulator
    (3) release the air from the strut
    (4) remove all fluid from the strut

24. Quick-disconnect fittings, in a hydraulic system, would most likely be found at the:
    (1) inlet side of the positive displacement pump
    (2) outlet side of the hydraulic reservoir
    (3) firewalls
    (4) negative end of an actuating cylinder

25. Porous paper filters in a hydraulic system should be:
    (1) cleaned by being washed in naptha and dried thoroughly
    (2) cleaned by turning the filter crank at least three complete turns
    (3) replaced after a specified time in service on the aircraft
    (4) washed in nepthane, slowly dried, and then dipped in vegetable fluid

26. You should use heavy paper or shim stock when:
    (1) installing a hydraulic accumulator
    (2) installing hydraulic seals over threads
    (3) scraping foreign material from hydraulic units
    (4) the accumulator center section is out of round

27. When using Skydrol fluid in hydraulic systems:
    (1) natural rubber seals should be used
    (2) synthetic (neoprene) rubber seals should be used
    (3) either natural rubber or synthetic seals are compatible with Skydrol
    (4) seals of a pressed fibre type should be used because of the deteriorating effect of Skydrol on either natural or synthetic rubber seals
    Note: Synthetic rubber is also known as Butyl rubber.

28. In the hydraulic system a grease retainer seal becomes hardened. You should soak it in:
    (1) oil
    (2) water
    (3) kerosene
    (4) gasoline

29. Hydraulic packing rings (also known as seals or gaskets) are of various shapes: (a) oval or round, and generally referred to as O rings, (b) chevron or V type, (c) cup-shaped or U type, and (d) square-shaped or square type. These packing rings are made from:
    (1) natural rubber
    (2) synthetic rubber (neoprene)
    (3) leather
    (4) all three of the above answers are correct

30. A particular hydraulic system is normally serviced with a vegetable-based fluid. By mistake, a mechanic services it with a mineral-based (red) hydraulic fluid. Upon discovering the mistake the mechanic should:
    (1) drain all lines and recondition all units that were contaminated by the mineral fluid
    (2) do nothing inasmuch as red mineral fluid is compatible with vegetable fluid
    (3) drain and flush the system with a weak thinner and service the system with vegetable-based fluid
    (4) complete the servicing and use red mineral-based fluid in the system from there on

31. Which of the following is used to install O rings over fittings?
    (1) paper or shim stock over the threads will give adequate protection
    (2) nothing, since O rings are made of tough material and can withstand being pulled or pushed over threaded portions of a fitting
    (3) any type of fabric material which is considered tear-proof
    (4) none of the above answers is correct

32. When a hydraulic system is serviced with Skydrol:
    (1) butyl seals should be used
    (2) caution must be observed because Skydrol is highly corrosive to metals
    (3) the Skydrol is relatively non-inflammable
    (4) all three of the above answers are correct

33. Synthetic rubber seals are compatible with mineral-based hydraulic fluid. Natural rubber seals are compatible with:
    (1) vegetable-mineral-based hydraulic fluid
    (2) mineral-based hydraulic fluid
    (3) alcohol-based hydraulic fluid
    (4) vegetable-based hydraulic fluid

34. A mechanic can determine the type of fluid used in an airplane hydraulic system by:

(1) draining a small quantity of fluid and placing it in a fluid-tester calculator

(2) checking the instructions printed on the decal found at the hydraulic reservoir tank

(3) checking the instructions printed on the decal found in the pilot's compartment

(4) checking the color, number, and type of seals used in the particular system

35. The hydraulic system unloading valve:
   (1) operates when the system is up to normal pressure
   (2) acts as a means of hydraulic system relief
   (3) operates at a time in which the hydraulic pumps are inoperative
   (4) all three of the above answers are correct

36. Which of the following in a hydraulic system will operate at the highest pressure value?
   (1) the system pressure regulator valve
   (2) a debooster system valve
   (3) the brake system relief valve
   (4) the hydraulic system relief valve

37. Regarding hydraulic lines installed in crew, baggage, or passenger compartments, and in entrance ways:
   (1) installations of this type are not permissible according to AC 43.13-1
   (2) installations of this type are permissible only if all surrounding materials are absolutely fireproof
   (3) they may be installed if properly protected from physical damage and properly vented
   (4) they may be installed if properly protected from physical damage, are properly vented, and are completely enclosed in an airtight capsule

38. Aircraft hydraulic systems utilize hydraulic fuses, which permit:
   (1) hydraulic fluid to flow only so long as there is no substantial pressure difference existing across the fuse
   (2) hydraulic fluid, when overheated, to bypass through the fluid cooler
   (3) the level of the reservoir to be constant through reserve fluid held in a standby tank
   (4) hydraulic fluid to flow so long as there is substantial pressure difference existing across the fuse

39. An engine-driven multi-stage reciprocating compressor is utilized in an aircraft pneumatic system. Which of the following would this system require?

(1) a sensitive and positive vacuum relief valve
(2) an approved oil separator
(3) a heavy-duty surge chamber
(4) a moisture separator

40. The hydraulic unit that permits normal flow in one direction and a restricted flow in the other direction is known as the:
(1) orifice sliding check valve
(2) orifice flow valve
(3) orifice check valve
(4) solenoid expander valve

41. The name of the hydraulic valve that opens the landing gear door prior to landing gear extension is:
(1) sequence valve
(2) gear-door time lag valve
(3) time lag extension valve
(4) reverse action ball-check valve

42. A common name for a sequence valve in a hydraulic system is:
(1) reverse check-out valve
(2) diaphragm-pressure relief valve
(3) time lag valve
(4) lag valve with a pop-out feature

43. A hydraulic system accumulator has been charged to 1,000 pounds per square inch (PSI) at a time when there is no hydraulic pressure in the system. The system is then charged to 3,000 PSI. Under these circumstances, the accumulator pressure should read (in PSI):
(1) 4,000
(2) 3,000
(3) 2,000
(4) 1,000

44. The hydraulic hand pump is plumbed into the:
(1) top of the hydraulic reservoir
(2) bottom of the nearest actuating piston
(3) bottom of the hydraulic reservoir at a standpipe port
(4) hydraulic reservoir vent line
Note: The suction line of the hand pump is joined to one port of the standpipe. The larger port in the standpipe is joined to the engine pump supply port.

45. The probable cause of severe chattering of an engine-driven hydraulic pump is:
(1) the ball check valve spring is too weak

(2) the check valve in the inlet side of the pump is stuck in an open position

(3) an air leak located in the inlet side of the hydraulic pump

(4) a stuck open main system relief valve

46. A hydraulic hand pump will draw fluid but will not pump pressure into the hydraulic system. The likely cause is:
(1) the check valve is not operating properly
(2) seals are leaking
(3) fluid is reverse flowing
(4) the engine driven pumps have cavitated

47. In aircraft hydraulics, acceptable fluids must have a low thermal expansion value. This is necessary to:
(1) reduce the possibility of fire to a minimum
(2) prevent the possibility of evaporation at high altitudes
(3) decrease the possibility of excessive pressures when the hydraulic units are exposed to high temperatures
(4) reduce the empty weight of the aircraft to an absolute minimum

48. A swaged-end hydraulic fluid hose has become damaged. You should:
(1) remove the damaged section and replace with a metal tubing and clamp combination
(2) properly replace the hose assembly
(3) replace the hose portion after removing the damaged ends
(4) remove the damaged section and replace with AN fittings and nipples

49. Which of the following hydraulic units convert hydraulic pressure into linear motion?
(1) scissor attachment fittings
(2) ball-check valves which have a horizontal motion during operation
(3) actuating cylinders incorporated in the hydraulic system
(4) hydraulic accumulators which are an integral part of the hydraulic system

50. What type of high-altitude multi-engine aircraft would *not* have a pressurized hydraulic reservoir?
(1) one that has a completely pressurized cockpit and passenger compartment
(2) any aircraft of this catagory equipped with multi-stage internal and external superchargers
(3) one which has the hydraulic system operating slowly on the piston type hydraulic pumps that incorporate shear pins
(4) a high-altitude multi-engine aircraft that has the hydraulic reservoir installed in a pressurized section of the cabin

51. The standpipe in a hydraulic system:
    (1) is in the accumulator
    (2) is adjacent to the hydraulic pumps
    (3) is at the debooster unit
    (4) projects into the fluid reservoir from the bottom of the tank

52. A bourdon tube is generally utilized in hydraulic systems. Which of the following best describes it?
    (1) it is used in the air-oil strut to prevent excessive fluid expansion
    (2) it performs the same functions as the spool-type regulator
    (3) it indicates the amount of pressure in the hydraulic fluid reservoir
    (4) it is the basic operating mechanism of many hydraulic pressure gages

53. The sight glass of a Freon air conditioning system is showing a low indication. Under this condition, you should:
    (1) immediately refill system, recheck the level, and place system back in service
    (2) continually operate system until the level stabilizes to a satisfactory level
    (3) check for leaks by operating the system
    (4) replenish the oil to the Freon and place system back in service

54. The jet pump in the cabin pressurization system of a jet aircraft consists of a venturi arrangement that uses bleed air to provide a source of low pressure to the control chambers of the outflow valve. In addition, a jet pump recirculates air from beneath the floor of cargo components. Another purpose of a jet pump in an air conditioning system in aircraft is to:
    (1) create a pressure on the compressor Freon
    (2) draw large amounts of air through the heat exchanger
    (3) pump extremely cold air to the mixing valve unit
    (4) create a pressure on the bleed air that goes to the turbine

55. Cabin pressurization of an aircraft consists of maintaining specified cabin altitude (generally 8,000 feet) to flight altitudes as specified in the operations manual of the aircraft and commensurate with government regulations. A pilot determines his flight rate-of-climb to arrive at a cruising altitude at a given time. He then sets the cabin rate-of-climb so that the cabin will be at, say, 8,000 feet when the aircraft reaches the cruising flight altitude of, say, 35,000 feet. During the flight climb, if the cabin rate-of-climb rises too rapidly:
    (1) there should be an adjustment of the inflow of air
    (2) there should be an adjustment of the outflow of air
    (3) the cabin pressure should immediately be equalized
    (4) do nothing as it will automatically adjust to a new rate-of-climb

56. Corrosion sets in and rusts the tanks if you allow oxygen bottles to remain with a low supply of oxygen in them. The leakage rate, after fully pressurizing an oxygen system, should not exceed:
    (1) 5% within a period of 24 hours
    (2) 3% within a 12 hour period
    (3) 1% within a 24 hour period
    (4) zero, as no leakage is permitted

57. In an air conditioning system, the mixing valve:
    (1) permits the mixing of cabin air and outside air
    (2) mixes flight deck air with passenger cabin air
    (3) limits the flow of cool, cold, and hot air
    (4) creates a condition of very low relative humidity

58. All of the Freon has escaped in an air conditioning system. Under these circumstances you should:
    (1) refill and return to service
    (2) evacuate, refill, and return to service
    (3) purge, refill, and return to service
    (4) evacuate, purge, evacuate, refill, and return to service

59. If the compressor fails, which of the following prevents loss of cabin pressurization?
    (1) air that is available from the secondary system
    (2) a check valve (one-way valve) in the air duct
    (3) the secondary cabin outflow valve
    (4) a duct that is controlled by the turbo-compresser

60. The thermostat in a small aircraft internal combustion heater system:
    (1) automatically turns the heater switch on and off as needed
    (2) automatically turns the igniter on and off as needed
    (3) meters the fuel as needed
    (4) cycles the fuel as needed

61. Cabin heating systems that utilize an exhaust heat exchanger as a source of heated air:
    (1) must be adequately checked for leaks due to the presence of carbon monoxide
    (2) should be checked frequently for leaks of exhaust gases
    (3) should be checked with a carbon monoxide testing instrument
    (4) all three of the above answers are correct

62. Which of the following actions should be taken if the cabin altitude warning device sounds in flight at 35,000 feet mean sea level (MSL)?
    (1) immediately check the warning device for possible malfunction
    (2) additional compressors should be turned ON

(3) immediately open the cabin pressure differential valves
(4) the passenger oxygen masks should be dumped

63. The outflow valve in the pressurization system relieves pressure if the cabin pressure exceeds the maximum allowable and the negative pressure relief valve relieves pressure if the pressure outside is greater than inside the cabin. The expansion turbine in the pressurization system cools the air flowing through it because the air:
(1) decreases in volume and decreases in pressure
(2) increases in volume and increases in pressure
(3) increases in volume and decreases in pressure
(4) none of the above answers is correct

64. Cabin pressurization is an important function of aircraft flying at high altitudes. In general, cabin pressurization differential is normally controlled by:
(1) adjusting the outflow valve at a constant pressure that varies only with engine RPM and MAP
(2) the utilization of constant volume superchargers and an automatically positioned cabin outflow valve
(3) the proper regulation of the butterfly valves located between the superchargers and the cabin
(4) utilization of cabin supercharger speed at fixed rates by a constant speed drive and without regard to flight altitude

65. The flow sequence in an air-cycle cooling system is: heat exchanger, compressor, heat exchanger, and turbine. The heat source for an air conditioning system during flight is the heat of compression. In the operation of the Freon system the Freon gas is compressed to a high temperature and pressure combination in the motor-driven two-stage centrifugal compressor. As it leaves the discharge of the compressor, the high temperature gas enters the condenser where it gives up heat to the ambient cooling air drawn by the condenser fan across the condenser surface. The removing of heat from the vapor is sufficient to condense the Freon gas to a liquid. The hot pressurized liquid Freon is then directed to the expansion valves. Thus, the liquid passing through the expansion valves experiences a reduction in pressure and temperature, and changes into a mixture of liquid gas. This mixture:
(1) enters the superchargers where it absorbs heat from the air, thus cooling the aircraft interior
(2) leaves the evaporators where it absorbs heat from the air, thus cooling the aircraft interior
(3) cools the aircraft interior after a final stage of supercharging the mixture
(4) enters the evaporators where it absorbs heat from the air, thus cooling the aircraft interior

66. When checking for oxygen leaks, you should use:
    (1) a high flash-point mineral solution
    (2) a soap solution specially prepared for this purpose
    (3) any suitable solution that will be compatible since oxygen is not inflammable
    (4) a kerosene solution since kerosene is not highly inflammable

67. Generally speaking, pure oxygen (the type used in breathing) may be used for welding purposes. What is added to the type of oxygen used normally for welding so that it can be used to meet crew members' oxygen requirements?
    (1) hydrogen
    (2) H_2O (water)
    (3) argon and helium
    (4) nothing—this type of oxygen cannot be used for breathing

68. The airspeed indicator uses both static and pitot (impact) pressure. If the main static source becomes disconnected within a pressurized cabin, the altimeter will indicate:
    (1) a high reading
    (2) no change from normal
    (3) cabin pressure
    (4) a low reading

69. When adjusted to the current or proper altimeter setting, the altimeter will indicate:
    (1) true altitude of the airplane
    (2) the approximate height of the airplane above MSL
    (3) the altitude of the airplane above field elevation
    (4) absolute altitude of the airplane

70. The main static source for the pitot static system becomes disconnected during flight. If the alternate source, located in the unpressurized cabin, is used:
    (1) a momentary descent will be indicated by the rate of climb indicator
    (2) the airspeed and altimeter indications will be too high
    (3) there will be no effect on the airspeed indicator because it senses pitot (impact) pressure only
    (4) the altimeter will indicate too low and the airspeed indicator will indicate too high

71. A vacuum system has a common intake. There should be which of the following?
    (1) a very small restrictor in the altimeter vacuum line
    (2) one main filter for all instruments

(3) one major filter for each of the instruments

(4) one filter for the vacuum manifold and one filter for each of the instruments

72. Which of the following instruments normally have operating range and limit markings?
    (1) directional gyro and airspeed indicator
    (2) cylinder head temperature gage (CHT) and altimeter
    (3) airspeed indicator and CHT
    (4) magnetic compass and rate of climb indicator

73. The operating principles of a thermocouple temperature indicating circuit is which of the following?
    (1) the circuit requires an outside source of direct current
    (2) there is alternator electrical means needed for thermocouple operation
    (3) two types of metal incorporated in the circuit (generally iron and constantin), when heated, create small unlike electrical currents that are transmitted through connecting wires to the receiver (instrument)
    (4) two similar types of metal (generally bronze) incorporated in the circuit, when heated, create small similar electrical currents that are transmitted through connecting wires to the receiver (instrument)

74. Most pressure gages are of which of the following types?
    (1) Bourdon tube
    (2) Magnesyn
    (3) Autosyn
    (4) Wheatstone

75. The face or dial of the airspeed indicator has several colored markings. They are: (a) the green arc, which indicates normal operating speed range; (b) the red line which indicates the never exceed speed; and (c) the yellow arc which indicates the operating caution range of speed. The white arc indicates the:
    (1) normal climbing speed
    (2) manuvering speed
    (3) flap operating speed
    (4) normal takeoff and landing speeds

76. If the static line in a pressurized cabin breaks, which of the following will occur?
    (1) only the altimeter will read low
    (2) only the airspeed indicator will read low
    (3) both altimeter and airspeed indicators will read high
    (4) both altimeter and airspeed indicators will read low

77. Because of its proximity to metal, electricity, etc., in the cabin of an aircraft, the magnetic compass should be "swung" frequently to determine which of the following errors?
    (1) variation
    (2) osculation
    (3) deviation
    (4) hysteresis

78. An instrument static system leak can be detected by observing the rate of change in the:
    (1) altimeter after suction has been applied to the static system to cause a prescribed equivalent altitude to be indicated
    (2) altimeter soon after pressure has been applied to the static system to create a prescribed equivalent altitude to be indicated
    (3) airspeed indicator after pressure has not been applied to the static system to cause a prescribed equal airspeed to be indicated
    (4) airspeed indicator after suction has not been applied to the static system to cause a prescribed equivalent airspeed value to be indicated

79. The primary purpose of the autopilot is to:
    (1) hold a predetermined heading of the aircraft under adverse wind conditions
    (2) relieve the pilot of undue stress during long periods of flight
    (3) aid the pilot when making the actual landing (not simulated) on the runway
    (4) compensate for changes in altitude, attitude, and heading under rough or turbulent conditions

80. During operation of an aircraft on autopilot, when the aircraft deviates from the altitude and heading in which the pilot has placed it when he engaged the autopilot, which of the following units send attitude signals?
    (1) servo motors
    (2) synchros in the gyros
    (3) gimbal rings
    (4) transponders

81. In high performance aircraft, which of the following regarding autopilot operation is most appropriate?
    (1) the autopilot generally can do a better job of flying under normal circumstances than can the human pilot
    (2) the autopilot can sense changes in heading, pitch, and trim faster than the human pilot generally speaking

(3) generally the autopilot can apply a more precise correction than the human pilot

(4) all three of the above answers are correct

82. In the basic autopilot operation, a vertical gyro, such as the remote gyro for the attitude indicator, provides a stable reference platform for pitch and roll axes. A directional gyro, for the heading indicator, provides a stable directional reference. To each of the three flight surfaces: elevator, aileron, and rudder, there are attached servo motors capable of moving the control surface. When the autopilot has been engaged by the pilot, synchros in the gyros signal:
    (1) the gimbal ring through a central coordinating computer to correct the deviation by movement of the control surface
    (2) the control surface needed to correct the deviation in a direct action
    (3) the servo motors through a central coordinating computer to correct the deviation by movement of the control surface
    (4) none of the above answers is correct

83. In the use of the autopilot a pilot should:
    (1) trim the aircraft properly before engaging the autopilot
    (2) pay attention to the aircraft trim during autopilot flight
    (3) know the limits of the autopilot and its affect on passenger comfort, especially during turbulent weather flight
    (4) all three of the above answers are correct

84. Which of the following units applies torque force to the control surface in an autopilot system?
    (1) the autopilot controller
    (2) the servo unit
    (3) a rate limiter
    (4) a gimbal ring

85. In the autopilot system, the position transmitter senses:
    (1) the actual rate of roll, pitch, and yaw
    (2) an indication of the static and dynamic surface pressure
    (3) the angular displacement of a control
    (4) any deviation from a selected heading

86. A jet pump may be utilized in a radio or electronics compartment. Its purpose is to:
    (1) prevent flashover by pressurizing the electrical units
    (2) adequately remove moisture or condensation from the area
    (3) adequately draw more air through the compartment for proper cooling of the equipment
    (4) to create a vacuum to aid the operation of the vacuum tubes

87. When installing radio equipment, it should be mounted:
   (1) by being supported by rigid grounding clamps
   (2) so as to be subjected to only ultimate accelerations
   (3) with adequate shock mounts between the aircraft structure and the equipment
   (4) by bolting direct to the aircraft primary structure

88. Which of the following will reduce engine noise in radio reception?
   (1) install proper shielding in the ignition system
   (2) dress the generator brushes by using emery cloth
   (3) install a growler in the generator circuit
   (4) reduce the secondary voltage output to the ignition distributor

89. Which of the following is correct in regards to Federal Communications Commission (FCC) regulations pertaining to the operation of two-way radio in aircraft?
   (1) each pilot (or other person) using two-way transmission and receiving must hold at least an FCC Radiotelephone Permit and have it on his person
   (2) each aircraft (with two-way radio) is assumed to be a mobile radio station and must have a proper FCC Radio Station License posted in the cockpit or flight deck area
   (3) any person approved for taxiing an aircraft on a radio controlled airport must hold at least a Restricted Radiotelephone Permit (FCC) and communicate with the control tower for permission to taxi or otherwise move the aircraft from one place to another on the airport
   (4) all three of the above answers are correct

90. An important consideration when installing radio equipment is:
   (1) that it will be as static-free as possible
   (2) the effect it will have on the weight and balance of the particular aircraft
   (3) that it will be as vibration-free as possible
   (4) all three of the above answers are correct

91. The preferred location of a VOR antenna on a small airplane is:
   (1) top forward of the cabin with the apex of the V pointing rearward
   (2) any conventional location so long as the structural considerations are satisfied
   (3) top forward of the cabin with the apex of the V pointing forward
   (4) top of the vertical stabilizer with the apex of the V pointing rearward

92. When installing a loop or DME (Distance Measuring Equipment) antenna on an aircraft, you should align it:
   (1) in such position that the "null" will be deleted
   (2) with the aircraft neutral axis

(3) as near the aircraft centerline as possible

(4) with the neutral-lateral axes

93. When making a loop-type antenna installation on the lower part of the fuselage, do not fasten the loop antenna to a primary structural member because:

(1) the unlike metals will set up an electrolytic action

(2) this type of installation will increase the corona and also the static in the radio reception.

(3) in the event of a landing with the landing gear retracted, the airplane may be severly damaged.

(4) this will tend to weaken the primary structure and cause a semi-monocoque structural reaction

94. Regarding fuel dump systems in large aircraft:

(1) fuel must discharge clear of any part of the aircraft structure

(2) the entire fuel dump system must be free of any possible fire hazard

(3) the fuel tanks are inter-connected to a common manifold

(4) all three of the above answers are correct

95. Flapper valves in large jet fuel tanks:

(1) set up a variable resistance within the fuel tank

(2) serve as a check valve, provide positive flow to the boost pump, and permit removal of a boost pump without draining the fuel tank

(3) permit sufficient negative pressure to form in the fuel tanks

(4) permit a positive pressure on top of the fuel to assure proper feed to the boost pump

96. Dirt, water, and other foreign matter normally drain to the sump in the bottom of the fuel tank. Ethylene dibromide is added to aviation gasoline to break up lead deposits that form in the combustion chambers of the engine cylinders. Ethylene dibromide also helps to suppress detonation when using takeoff engine power. Fuel strainers are generally located in a fuel system:

(1) between the fuel tank outlet and the fuel pump

(2) in the outlet side of the carburetor

(3) between the vapor vent line and the tank

(4) between the engine fuel pump and the gascolater

97. Aircraft should always be defueled outside in the open air. In fuel systems that have cross-feeding, the cross-feed lines can be utilized to:

(1) balance out the fuel load

(2) properly transfer fuel from one tank to another

(3) supply the fuel from any tank to one engine

(4) all three of the above answers are correct

98. To repair an integral fuel tank you should use:
    (1) rivets
    (2) bronzing
    (3) welding
    (4) soldering

99. Single-point refueling has which of the following primary advantage?
    (1) vapor elimination during the refueling operations
    (2) reduces the otherwise larger number of personnel
    (3) reduces the time of refueling
    (4) eliminates completely potential fire hazards

100. Inter-connected fuel tanks are vented to:
    (1) eliminate fuel vapors
    (2) prevent fuel expansion at high atmospheric temperature
    (3) prevent condensation (moisture) forming in the tanks
    (4) permit equalization of fuel tank pressures

101. Fuel systems that have probes in each tank are described as:
    (1) electro-mechanical
    (2) electrical
    (3) manual
    (4) inductor-reactance

102. An aircraft engine fuel system incorporates emergency valves. If the valves are in the shut-off position, the:
    (1) warning light (fire) in the flight deck will be ON
    (2) fuel will be shut off to the particular engine
    (3) fire warning system will be de-activated
    (4) fuel will be drained from the carburetor fuel chambers

103. Two inter-connected fuel tanks and one valve are incorporated in a gravity type fuel system. A requirement, in this case, is that:
    (1) all outlets must be of the same diameter
    (2) both tanks must be serviced with the same quantity of fuel
    (3) both fuel tanks must be properly vented to each other
    (4) there must be incorporated an automatic valve to close whenever either tank is empty

104. An aircraft is equipped with a manual fuel selector valve on the flight deck. In this case the valve should be:
    (1) so installed that loads, etc., are transmitted to the lines rather than to the valve itself

(2) embossed with definite index positions or stops so the flight crew can change fuel tanks, etc., under all types of lighting conditions, even in total darkness

(3) pre-tested under adequate pressure at each 50 hours of flight time

(4) always installed on the left side of the cockpit (or flight deck) so that it is within easy reach of the pilot in command

105. The primary advantage of integral fuel tanks is reduced:
    (1) weight
    (2) capacity
    (3) fuel leakage
    (4) fire hazard

106. Which of the following precautions should *not* be taken when a mechanic is to work inside a fuel cell?
    (1) before entering the fuel cell touch a static discharge plate
    (2) another person should stand by outside the fuel cell
    (3) defuel and purge the fuel cell inside an air-conditioned hangar
    (4) the fuel cell should be purged continually while the work is in progress

107. When defueling a swept-back wing, the outboard cells should be defueled first. Fuel outlet finger screens installed in the bottom of a fuel tank should be:
    (1) the same size or greater than the diameter of the fuel tank outlet
    (2) double the size of the fuel tank outlet
    (3) only equal in size to the fuel tank outlet
    (4) the same size as other screens in the particular fuel system

108. Fuel boost pumps are an aid in the prevention of vapor lock in the fuel lines. Fuel tanks are required to withstand an internal pressure of at least 3.5 PSI and have an expansion space of at least 2% of the tank capacity. Flux should be removed from a fuel tank, after repair by welding, by:
    (1) flushing the tank with 10% silver nitrate for at least two hours
    (2) light sanding, followed by washing thoroughly with nitric acid
    (3) washing both inside and outside with hot water, follow with an acid bath, then thoroughly rinse with hot water
    (4) using a solution of boric acid to rinse the tank inside and outside

109. Baffle plates installed in fuel tanks tend to decrease the surging (or sloshing) of the fuel while in flight. This reduces the movement of the fuel and aids airplane stability. Which of the following markings should be on or near the fuel filler openings?
    (1) type of seals used throughout the fuel system

(2) blue marking for low-octane fuel, red marking for high-octane fuel, or a white marking for jet fuel

(3) the type of fuel to be used and the capacity of the fuel tank

(4) the type of fuel to be used and the tank pressure limits

110. Vapor lock (excessive fuel evaporation) in the fuel lines is least likely to occur:
   (1) at sharp bends in the fuel lines
   (2) at a steep rise or curvature in the fuel lines caused by obstructions in the aircraft structure
   (3) between the engine fuel pump and the carburetor
   (4) between the main fuel strainer and the engine fuel pump

111. The major cause of vapor lock (restriction of fuel flow) is:
   (1) low fuel pressure and high fuel temperature in one or more sections of the fuel system
   (2) fuel that is exceptionally susceptible to vaporization and excessive turbulence
   (3) excessive heat in close proximity to any section of the fuel system
   (4) all three of the above answers are correct

112. In aircraft operation, aromatic fuels:
   (1) should not be used because of their high potential heat value
   (2) require that pure rubber fuel hose should not be used in the fuel system
   (3) may be used only in engines that are compounded
   (4) may be used only in engines that have power recovery turbines (PRT's)

113. In the event an error has been made in the fueling of a turbojet aircraft, it can normally be de-fueled by means of the:
   (1) dump chutes
   (2) fuel cross-feed lines
   (3) under-wing filler points
   (4) over-wing filler points

114. In a typical turbojet airplane, fuel is supplied under pressure directly from each fuel tank to the corresponding engine by the main fuel tank boost pumps. Regarding fuel systems on jet aircraft, which of the following is most appropriate?
   (1) fuel may be supplied to the engine from the fuel manifold by opening the fuel manifold valve and shutting off or over-riding the main tank boost pump to that engine
   (2) an independent tank-to-engine fuel feed system is used for each engine and is interconnected by a fuel manifold, and fuel can be supplied to any or all of the engines through this manifold

(3) fuel is supplied to the fuel manifold from the center tank or from any main tank, but the reserve tanks are not connected to the fuel manifold and check valves prevent the flow of fuel from the manifold into the fuel tanks

(4) all three of the above answers are correct

115. Excess fuel from the engine-driven fuel pump:
(1) returns to the main fuel tank
(2) is returned to the outlet port of the pump
(3) is temporarily stored in the sylphon bellows
(4) is returned to the inlet side of the pump

116. The purpose of a ratiometer is to:
(1) indicate the level of the fuel in the tanks and is electrically operated
(2) determine the ratio between the primary and secondary output of the ignition system
(3) indicate the fuel-air ratio of the fuel being burned in the combustion chamber of the engine cylinders
(4) determine the density of the fuel-air ratio in the carburetor float chamber

117. Potassium dichromate is used in aluminum fuel tanks to:
(1) counteract corrosion
(2) aid in fuel filtering
(3) counteract internal fuel pressure
(4) add to the fuel's capacity for better vaporizing

118. The fuel booster pump is generally which of the following types?
(1) vane
(2) gerotor
(3) centrifugal
(4) gear

119. The engine-driven fuel pump is generally which of the following types?
(1) vane
(2) gerotor
(3) centrifugal
(4) gear

120. An electrical centrifugal boost pump located in the fuel tanks creates fuel flow to the engine-driven sliding vane fuel pump. The fuel then enters the pressure carburetor (via appropriate screen or strainers) where the vapor vent float is located. Assuming the vapor vent float leaks, fills with fuel, and sinks:
(1) the engine will not be effected due to the internal safety device

    (2) vapor will continue to be vented back to the fuel tank with no apparent malfunction of the engine

    (3) fuel will be syphoned back to the fuel tank with a possible malfunction in engine operation

    (4) the fuel flow to the poppet valve will be completely shut off

121. Which of the following applies to aircraft fuel systems?
    (1) there must be fuel shutoff valves to at least two of the engines on a four-engine aircraft
    (2) all aircraft must have a fuel shutoff valve to each engine
    (3) all aircraft with more than one engine must have a fuel shutoff valve to each engine
    (4) all multi-engine aircraft must have a fuel shutoff valve only to the engine that is started first

122. Position (navigation) lights are connected in parallel with each other and in series with the single pole single throw (SPST) two position switch. Regarding navigation light colors, which of the following applies:
    (1) right (starboard) is green
    (2) left (port) is red
    (3) tail (aft) is white
    (4) all three of the above answers are correct

123. The field strength of a magnetic core can be increased by:
    (1) an increase in coil windings and an increase in the applied amperes
    (2) a decrease in coil windings and a decrease in the applied amperes
    (3) an increase in coil windings and a decrease in the applied amperes
    (4) a decrease in coil windings and an increase in the applied amperes

124. Starter motors develop a high torque at relatively low RPM and are thus very suitable to incorporate in the starting system of an engine. The type of motor most suitable for operating a landing gear is:
    (1) a compound wound motor
    (2) a shunt or parallel wound motor
    (3) a series wound motor
    (4) any type of motor that is classified as an induction motor

125. Rotating beacon (anti-collision) lights require relatively high amperage for operation. When wiring this type of light on an aircraft you should:
    (1) install it in a separate circuit and switch
    (2) install it as close to the bus as possible
    (3) install it entirely separate from the navigation light circuit and switch
    (4) all three of the above answers are correct

126. The commutator in a DC electrical motor:
     (1) should be cleaned or "dressed" with course emery cloth
     (3) basically converts field alternating current to direct current for charging batteries, etc.
     (3) continually switches the input current to new sections of the armature in order that the top of the armature is always the north pole
     (4) makes possible the operation of the (TR) transmitter-rectifier unit

127. Electrical cable terminals, when properly installed, should develop 100% of the electric cable strength. Turn or "dress" the commutator surface of an electrical motor or generator with No. 000 sandpaper. When connecting electrical wire terminals to a stud:
     (1) the maximum number of terminals per stud is six
     (2) use both plain and lock washers in the installation
     (3) the maximum number of terminals per stud is five
     (4) after installation, check for possible terminal movement

128. An aircraft is equipped with a 115 volt, 400 cycle alternating current electrical system. Direct current is obtained for battery charging:
     (1) by incorporating an inverter in the system between the alternator and the battery
     (2) by the utilization of a dynamotor
     (3) by the incorporation of a rectifier-transformer unit between the alternator and the battery
     (4) by the use of a DC bus-tie breaker

129. Generator brushes should be seated, after generator overhaul, by using:
     (1) emery cloth
     (2) any grade of steel wool
     (3) only crocus cloth
     (4) a good grade of fine sandpaper

130. Which of the following is the effect of counter electromotive force (EMF) in an aircraft engine electrical starter?
     (1) opposes normal EMF and decreases the current (amperes) at high speed
     (2) is much greater in the lower speed range
     (3) opposes normal EMF and creates more current at high speed
     (4) adds greatly to the EMF in order to give current at the higher speeds

131. In aircraft electrical systems, the advantage of using alternating current rather than direct current is which of the following?
     (1) AC impedance can be controlled much easier than DC impedance
     (2) AC is more responsive and easier to step up or step down the voltage
     (3) AC carries much less reluctance in the circuits than DC

(4) instead of using fuses as safety devices, AC permits the use of circuit breakers and circuit protectors

132. When routing wire bundles through aircraft structures, they can be protected by the use of:
(1) aircraft electrical tape of the friction type
(2) approved safety wire of the brass type
(3) clamps that are rubber lined
(4) caul blocks or gusset plates

133. Some electrical motors utilized in aircraft are equipped with a magnetic brake. The brake is applied by springs and released by:
(1) hydro-mechanical action
(2) mechanical linkage
(3) confined hydraulic pressure
(4) a solenoid coil

134. In a retractable landing gear system, lights indicate the position of the landing gear. Which of the following are correct in regards to indicating landing gear positions?
(1) no light–gear is up, green light–gear is down, and red light–gear is in unsafe position
(2) no light–gear is in unsafe position, green light–gear is down, and red light–gear is up
(3) green light–gear is down, red light–gear is in unsafe position, and no light–gear warning is ON
(4) red light–gear is in unsafe position, green light–gear is down, and no light–gear warning is OFF

135. A major characteristic of a thermocouple is that:
(1) it requires a special circuit from the bus bar
(2) the autosyn-selsyn principle is utilized
(3) it does not require any source of outside electricity for its operation
(4) its source of electricity must be derived from a rectifier-transformer unit

136. On removal of the end cap of a generator, small particles of solder are found on the generator face plate. This condition indicates:
(1) a closed generator field circuit
(2) an open generator field circuit
(3) excessive shaft bearing wear
(4) the generator armature has shorted out

137. The maximum safe current in the output of a generator is controlled by a:

    (1) governor relay
    (2) RCR (reverse current relay)
    (3) current limiter
    (4) vibrating or carbon pile voltage regulator

138. A growler is used for which of the following purposes?
    (1) reduces generator ripple and generator reaction
    (2) supplements the carbon pile type of voltage regulator
    (3) testing the condition of an armature
    (4) checking field windings for polarity

139. An electrical motor that incorporates a magnetic brake would most likely be used in which of the following units?
    (1) engine starter motor
    (2) rotating beacon (anti-collision light) motor
    (3) landing gear anti-skid motor
    (4) landing gear motor

140. The voltage regulator is properly adjusted. However, in this 30 volt system the voltmeter is reading 10 to 15 volts. The most probable malfunction or problem point is in the:
    (1) inverter-rectifier circuit
    (2) storage battery circuit
    (3) generator circuit
    (4) reverse current relay circuit

141. Double field windings are found in many electric motors. This permits:
    (1) control of the motor speed
    (2) reversing the direction of rotation of the motor
    (3) a great reduction in motor ripple
    (4) a reversal of the reverse current relay output

142. Copper is considered to be the best conductor of electricity since it allows easy movement of electrons. When increasing the size of the plates of a given storage battery, the voltage will remain the same but the ampere-hour capacity of the battery will increase proportionately; i.e. double the size of the plates and the ampere-hour capacity will double. A generator actually creates AC but it must put out DC to keep the DC storage battery up to proper charge. This conversion of AC to DC is accomplished by:
    (1) the generator commutator
    (2) a battery-connected inverter
    (3) two groups of generator interpoles
    (4) a battery-connected rectifier-transformer combination

143. The power factor in electrical circuits is the ratio of the true power to the apparent power. The power factor is an indication of the:
    (1) efficiency of an aircraft electrical circuit
    (2) resistance of a DC circuit
    (3) EMF of an AC circuit
    (4) EMF of a DC circuit

144. Starting an aircraft engine requires a motor that offers a wide range of torque and speed. Such a motor (actually used in aircraft) is of the:
    (1) series-wound DC type
    (2) shunt-wound AC type
    (3) parallel-shunt wound type
    (4) series-wound alternating type

145. The frequency of pulsations increases with an increase in the number of coils in a DC generator and produces a slight tremor called commutator ripple. This condition:
    (1) has an adverse effect on the requirements of an electrical system
    (2) does not cause interference in a radio receiver
    (3) is reduced or eliminated by connecting a filter across the generator terminals
    (4) may cause a mechanical breakdown of the generator

146. The use of interpoles (commutating poles) makes it possible to have sparkless commutation. This makes it unnecessary to use the method of shifting the brushes on the commutator. Interpoles:
    (1) improve generator output
    (2) prolong the life of the brushes and armature
    (3) reduce radio interference
    (4) all three of the above answers are correct

147. When the generator speed is increased, the carbon pile of the voltage regulator:
    (1) is spread and the resistance is increased
    (2) is drawn together and the resistance is decreased
    (3) remains at a given spacing with a constant resistance
    (4) none of the above answers is correct

148. An aircraft is equipped with a four-pole series generator. The generator is 75% efficient and it furnishes 100 amperes to its load (generator draw) at 5,000 RPM. Under these circumstances, how much current (amperes) will flow through each of the field windings?
    (1) 100

(2) 50
(3) 40
(4) 25

149. A starter motor is one that must be able to create a high initial torque. Starter motors are wound in:
(1) parallel
(2) shunt
(3) series
(4) series-parallel

150. The points stick open on a generator voltage regulator. This will cause:
(1) high voltage output
(2) low voltage output
(3) no malfunction of the generator
(4) a high residual voltage output

151. During the operation of a four-engine aircraft, which of the following generator controls are available to the pilot?
(1) each of the four generators has a switch that is in series with each generator field
(2) one master field switch controls the four generator fields that are in parallel with each other
(3) there is an individual switch in series with the armature of each generator
(4) there are four switches, each in parallel with the reverse current relay points

152. During flight, when more electrical power is needed, the generator voltage will:
(1) remain at a constant value; however, the current flow will increase
(2) automatically decrease because of the greater current flow
(3) remain at a constant value; the current will also remain at a constant value
(4) increase at a steady value in order to restrict the current flow

153. Which of the following forces (voltage) causes the points of a reverse current relay to close?
(1) residual
(2) battery
(3) generator
(4) magneto

154. The vibrator-type voltage regulator has:
(1) few turns of heavy wire
(2) many turns of fine wire

(3) a series of carbon pile conductors hooked up in parallel

(4) carbon pile conductors hooked up in series

155. The voltage regulator has a broken wire in the coil. Under this circumstance you will obtain:
   (1) normal voltage
   (2) a 50% reduction in voltage
   (3) zero voltage
   (4) a gradually decreasing voltage

156. If oil is found in the generator you should:
   (1) replace the armature
   (2) replace the brushes
   (3) check the engine oil seal for leakage
   (4) do nothing; this is a normal condition

157. All resettable circuit breakers:
   (1) should open the circuit irrespective of the position of the operating control when an overload exists
   (2) should open the circuit irrespective of the position of the operating control when a circuit fault exists
   (3) are referred to as *trip-free* circuit breakers
   (4) all three of the above are correct

158. Circuit breakers are located:
   (1) generally far from the power source
   (2) close to the power source
   (3) at any location compatible to the particular circuit
   (4) so as not to be exposed to heat of any kind

159. The type of control switch used for a retractable landing light is described as:
   (1) single-pole double-throw (SPDT)
   (2) double-pole double-throw (DPDT)
   (3) double-pole single-throw (DPST)
   (4) single-pole single-throw (SPST)

160. In an aircraft electrical circuit an SPST switch is described as a:
   (1) single-pole double-throw
   (2) single-pole switch-throw
   (3) switch-pole single-throw
   (4) single-pole single throw

161. Navigation lights are operated by a switch marked ON and OFF. This type of switch is best described as an:
   (1) SPDT

(2) DPST
(3) SPTT
(4) SPST

162. Multiple splices in a cable bundle:
    (1) are not permissible in accepted repair procedures
    (2) should be staggered along the run of the bundle
    (3) should be concentrated in one general location in the bundle
    (4) should not be closer than 15 inches from each other

163. You are assigned an electrical job that calls for attaching six cables
    to the bus bar. You should connect:
    (1) all six cables to one stud
    (2) one cable to the next stud since five cables on one stud is the
        maximum allowable
    (3) three cables to one stud and three cables to the next stud, properly
        connected by a metallic strip
    (4) all six in any combination since there is no limit to the number
        of cables that may be placed on one stud

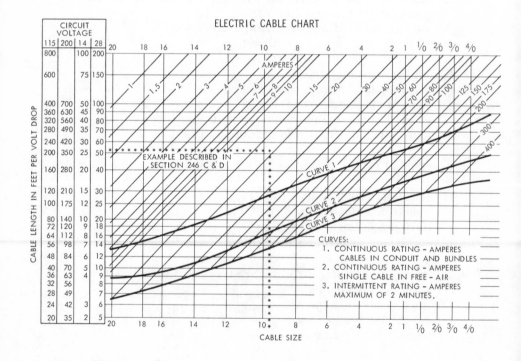

ELECTRIC CABLE CHART

164. A certain 28-volt flap motor has an output power of 1,742 watts and requires 62.2 amperes, operates intermittently, is located 50 feet from the power source, and has no return wire. An allowable voltage drop of two volts is permitted in a 28-volt system. Thus the cable length per volt drop is 25 feet. With this length and the 62.2 amperes required, the electrical cable chart shows that the minimum copper electrical cable size which can be used is:
    (1) #8
    (2) #6
    (3) #4
    (4) #2

165. In order to eliminate charges of static electricity throughout the aircraft structure, bonding jumpers are used to interconnect parts of the structure. These jumpers should be of the same metal as that which they are interconnecting; they should be clamped and as short as practicable. They should be installed in such manner that the resistance (ohm) of each connection does not exceed:
    (1) .672
    (2) .003
    (3) .033
    (4) .333

166. The one-wire aircraft electrical system differs from the two-wire system in what way?
    (1) one uses a DC source and the other an AC source
    (2) one is for large and the other for small aircraft
    (3) one uses low and the other high voltage
    (4) the manner in which the circuit is grounded back to the source

167. The term "spaghetti" in electrical installations describes:
    (1) the specific types of cord used to tie wires into bundles in most general types of electrical conduit
    (2) the wire bundles themselves
    (3) the plastic tubing used to cover electrical wires and terminals
    (4) a small type of bonding jumper

168. Generally speaking, electrical cables should not be spliced but replaced. However, if an electrical splice must be made using a soldered joint, which of the following must be taken into consideration?
    (1) the soldering should be accomplished by the cold solder method
    (2) tinning of both wires is a necessity
    (3) the finished splice must be at least 50% as strong as the cable
    (4) the splice must be mechanically sound

169. An electrical copper cable is to be installed in a length of 100 feet. It will carry 20 amperes and the circuit has a 1-volt drop. According to the cable chart, what sizes of copper and aluminum cable can be used and for what purposes?
    (1) #4 copper cable, #6 aluminum cable, and all three purposes listed under curves 1, 2, and 3
    (2) #6 copper cable, #4 aluminum cable, and all three purposes listed under curves 1, 2, and 3
    (3) #8 copper cable, #6 aluminum cable, and all three purposes listed under curves 1, 2, and 3
    (4) #4 copper cable, #2 aluminum cable, and all three purposes listed under curves 1, 2, and 3

170. Installation length of a copper cable is 75 feet, voltage drop is 3, and the circuit is to carry 50 amperes. According to the cable chart, what size copper cable is required, what size aluminum cable, and for what purpose can the cable be used?
    (1) #6 copper cable, #4 aluminum cable, and for purposes under curves 2 and 3
    (2) #10 copper cable, #8 aluminum cable, and for purposes under curves 2 and 3
    (3) #8 copper cable, #6 aluminum cable, and for purposes under curves 2 and 3
    (4) none of the above answers is correct

171. Which of the following is most appropriate regarding the use of electrical cables and the electrical cable charts?
    (1) if the area of a cable is reduced by one-half and the length is then doubled, the resistance of the replaced cable will be four times as much as the original cable
    (2) if the diameter of a cable is reduced by one-half and the length is then doubled, the resistance of the replaced cable will be eight times as much as the original cable
    (3) you plan to use a #8 copper cable, but if #8 copper cable is not available you could use #6 aluminum
    (4) all three of the above answers are correct

172. The generator may burn out if the circuit is too heavily loaded. The excessive current may generate enough heat to burn off the insulation, melt soldered connections, and possibly start a fire. The device that prevents this is known as a:
    (1) current limiter
    (2) current limitator
    (3) current regulator
    (4) all three of the above answers are correct

173. The strength of a solenoid depends on the:
  (1) type of core
  (2) size of core
  (3) number of turns
  (4) applied voltage

174. The current flows through the coil in the solenoid switch:
  (1) only until the points make contact
  (2) continuously when the master switch is ON.
  (3) until just after the master switch is ON.
  (4) when the appropriate cockpit switch is ON.

175. Aircraft circuits use AC, but storage batteries create DC. Which of the following is used to convert battery DC current to airplane AC current?
  (1) galvonmeter
  (2) rectifier
  (3) inverter
  (4) autosyn changeover segment

176. Landing lights are designed and operated to draw a very high current during operation. They are:
  (1) connected directly to the bus bar
  (2) remotely controlled by a cockpit switch that operates a solenoid relay switch which closes the circuit from the battery to the light
  (3) operated by the same switch and relay when more than one light is installed
  (4) connected indirectly to the bus bar

177. A generator is dependent on its own output to energize the field windings to build up a magnetic field for the generation of electrical energy. Considering this, how does a generator create electrical energy immediately upon initial rotation?
  (1) by the normal retention of a small amount of residual magnetism in the field
  (2) through temporary use of the DC battery current
  (3) by static electricity created through the rotating armature
  (4) through the use of current temporarily retained in a condenser

178. During flight, all the gear-down-lock indicator lights burn out. How would this affect the gear-warning-horn system?
  (1) the system would become inoperative
  (2) it would have no effect
  (3) the entire gear warning system would be rendered inoperative
  (4) all three of the above answers are correct

179. To fasten deicer boots to aircraft surfaces you should use:
    (1) clamps
    (2) rivets
    (3) rivnuts
    (4) small hex-head bolts

180. On a typical large type aircraft, an anti-skid system is provided to prevent excessive skidding during heavy braking. A skid in any main gear wheel will cause a flywheel detector to energize an anti-skid valve, releasing brake pressure from the affected tandem pair of wheels. This action is indicated by kickback felt through rudder pedals and by one of four brakes-released indicators on the overhead panel. Regarding the brake anti-skid system:
    (1) protection is provided for a locked wheel by time delay circuits which extend the brake-released time by approximately 0.7 seconds
    (2) the time delay circuits, energized when the landing gear is extended (as on an approach), prevent the brakes from being applied until 0.7 seconds after touchdown
    (3) rapid wheel deceleration detection is an important function of the anti-skid system
    (4) all three of the above answers are correct

181. In an aircraft anti-skid system, the electrical solenoid does which of the following?
    (1) allows less brake pressure by allowing a shuttling of the brake booster
    (2) relieves the pressure to the brakes of the affected wheels
    (3) immediately creates a transfer of pressure to the unaffected wheel brakes
    (4) temporarily traps a given amount of pressure in the affected brakes

182. Which of the following applies to the installation of deicer boots?
    (1) the surface to which the deicer boot is to be installed should be cleaned thoroughly (remove paint where necessary)
    (2) talcum powder should be applied to the boot prior to installation
    (3) rivnuts or glue is an approved method of installing deicer boots
    (4) all three of the above answers are correct

183. Deicer boots installed on the leading edges of wings are inflated by which of the following types of pumps?
    (1) electrical centrifugal boost pump in conjunction with a gerotor pump
    (2) a piston type pump whose output can be controlled by shifting the pump housing
    (3) a gear type pump that utilizes floating vanes
    (4) a vane type of pump

184. A good cleaning agent for deicer boots is:
   (1) toluene thinner
   (2) thinned turpentine
   (3) hot linseed oil
   (4) soap and water

185. Turbojet aircraft wings (fixed-wing type) are generally protected from ice formation by the use of:
   (1) pneumatically controlled expandable boots installed on the leading edges
   (2) hot air bleed emanating from the engine compressor section that flows to the wing leading edges
   (3) synthetic rubber boots that are electrically heated
   (4) hot air bled from the heaters that are located near the wing area

186. When heat is used as a means of preventing ice from forming on the aircraft leading edges, which of the following is correct?
   (1) heat is applied to the leading edges at all times during flight
   (2) heat is applied only when the outside air temperature is below the freezing point, either 32°F. or 0°C.
   (3) heat is applied only when the outside air has a relative humidity of at least 75%
   (4) heat is generally applied only when the aircraft is flying under known possible icing conditions as determined by the flight crew

187. When wing deicer boots are utilized, the purpose of an air-oil separator is to:
   (1) prevent the deicer boot material from deteriorating
   (2) eliminate oil from the vacuum pump that supplies the compressed air for operating the deicer boots
   (3) prevent oil from spreading out through the vacuum system
   (4) aid the operation of the deicer boot movement by injecting fluid into the deicer boot channels

188. Large jet type aircraft fire extinguishing systems are charged with Freon and nitrogen. Built-in fire extinguishing systems for piston-type (recip- rocating engines) are usually charged with:
   (1) bicarbonate of soda mixture
   (2) a thin solution of carbon tetrachloride
   (3) Freon and nitrogen
   (4) carbon dioxide and nitrogen

189. Colors for fluid line identification in aircraft are: (a) lubrication sys- tem—yellow, (b) fuel system—red, (c) breathing oxygen—green, (d) hydraulic system—yellow-blue, (e) deicing system—grey, and (f) air

conditioning system—brown-grey. Fire protection systems (fire extinguisher lines) are color coded with which of the following colors?
(1) blue-orange
(2) brown-orange
(3) brown
(4) yellow-green

190. While starting an engine, an induction system fire occurs. You should do which of the following?
(1) discontinue engine rotation with the starter immediately
(2) lean out the mixture and extinguish the fire by repeated backfirings
(3) immediately stop rotation of the engine and energize the engine fire extinguishing system
(4) continue rotation of the engine with the starter

191. A loop-type fire warning (signaling) device is based on the closed loop principle and under some circumstances can give a false fire warning. A false fire warning could occur due to which of the following conditions?
(1) moisture condensing on the exterior of the loop
(2) capacitance that has escaped from nearby electrical condensers
(3) bends, a crushed sensing unit device, or kinks
(4) use of lead wires of excessive strong capacity

192. The typical loop fire warning unit operates by which of the following?
(1) pressure exerted in the lines due to heated gases
(2) the major terminals coming to a temperature that causes melting
(3) change of resistance of the ceramic material in the tube when exposed to excessive heat
(4) the same principle that energizes the typical electric thermocouple

193. During aircraft operation, a wheel brake fire occurs. Which of the following should be used to extinguish this type of fire?
(1) dry chemicals
(2) moist chemicals
(3) foam similar as that used to coat the runway in case of an aircraft wheels-up landing
(4) toluene mixed with proper proportions of powdered bicarbonate of soda

194. You are instructed to check a fire extinguisher bottle for full charge. You can determine the amount of contents by:
(1) inspection of the pressure-specific gravity meter
(2) checking the height of the contents by looking through the sight gage

(3) the use of a standard dip stick

(4) weighing the bottle

195. In a typical large aircraft, smoke detectors (or smoke indication systems) are installed to indicate the presence of smoke in the main compartment or the electronic rack. The indicator, located on the system operator's center panel, consists of seven windows, a shrouded bulb, a detect light, and a "push to detect" button. When the button is depressed:

(1) the shrouded bulb illuminates but is normally invisible in the windows

(2) if any smoke is present, the smoke particles will reflect the light and the appropriate window will appear to illuminate

(3) the detect light is an indication that the shrouded bulb is serviceable; the air is continually routed through the indicator and only a momentary depression of the button is necessary to detect smoke

(4) all three of the above answers are correct

196. Which of the following is correct regarding a CO_2 bottle?

(1) a manual discharge is indicated by a red disc

(2) a thermal discharge is indicated by a yellow disc

(3) weighing the CO_2 bottle would not indicate whether or not it has been discharged

(4) a manual discharge is indicated by a yellow disc

THE CORRECT ANSWERS TO THE AIRFRAME
GENERAL EXAMINATION

1. (3)	40. (4)	79. (4)	118. (4)
2. (4)	41. (4)	80. (4)	119. (3)
3. (1)	42. (4)	81. (4)	120. (4)
4. (4)	43. (2)	82. (3)	121. (3)
5. (4)	44. (4)	83. (4)	122. (2)
6. (2)	45. (3)	84. (2)	123. (4)
7. (1)	46. (2)	85. (3)	124. (4)
8. (3)	47. (1)	86. (3)	125. (3)
9. (4)	48. (4)	87. (1)	126. (1)
10. (3)	49. (1)	88. (4)	127. (3)
11. (3)	50. (3)	89. (3)	128. (1)
12. (2)	51. (4)	90. (4)	129. (3)
13. (1)	52. (3)	91. (2)	130. (4)
14. (2)	53. (4)	92. (4)	131. (3)
15. (4)	54. (3)	93. (3)	132. (3)
16. (3)	55. (2)	94. (4)	133. (2)
17. (4)	56. (3)	95. (4)	134. (4)
18. (3)	57. (3)	96. (2)	135. (4)
19. (4)	58. (3)	97. (2)	136. (3)
20. (4)	59. (3)	98. (1)	137. (2)
21. (3)	60. (2)	99. (3)	138. (2)
22. (2)	61. (4)	100. (4)	139. (2)
23. (4)	62. (2)	101. (1)	140. (4)
24. (1)	63. (1)	102. (4)	141. (2)
25. (2)	64. (3)	103. (4)	142. (3)
26. (4)	65. (3)	104. (2)	143. (4)
27. (2)	66. (3)	105. (1)	144. (4)
28. (3)	67. (2)	106. (2)	145. (3)
29. (1)	68. (4)	107. (4)	146. (2)
30. (2)	69. (2)	108. (3)	147. (1)
31. (4)	70. (4)	109. (2)	148. (3)
32. (1)	71. (3)	110. (3)	149. (2)
22. (4)	72. (3)	111. (3)	150. (4)
34. (1)	73. (4)	112. (2)	151. (2)
35. (3)	74. (2)	113. (2)	152. (3)
36. (3)	75. (4)	114. (4)	153. (1)
37. (1)	76. (1)	115. (4)	154. (3)
38. (4)	77. (3)	116. (1)	155. (2)
39. (3)	78. (3)	117. (1)	156. (3)

157. (4)	199. (3)	241. (1)
158. (3)	200. (1)	242. (4)
159. (1)	201. (2)	243. (4)
160. (3)	202. (3)	244. (2)
161. (4)	203. (1)	245. (3)
162. (4)	204. (4)	246. (4)
163. (3)	205. (1)	247. (4)
164. (3)	206. (3)	248. (3)
165. (3)	207. (4)	249. (4)
166. (2)	208. (2)	250. (4)
167. (4)	209. (4)	251. (2)
168. (4)	210. (3)	252. (3)
169. (4)	211. (4)	253. (4)
170. (3)	212. (1)	254. (2)
171. (4)	213. (2)	255. (4)
172. (4)	214. (3)	256. (3)
173. (3)	215. (1)	257. (3)
174. (4)	216. (3)	258. (4)
175. (1)	217. (2)	259. (3)
176. (3)	218. (4)	260. (4)
177. (2)	219. (4)	261. (2)
178. (3)	220. (4)	262. (1)
179. (4)	221. (3)	263. (2)
180. (1)	222. (1)	264. (4)
181. (2)	223. (4)	265. (3)
182. (3)	224. (1)	266. (4)
183. (4)	225. (4)	267. (2)
184. (3)	226. (3)	268. (3)
185. (2)	227. (4)	269. (1)
186. (1)	228. (2)	270. (4)
187. (4)	229. (1)	271. (3)
188. (1)	230. (2)	272. (3)
189. (2)	231. (4)	273. (2)
190. (4)	232. (1)	274. (3)
191. (3)	233. (3)	275. (2)
192. (3)	234. (2)	276. (4)
193. (1)	235. (3)	277. (4)
194. (2)	236. (4)	
195. (3)	237. (1)	
196. (4)	238. (4)	
197. (3)	239. (1)	
198. (2)	240. (4)	

THE CORRECT ANSWERS TO THE AIRFRAME STRUCTURES EXAMINATION

1. (4)	40. (2)	79. (3)	118. (3)
2. (2)	41. (3)	80. (3)	119. (2)
3. (2)	42. (2)	81. (2)	120. (4)
4. (1)	43. (1)	82. (2)	121. (3)
5. (4)	44. (4)	83. (1)	122. (3)
6. (2)	45. (4)	84. (4)	123. (4)
7. (1)	46. (3)	85. (4)	124. (1)
8. (3)	47. (4)	86. (2)	125. (2)
9. (1)	48. (1)	87. (3)	126. (3)
10. (2)	49. (4)	88. (1)	127. (3)
11. (3)	50. (1)	89. (2)	128. (2)
12. (3)	51. (1)	90. (4)	129. (4)
13. (4)	52. (3)	91. (3)	130. (4)
14. (4)	53. (1)	92. (1)	131. (1)
15. (4)	54. (3)	93. (1)	132. (4)
16. (3)	55. (1)	94. (1)	133. (1)
17. (3)	56. (2)	95. (4)	134. (4)
18. (3)	57. (4)	96. (1)	135. (4)
19. (4)	58. (2)	97. (1)	136. (4)
20. (2)	59. (3)	98. (4)	137. (4)
21. (1)	60. (2)	99. (3)	138. (3)
22. (3)	61. (2)	100. (1)	139. (4)
23. (4)	62. (1)	101. (4)	140. (1)
24. (2)	63. (4)	102. (2)	141. (1)
25. (3)	64. (1)	103. (1)	142. (2)
26. (2)	65. (1)	104. (2)	143. (1)
27. (4)	66. (4)	105. (3)	144. (3)
28. (3)	67. (3)	106. (1)	145. (1)
29. (4)	68. (3)	107. (4)	146. (3)
30. (2)	69. (2)	108. (4)	147. (1)
31. (2)	70. (1)	109. (3)	148. (2)
32. (1)	71. (3)	110. (1)	149. (4)
33. (1)	72. (1)	111. (2)	150. (3)
34. (4)	73. (4)	112. (4)	151. (1)
35. (3)	74. (3)	113. (4)	152. (2)
36. (4)	75. (4)	114. (4)	153. (3)
37. (1)	76. (3)	115. (2)	154. (4)
38. (4)	77. (4)	116. (1)	155. (3)
39. (4)	78. (4)	117. (1)	156. (2)

157. (4)	178. (1)
158. (2)	179. (4)
159. (1)	180. (3)
160. (3)	181. (2)
161. (4)	182. (4)
162. (4)	183. (1)
163. (2)	184. (4)
164. (1)	185. (4)
165. (4)	186. (3)
166. (4)	187. (4)
167. (1)	188. (2)
168. (3)	189. (3)
169. (3)	190. (1)
170. (4)	191. (3)
171. (2)	192. (2)
172. (3)	193. (4)
173. (3)	194. (2)
174. (4)	195. (2)
175. (2)	196. (4)
176. (1)	197. (3)
177. (4)	

THE CORRECT ANSWERS TO THE AIRFRAME SYSTEMS AND COMPONENTS EXAMINATION

1. (4)	40. (3)	79. (2)	118. (3)
2. (1)	41. (1)	80. (2)	119. (1)
3. (2)	42. (3)	81. (4)	120. (3)
4. (1)	43. (2)	82. (3)	121. (2)
5. (3)	44. (3)	83. (4)	122. (4)
6. (3)	45. (3)	84. (2)	123. (1)
7. (4)	46. (1)	85. (3)	124. (3)
8. (3)	47. (3)	86. (3)	125. (4)
9. (3)	48. (2)	87. (3)	126. (3)
10. (1)	49. (3)	88. (1)	127. (4)
11. (3)	50. (4)	89. (4)	128. (3)
12. (2)	51. (4)	90. (4)	129. (4)
13. (3)	52. (4)	91. (3)	130. (1)
14. (2)	53. (3)	92. (3)	131. (2)
15. (2)	54. (2)	93. (3)	132. (3)
16. (2)	55. (2)	94. (4)	133. (4)
17. (2)	56. (3)	95. (2)	134. (1)
18. (2)	57. (3)	96. (1)	135. (3)
19. (2)	58. (4)	97. (4)	136. (4)
20. (1)	59. (2)	98. (1)	137. (3)
21. (4)	60. (4)	99. (3)	138. (3)
22. (3)	61. (4)	100. (4)	139. (4)
23. (1)	62. (4)	101. (2)	140. (3)
24. (3)	63. (3)	102. (2)	141. (2)
25. (3)	64. (2)	103. (3)	142. (1)
26. (2)	65. (4)	104. (2)	143. (1)
27. (2)	66. (2)	105. (1)	144. (1)
28. (4)	67. (4)	106. (3)	145. (3)
29. (4)	68. (4)	107. (1)	146. (4)
30. (1)	69. (2)	108. (3)	147. (1)
31. (1)	70. (2)	109. (3)	148. (1)
32. (4)	71. (4)	110. (3)	149. (3)
33. (4)	72. (3)	111. (4)	150. (2)
34. (2)	73. (3)	112. (2)	151. (1)
35. (4)	74. (1)	113. (3)	152. (1)
36. (4)	75. (3)	114. (4)	153. (3)
37. (3)	76. (4)	115. (4)	154. (2)
38. (1)	77. (3)	116. (1)	155. (3)
39. (4)	78. (1)	117. (1)	156. (3)

157. (4)
158. (2)
159. (1)
160. (4)
161. (4)
162. (2)
163. (3)
164. (2)
165. (2)
166. (4)
167. (3)
168. (4)
169. (2)
170. (3)
171. (4)
172. (4)
173. (4)
174. (4)
175. (3)
176. (2)

177. (1)
178. (2)
179. (3)
180. (4)
181. (2)
182. (4)
183. (4)
184. (4)
185. (2)
186. (4)
187. (1)
188. (4)
189. (3)
190. (4)
191. (3)
192. (3)
193. (1)
194. (4)
195. (4)
196. (4)

INDEX

You may also be interested in . . .

AIRFRAME & POWERPLANT MECHANICS MANUAL: to prepare you for your Powerplant Rating now that you have the Airframe. Nearly 3,000 exam questions (with correct answers), about half of them on the Powerplant Rating. (The material on the Airframe has been superseded by this new manual in your hands.) $7.50

FLIGHT ENGINEERS MANUAL: text, airplane operating manuals, FAR's, and typical written examinations for the Flight Engineer Certificate, including Turbojet, Turboprop, and Reciprocating-Engine Class Ratings. The only complete guide available. $8.75

THE NEW PRIVATE PILOT: if you're interested in acquiring your basic pilot's license, this is the manual that will guide you to it. With clear, simple, concise text, plus two sample written exams. More than 150,000 copies in circulation. All necessary maps and reference material included. $6.75

Write for a free complete catalog describing all current Zweng Airman Training Manuals.

PAN AMERICAN NAVIGATION SERVICE, INC.
12021 VENTURA BLVD., NORTH HOLLYWOOD, CALIF. 91604